Emerging Trends in Energy and Environmental Design Integrating New Services and Tools for Smart Cities and Smart Buildings

Emerging Trends in Energy and Environmental Design Integrating New Services and Tools for Smart Cities and Smart Buildings

Guest Editors

Saveria Olga Murielle Boulanger
Andrea Boeri
Gonçalo Mendes

MDPI
Basel • Beijing • Wuhan • Barcelona • Belgrade • Novi Sad • Cluj • Manchester

Guest Editors
Saveria Olga Murielle Boulanger
Department of Architecture
University of Bologna
Bologna
Italy

Andrea Boeri
Department of Architecture
University of Bologna
Bologna
Italy

Gonçalo Mendes
Laboratory of Electricity Markets and Power Systems
Lappeenranta University of Technology (LUT)
Lappeenranta
Finland

Editorial Office
MDPI AG
Grosspeteranlage 5
4052 Basel, Switzerland

This is a reprint of the Special Issue, published open access by the journal *Energies* (ISSN 1996-1073), freely accessible at: https://www.mdpi.com/journal/energies/special_issues/KX594XBL6H.

For citation purposes, cite each article independently as indicated on the article page online and as indicated below:

Lastname, A.A.; Lastname, B.B. Article Title. *Journal Name* **Year**, *Volume Number*, Page Range.

ISBN 978-3-7258-6412-6 (Hbk)
ISBN 978-3-7258-6413-3 (PDF)
https://doi.org/10.3390/books978-3-7258-6413-3

Cover image courtesy of Saveria Olga Murielle Boulanger

Contents

Preface . . . . **vii**

Lia Marchi and Jacopo Gaspari
Energy Conservation at Home: A Critical Review on the Role of End-User Behavior
Reprinted from: *Energies* **2023**, *16*, 7596, https://doi.org/10.3390/en16227596 . . . . **1**

Sebastiano Anselmo and Maria Ferrara
Trends and Evolution of the GIS-Based Photovoltaic Potential Calculation
Reprinted from: *Energies* **2023**, *16*, 7760, https://doi.org/10.3390/en16237760 . . . . **23**

Roberto Bosco, Savino Giacobbe, Salvatore Losco, Louise Anna Mozingo and Renata Valente
Environmental Design on Site-Specific Energy Solidary Communities around Public High Schools in the Metropolitan Area of Naples (Italy)
Reprinted from: *Energies* **2024**, *17*, 2247, https://doi.org/10.3390/en17102247 . . . . **50**

Saveria Olga Murielle Boulanger
Smart City and Energy: A Bibliometric Review of the Smart City and Smart Energy Concept from the Perspective of the Bioclimatic Approach
Reprinted from: *Energies* **2024**, *17*, 2486, https://doi.org/10.3390/en17112486 . . . . **76**

Mariana Januário, Ricardo Gomes, Patrícia Baptista and Paulo Ferrão
Integrated Energy and Environmental Modeling to Design Cost-Effective Building Solutions at a Regional Level
Reprinted from: *Energies* **2024**, *17*, 5730, https://doi.org/10.3390/en17225730 . . . . **97**

António J. Arsénio Costa and Hugo Morais
Power Quality Control Using Superconducting Magnetic Energy Storage in Power Systems with High Penetration of Renewables: A Review of Systems and Applications
Reprinted from: *Energies* **2024**, *17*, 6028, https://doi.org/10.3390/en17236028 . . . . **130**

Francesco Montana, Maurizio Cellura, Maria Luisa Di Silvestre, Sonia Longo, Le Quyen Luu, Eleonora Riva Sanseverino and Giuseppe Sciumè
Assessing Critical Raw Materials and Their Supply Risk in Energy Technologies—A Literature Review
Reprinted from: *Energies* **2025**, *18*, 86, https://doi.org/10.3390/en18010086 . . . . **161**

Negar Mohtashami, Nils Sauer, Rita Streblow and Dirk Müller
Comparative Analysis of Smart Building Solutions in Europe: Technological Advancements and Market Strategies
Reprinted from: *Energies* **2025**, *18*, 682, https://doi.org/10.3390/en18030682 . . . . **175**

Preface

The ongoing climate crisis, the instability of global energy systems, and broader social and political changes are increasing the complexity of urban sustainability challenges. Despite international commitments and growing awareness, cities are still not on track to meet the mitigation and adaptation targets outlined in the latest IPCC and SDG reports. In this context, the built environment emerges as both a critical issue and a key area for intervention. Energy and environmental design can contribute to more robust and adaptive urban systems, especially when informed by site-specific conditions and supported by inclusive processes.

This volume brings together eight contributions that examine sustainable energy in urban settings from multiple disciplinary perspectives. The papers cover a range of topics, including patterns of energy use in households and the influence of behavioral factors; the development of user-centered tools for improving energy performance; and the use of GIS-based methods for estimating solar energy potential in cities. Other articles address the planning and implementation of renewable energy communities at the neighborhood level, and critically assess prevailing Smart City frameworks, proposing a broader integration of environmental design approaches. Further studies focus on strategies for building-level decarbonization, the implications of critical raw materials for energy infrastructure planning, and a comparative analysis of smart building technologies across European markets. Together, these studies highlight how energy transition processes intersect with technical, social, and institutional dimensions. Rather than offering a single recipe, they emphasize the importance of integrating knowledge across scales, from individual buildings to urban districts, and across domains, from energy systems to user practices. The Volume may offer useful insights for researchers, practitioners, and policymakers working on sustainable urban development.

We would like to thank all the authors for their contributions, the reviewers for their time and expertise in supporting the quality of this collection, and the editorial staff for their support.

Saveria Olga Murielle Boulanger, Andrea Boeri, and Gonçalo Mendes
Guest Editors

energies

MDPI

Review

Energy Conservation at Home: A Critical Review on the Role of End-User Behavior

Lia Marchi and Jacopo Gaspari *

Department of Architecture, University of Bologna, 40136 Bologna, Italy; lia.marchi3@unibo.it
* Correspondence: jacopo.gaspari@unibo.it

Abstract: The recent literature emphasizes the significance of occupants' behavior in shaping home energy demand. Several policies have been defined and tools and technologies have been developed to raise people's awareness and encourage energy-saving practices at home, but households' energy demand keeps rising. The thesis is that the fundamentals on this topic are still unclear and that available tools, strategies and measures should be approached in a more integrated way, as they are not now effective enough to encourage energy savings. How these could be successfully combined is still a major knowledge gap. Thus, this article proposes a critical review of the literature to discuss the potential role of end users in energy conservation at home, preparing the ground for truly effective engagement strategies and tools to encourage behavioral change. To that end, a systematic literature review is performed, including over 130 relevant articles. According to the critical interpretation of their content, after years of technologically driven strategies, the most promising approaches capable of overcoming the intention–action gap are those more user-centered. However, relying solely on the social aspect is not effective. Synergistic integration of the two main clusters of studies has been identified as a promising field of research for the future.

Keywords: smart metering; behavioral implications; energy efficiency; energy savings; intention–action gap

1. Introduction

Reducing the energy demand of the building sector is key to cutting carbon emissions, tackling climate change, and, thus, achieving sustainable development at the global level [1,2]. Several United Nations Sustainable Development Goals (SDGs) are indeed connected with energy use in the built environment, from Goal 7 'Ensure access to affordable, reliable, sustainable and modern energy for all' to Goal 11 'Make cities and human settlements inclusive, safe, resilient and sustainable' and Goal 13 'Take urgent action to combat climate change and its impacts' [3,4]. Over 17% of the SDG targets are directly dependent on construction, and 27% are indirectly dependent on this sector's activities [5], most of which are highly energy-consuming.

Indeed, according to the latest estimations, over one-third of global energy consumption and emissions are attributed to the building sector, which includes energy used to build, heat, cool, and light either living or working spaces, including equipment and appliances [6]. Space and water heating account for about half of these, which largely depend on building energy performance and system efficiency [6–8]. The type of fuel energy also matters in this regard: even though the market share of fossil fuel-based heating systems is declining, gas boilers continue to be the most sold and used in a variety of global markets, and cleaner alternatives are scarcely increasing [9]. Therefore, at least in Europe, significant efforts have been made in the last decade to achieve extensive renovation of the building stock by reducing the envelope thermal flows, upgrading heating and cooling systems, and shifting to renewable energy sources [2,6,10–14].

In recent decades, in addition to technical factors, increasing attention has been paid to social factors relevant to buildings' energy efficiency, in particular in relation to occupants' behavior. For example, one of the key elements of the Clean Energy for All Europeans package is to empower end users to change their passive behavior and make their home energy demands more flexible [15]. This is estimated to potentially contribute a 4% to 30% saving in home energy consumption [16].

One of the biggest energy and carbon-saving behavioral shifts in buildings deals with lowering heating/raising cooling set points or introducing smart and/or user-centered technologies to manage systems and appliances' schedules. However, their implementation and effectiveness are highly dependent on personal circumstances that are difficult to predict and are determined by a series of technical, economic, and social factors that still require a significant amount of work [17].

Rising energy prices and growing inflation rates after the COVID-19 pandemic and the Russian–Ukrainian conflict have contributed somewhat to raising awareness of this topic at all levels. Lockdown events also have led to diverse home occupancy patterns—typically for longer periods during the day. Consequently, changes in energy consumption have emerged. Energy use in buildings can indeed affect households' economic stability, especially that of the most economically vulnerable consumers who struggle daily to cover basic living expenses [18–21].

Consumers have become more interested in both flexible energy contracts and retrofit interventions. Discontinuous energy demand during COVID-19, and later gas shortages from Eastern Europe, have driven utility organizations towards new demand response (DR) programs and clean energy sources. These, combined with the overarching goal of a just and clean energy transition, have pushed policymakers to more seriously consider households' behavior, energy nudging, and economic incentives for deep retrofitting. At the European level, for example, the program Repower EU [22] includes the EU 'Save Energy', plan which stresses the importance of short-term and long-term measures to reduce the energy consumption of both households and industries by pushing voluntary habit changes and structural retrofitting measures in combination: changes to our lifestyles and behavior can help significantly lower our energy consumption. Choosing to reduce heating temperatures, [...] use household appliances and air-conditioning more efficiently and switch off the lights can deliver substantial, short-term savings [23].

As buildings become smarter and more energy efficient (towards net zero or positive energy), it has been estimated that the role of end-users in shaping the energy demand will also increase [24]. Accordingly, the topic has been progressively researched and considered in a variety of sectors [25–27]. Energy conservation options and user energy profiles are increasingly studied to drive the development of policies, strategies, and tools more effectively. However, great uncertainty on the actual role and weight of home occupants in energy conservation is still observed in the literature, which partly depends on what behaviors are considered (e.g., only those dependent on routines, or also one-shot investments in energy-saving appliances) [28] and partly on the fact that the determinants of user behavioral change are still unclear. Also, the effectiveness of energy monitoring and management tools, which have been long considered as a solution to this issue, is now an open debate [29–32].

Overall, it seems that the main research gap is how to make available means and tools work together effectively, as the topic is so complex that environmental, economic, social and institutional factors are all relevant and should be considered together. However, it is still uncertain which drivers and leverages of the system are more promising to be studied further because most studies tend to focus on one aspect at a time.

This paper presents a critical State-of-the-Art review to investigate the role of users' behavior in shaping home energy demand and the opportunities for energy savings that are recurrent in the literature, along with strategies that might encourage users to adopt such behaviors. In addition, possible gaps to be filled by future research, intersections between the diverse categories of end-users contributions, and frontiers of the topic will be identified.

Future research could build on this knowledge and evolve practical strategies for accelerating the energy transition at home and thus contribute to global sustainability targets.

2. Materials and Methods

According to the research workflow presented in Figure 1, a systematic literature review was performed in Web of Science (WoS) and Scopus (stage 1). Science Direct, Research Gate, and Google Scholar were also considered, but then not included as the first is very similar to Scopus; the second is a voluntary repository which might not be fully updated; and the third is not properly structured with filters to scan and select the wide number of expected records. The two selected electronic databases are widely used in bibliometric analysis and considered highly authoritative and credible in academia as they include peer-reviewed records and are managed by third-party entities [33–36].

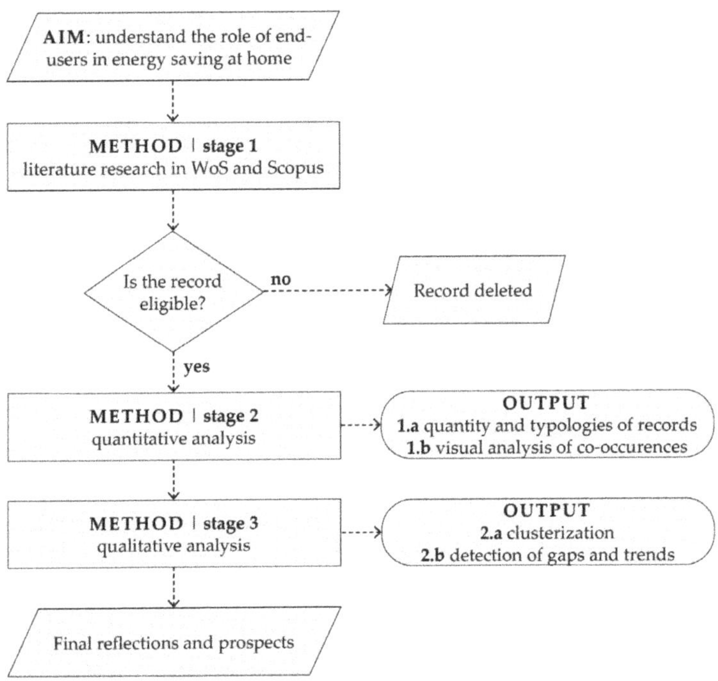

Figure 1. Literature research workflow.

Articles were searched by title and keywords. A 3W criterion was used to build the keyword string:

- *What*: Energy AND (conservation OR saving* OR reduction OR consumption OR use* OR behavio*)
- *Who*: User* OR Occupant* OR Household* OR Tenant*
- *Where*: Home* OR House* OR Resident*

Accordingly, the complete search string in each database is: (Energy AND (conservation OR saving* OR reduction OR consumption OR use* OR behavio*)) AND (User* OR Occupant* OR Household* OR Tenant*) AND (Home* OR House* OR Resident*).

The following filters were applied:

- Years: 2013–2023, assumed as a reasonable time during which the issue has evolved considering the recent energy challenges.
- Language: English

- Types of work: reviews, articles, and book/book chapters, to ensure the quality of contribution and homogeneity of data.
- Research area in WoS: Engineering; Construction Building Technology; Energy Fuels; Environmental Science Ecology; Architecture; Urban Studies; Behavioral Sciences. Subject area in Scopus: Engineering; Energy; Environmental Science; Social Sciences.

The search was conducted between May and August 2023, with regular updates (the last search was on 29 August 2023). Once the search was completed, records were screened according to the Preferred Reporting Items for Systematic reviews and Meta-Analyses (PRISMA) diagram [37], as shown in Figure 2.

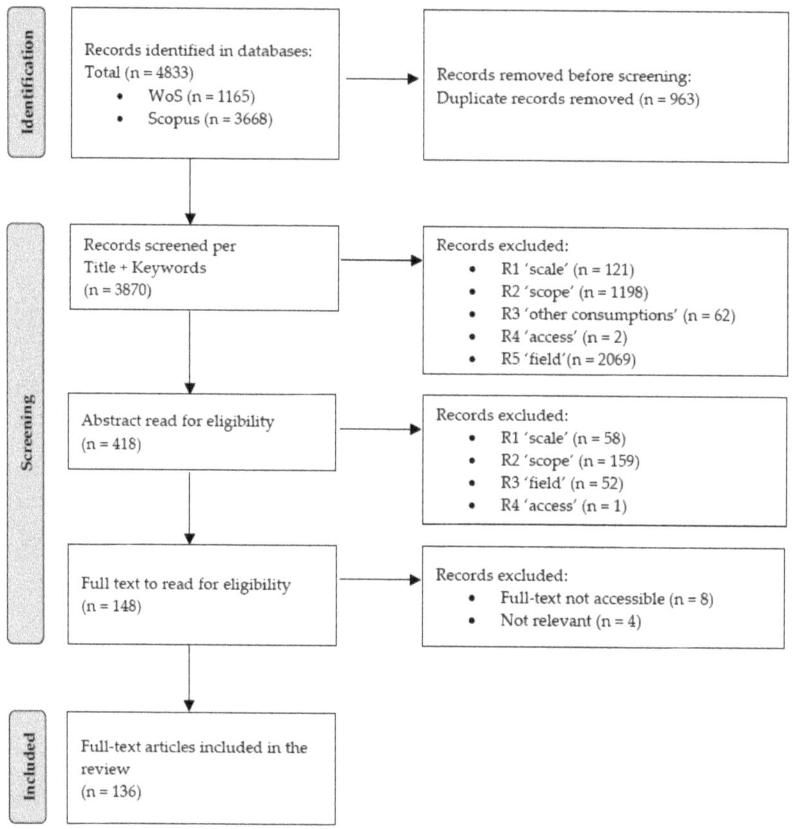

Figure 2. PRISMA workflow with the number of records processed per stage.

The identified 4833 records were imported into an Excel file and duplications excluded through automatic detection by title (963). Accordingly, 3870 records were selected for the screening process.

The first consistency screening was by title and keywords. In the Excel file, the records were categorized as "consistent" or "not consistent" by the following reasons for exclusion:

- Reason 1, the scale of the study is too large (e.g., city or regional level) or limited (e.g., specific appliances).
- Reason 2, the study is out of scope, e.g., focused on the Internet of Things, fuel and energy systems, building energy retrofitting, green purchasing, energy literacy, energy poverty, etc.
- Reason 3, the study addresses households' consumption other than domestic energy use (e.g., transport, food, clothing).

- Reason 4, content not retrieved, including abstract and keywords.
- Reason 5, the content is from another field/discipline.

At the end of the first screening phase, 418 records were selected for the following stage. The second screening phase consisted of reading the abstract to check for other excluding reasons. The identified ones were nearly the same as in the previous step:

- Reason 1, the scale of the study is too large or limited.
- Reason 2, the study is out of scope.
- Reason 3, the content is from another field/discipline.
- Reason 4, the abstract is not retrieved.

The third phase of screening regarded the full text of the resulting 148 records, 8 of which were discarded as the file was not fully accessible and 4 were deemed as not relevant to this study. A total of 136 records were included in this study.

The second stage of the review consisted of the quantitative analysis of the selected records (output 1.a), per time distribution, source, and type of document. Then, the open-source tool VOSviewer was used to find the co-occurrence of terms (output 1.b). The tool is indeed particularly useful to identify and visualize trends, patterns, and clusters of recurrent topics in bibliometric networks [38], which can be of help for the following stage. A Research Information System file (.ris) was exported from Mendeley (reference software, version Desktop 1.19.8) with titles and abstracts and used in VOSviewer v. 1.6.19 to obtain maps of terms. The counting method was set as binary, which means that a term was accounted for once in each document. The minimum number of occurrences in the database was set to 5 to register only frequent terms.

The authors' critical interpretation of clusters, trends, and gaps—which was derived from reading articles—was combined with this visual finding to form the basis of the qualitative analysis of records (outputs 2.1 and 2.2).

3. Results

Quantitative Analysis

The first output of the quantitative analysis is the time distribution of detected records for both stage 1 (including duplicates, divided per database) and stage 3 (articles included in the review) (Figures 3 and 4). In addition, top-ranking journals where the included articles were published are shown in Table 1.

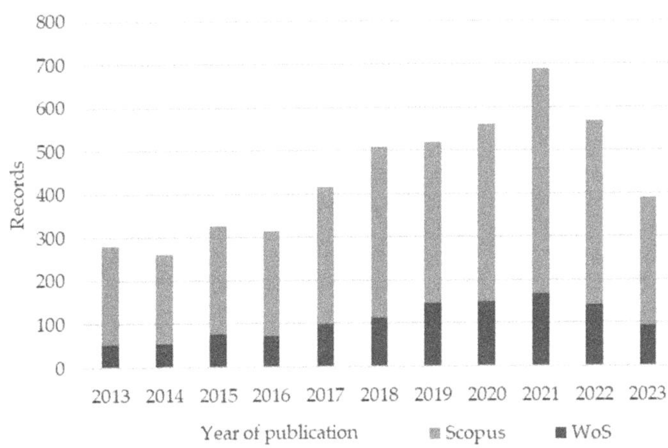

Figure 3. Time distribution of records per database, duplicates included.

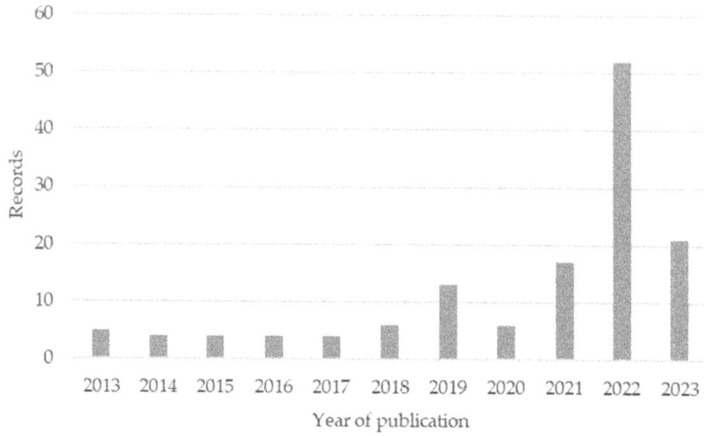

Figure 4. Time distribution of articles included in the review.

Table 1. Top-ranking journals where selected articles are published.

Source Title	Number of Articles
Energy Policy	13
Energies	12
Energy and Buildings	10
Energy Efficiency	10
Renewable and Sustainable Energy Reviews	6
Journal of Cleaner Production	5
Energy Research and Social Science	5
Sustainability	4
Energy	4
Sustainable Cities and Society	4
Energy Economics	4

The graphs show a significant increase in records during the last 3 years (2021–2023), with a peak in 2022 which demonstrates the increasing interest the topic has gained in several scientific and nonscientific sectors. This can be partly explained by contextual factors such as the rise in energy prices, while the 2023 apparent decrease might be related to the time period considered for this review (the end of August).

As for the document type, 122 are 'Articles', 13 are 'Reviews', and 1 is a 'Book chapter'. The first map created in VOSviewer was made by selecting the most relevant terms, which are obtained by default as 60% of the total co-occurrences detected. So, 113 terms were included in the map shown in Figure 5, where the software has grouped the items into four clusters (red, green, blue, and yellow).

The first two groups are significantly bigger than the others and contain the top five cited terms, which are: use (50 occurrences), model (34), efficiency (31), user (29), and role (29). The distance between the terms here represents the thematic distance in the research topics: while red and green clusters are quite homogeneous and compact, the other two groups contain quite scattered items suggesting that they are more heterogeneous.

The red cluster refers to terms pertaining more to the technological/physical dimension of energy conservation, such as: 'use', 'feedback', 'efficiency', and 'system'. However, a strong user-center approach emerges in this group by terms such as 'user', 'preference', and 'profile'. The green cluster is more related to social sciences/intangible aspects and behavioral theories, being the mainstream items 'model', 'role', 'attitude', 'determinant', and 'intention'. The blue cluster deals more with methodological aspects of the research in this field, with terms such as 'question', 'barrier', 'target', and 'subject', while the yellow one contains the remaining and less related terms.

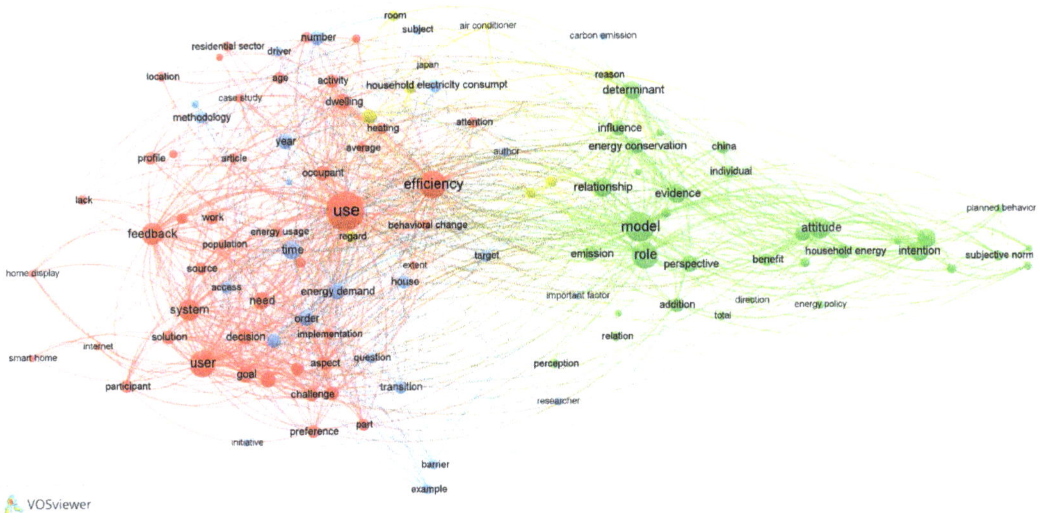

Figure 5. Co-occurrence of terms in the selected records, grouped per cluster and linked to each other. Diagram processed through VOSviewer.

Figure 6 visualizes the same clusters and terms in another way, giving more emphasis to the time distribution of the terms in recent years.

Figure 6. Visualization of recurrent terms as hot topics (in red) during the last 4 years. Diagram processed through VOSviewer.

Figure 7 focuses on the term 'use' (the most recurrent) and its over 110 links (represented by the curved lines). As the term is quite general, in this diagram it is more interesting to observe the connection between the item and the other ones.

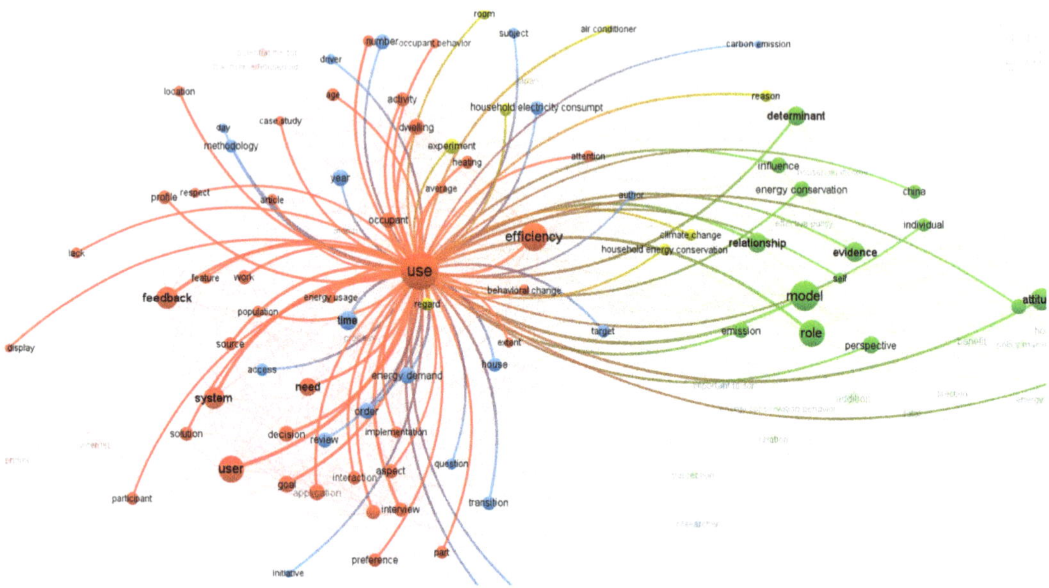

Figure 7. Extract of the network map focused on the term 'use' and its links. Diagram processed through VOSviewer.

4. Discussion

The screening of records in the two selected databases demonstrated that the topic is highly relevant. It has been progressively and increasingly investigated in the literature in attempts to address the environmental, societal, and economic concerns associated with energy conservation in the residential sector. The time distribution of records clearly illustrates that the research interest has been growing towards the topic, with a peak in 2021 (Figure 3, records not screened) probably due to changes in residential energy consumption patterns induced by the COVID-19 pandemic and the increasing harshness of climate change effects in cities.

From the over 3600 records identified at first, only 136 were selected in the end for the full-text reading (screening step 3). However, before discussing the latter, it is worth briefly reflecting on the discarded records. Although some were removed as not consistent, most of them were related to the general challenge of household energy transition and can help to frame the general discussion.

The most frequent and relevant research lines observed in the deleted records were:

Technical aspects

- Studies on information and communication technology (ICT) or the Internet of Things (IoT) for the management of either energy networks or smart homes.
- Studies on the impact of the primary energy type on the energy demand or, on a larger scale, on the national energy mix/energy market.
- Studies on the energy sources and network efficiency.
- Studies on the effect of photovoltaics and energy storage systems on the energy demand at home.
- Studies on the impact of urban morphology and/or pattern on residential energy use.
- Studies on the efficiency of cooling/heating systems.

- Studies on statistical methods for energy use trend predictions
- Studies on the efficiency of cooling/heating systems and/or the increased use of cooling to cope with global warming effects.
- Studies on statistical methods for energy use trend predictions.
- Studies on the performance gap between energy efficiency simulations of buildings and real consumption due to the human factor (occupancy patterns).
- Studies on other energy and carbon-saving practices for households, including food, mobility, and clothing choices.

Economic aspects

- Studies on the impact of the energy market or energy price on energy consumption and home system choices.
- Studies on the optimization of demand response programs and/or their effect on energy bills and user comfort.
- Studies on home appliances green purchasing and/or household energy-saving options (HESO), including preferences and motivation for their adoption.

Social aspects

- Studies on the role of energy consumption on the occupants' comfort and living standards.
- Studies on the human factors that may influence energy or fuel poverty.
- Studies on the preference, diffusion, or acceptance of electric vehicles rechargeable at home.
- Studies on the change induced by the COVID-19 pandemic in occupancy patterns on the energy demand.

Although it includes discarded records (due to scale, scope, etc.), this list suggests that when searching for households' energy saving, the mainstream topics are related to the technical dimension of the issue, from the efficiency of the energy network to the improvement in the technical performances (thermal insulation, system upgrading or shift, etc.). In addition, a significant number of discarded records were related to the economic dimension of households' energy transition, implying that the problem requires not only multidisciplinary but also multi-scale approaches and solutions across the market. Even though the subjects of the discarded studies were diverse, the marketing strategies and behavioral sciences issues identified suggest further investigation could help point out whether there is a spillover potential for approaches and tools used in different sectors.

As for the selected records, the observations of the VOSviewer maps (Figures 5–7) solicit some relevant reflections. It strongly emerges, as confirmed by the literature investigation, that there is a limited connection between the topics social and technical dimensions in existing studies. This has largely been identified as a significant barrier to effective advancements in the field, as well as a clear prevalence of technologically driven approaches to users' energy savings at home. Nonetheless, the hot topic visualization in Figure 6 shows a slightly promising shift in recent years toward social aspects. Terms such as 'solution' seem now to be outdated as it was largely proven to be ineffective in solving such a complex and multifaceted issue, that is the adoption of energy savings in the residential sector. An abundance of terms can be observed, such as 'feedback', 'preference', and 'role', supported by 'smart home' and 'internet'.

These general reflections are strengthened by a detailed analysis of the records, which is addressed in the thematic discussion that follows and whose main thematic clusters are displayed in Table 2.

Table 2. Distribution of records per thematic cluster, corresponding to the breakdown of the discussion sub-sections.

Clusters	Relevant Records
Chance for savings and role of users' behavior	
Energy saving practices and behaviors	Bouktif et al., 2022 [39]; Karahan et al., 2021 [40]; Khani et al., 2021 [41]
Energy price and energy savings link	Nahiduzzaman et al., 2023 [42]; Sloot et al., 2022 [43]; Al Mamun et al., 2022 [44]; Xu et al., 2021 [45]; Wang et al., 2021 [46]; Deumling et al., 2019 [47]
Efficiency investments and curtailment practices	Matsumoto et al., 2022 [48]; Never et al., 2022 [49]; Gajdzik et al., 2023 [50]
Energy efficiency paradox	Wester 2022 [51]; Aydın et al., 2023 [52]; Moeller et al., 2022 [53]
How to mobilize the adoption of energy conservation practices	
Intention–action gap	Gaspar et al., 2017 [54]; Yue et al., 2023 [55]; Zhang et al., 2021 [56]; Frederiks et al., 2015 [57]
Determinants of energy behavior	Al Mamun et al., 2022 [44]; Hori et al., 2013 [58]; Belaïd et al., 2016 [59]; Stikvoort et al., 2018 [60]; Hagejärd et al., 2023 [61]; Jakučionytė-Skodienė et al., 2023 [62]; Han et al., 2022 [63]; Owusu-Manu et al., 2022 [64]; Duong et al., 2022 [65]; Chen et al., 2022 [66]; Yixuan et al., 2022 [67]; Żywiołek et al., 2021 [68]; Wang et al., 2021 [69]
Nudging and behavioral interventions	Streimikiene 2023 [70]; Sudarmaji et al., 2022 [71]; Williams et al., 2022 [72]; Ruokamo et al., 2022 [73]; Henry et al., 2019 [74]; Sohre et al., 2022 [26]
Impact of social factors and comparison on savings	Fraser 2023 [75]; Bohdanowicz et al., 2021 [76]; Vassileva et al., 2014 [77]; Mukai et al., 2022 [78]; Hafner et al., 2019 [79]; Schneider et al., 2023 [80]; Jorgensen et al., 2020 [81]
Effect of smart energy management devices and apps	Chen et al., 2022 [66]; Caldera et al., 2023 [82]; Mostafa et al., 2022 [83]; Hess et al., 2022 [84]; Wood et al., 2019 [85]
Effectiveness of feedback and user engagement	Hagejärd et al., 2023 [61]; LaMarche et al., 2014 [86]; Pombeiro et al., 2019 [87]; Kendel et al., 2017 [88]; Nilsson et al., 2018 [89]; Geelen et al., 2019 [90]; Dalvi et al., 2016 [31]; Straub et al., 2018 [91]; Paneru et al., 2023 [92]; Garg et al., 2023 [93]; Madsen et al., 2023 [94]; Kim et al., 2023 [95]; Mataloto et al., 2023 [96]; Al-Kababji et al., 2022 [97]; Chatzigeorgiou et al., 2021 [98]; Song et al., 2020 [99]; Csoknyai et al., 2019 [100]; Méndez et al., 2022 [101]; Méndez et al., 2023 [102]; Papineau et al., 2022 [103]; Koasidis et al., [104]
Contextual and user differences in energy savings	
Geographic impact on energy patterns	Johansson et al., 2019 [105]; Iwata et al., 2015 [106]; Long et al., 2018 [107]
Socio-demographic features and dwelling characteristics impact	Karahan et al., 2021 [40]; Chen et al., 2022 [66]; Sudarmaji et al., 2022 [71]; Zhao et al., 2019 [108]; Spandagos et al., 2020 [109]; Kumar et al., 2023 [110]; Lei et al., 2022 [111]; Sen et al., 2022 [112]; Wang et al., 2022 [113]; Jareemit et al., 2019 [114]
Age factor in energy use	Pais-Magalhães et al., 2022 [115]; Dai et al., 2021 [116]; Lv et al., 2022 [117]
Income level and energy consumption	Vassileva et al., 2014 [77]; Malama et al., 2015 [118]; Podgornik et al., 2016 [119]; Romero-Jordán et al., 2022 [120,121]; Kaplowitz et al., 2022 [121]; Matthies et al., 2022 [122]; Perez-Bezos et al., 2023 [123]; Godoy-Shimizu et al., [124]
Genre or ethnicity effect on saving attitudes	Jareemit et al., 2019 [114]; Shrestha et al., 2021 [125]; Leslie et al., 2022 [126]
Ownership status and energy use	Taneja and Mandys 2022 [127]; Boudet at al. [128]
Household archetypes and related energy use profiles	Gaspar et al., 2017 [54]; Bedir et al., 2017 [129]; Ben et al., 2018 [130]; Ortiz et al., 2019 [131]; Mi et al., 2021 [132]; Akbari et al., 2021 [133]; Lu et al., 2022 [134]; Heinrich et al., 2022 [135]; Chen et al., 2022 [136]
Intersection with environmental and societal issues	
Energy consumption and environmental awareness nexus	Żywiołek et al., 2021 [68]; Rosak-Szyrocka et al., 2022 [137]; Kaplowitz et al., 2022 [121]; Kopsakangas-Savolainen et al., 2013 [138]; Sapci et al., 2014 [139]; Zhang et al., 2021 [140]; Jaciow et al., 2022 [141]; Chen et al., 2023 [142]; Du et al., [143]; Lam et al., [144]
COVID-19 effects on energy consumption patterns	Mataloto et al., 2023 [96]; Ueno 2022 [145]; Khalil et al., 2022 [146]; Balest et al., 2022 [147]
Research methods, gaps, and frontiers	
Literature reviews and meta-studies	Bouktif et al., 2022 [39]; Krishnan et al., 2022 [148]; Composto et al., 2022 [149]; Hu et al., 2022 [150]
Survey as research tool	Gaspari et al., 2021 [28]; Zhang et al., 2021 [56]; Hori et al., 2013 [58]; Johansson et al., 2019 [105]; Iwata et al., 2015 [106]; Zhao et al., 2019 [108]; Godoy-Shimizu et al., [124]; Kopsakangas-Savolainen et al., 2013 [138]; Chen et al., 2023 [142]; Venkatesh et al., 2020 [151]
Studies on technology impact on savings	Zhang et al., 2021 [140]; Kim et al., 2022 [152]; Andrade et al., 2022 [153]; Bastida et al., 2019 [154]; Qin et al., 2022 [155]; Pothitou et al., 2017 [156]
Studies that apply behavioral theories	Wang et al., 2018 [157]; Xu et al., 2021 [158]; Yue et al., 2019 [159]; Qalati et al., 2022 [160]; Webb et al., 2013 [161]; Conradie et al., 2023 [162]; Le-Anh et al., 2023 [163]; Nguyen et al., 2022 [164]; Fatoki 2022 [165]
Need for an integrated approach	Gaspari et al., 2021 [28]; Dietz et al., 2013 [166]; Gram-Hanssen 2013 [167]; Savvidou et al., 2020 [168]

4.1. Chance for Savings and Role of Users' Behavior

Nowadays, the majority of researchers in the field agree that occupants can strongly affect the energy demand, either acting on the physical features of the building or changing energy use practices [39,40]. Accordingly, energy-saving opportunities emerging from the selected articles can be grouped into two main clusters:

- Investments to reduce operational energy use (e.g., system replacements/upgrades, envelope thermal retrofit, replacement of appliances with new efficient ones).
- Change in energy consumption routine (i.e., conservation) or one-shot energy-saving actions. The change can involve either a different time or spatial use of energy [41].

As for the first category, the discussion can be extended on a larger scale to the shift in energy sources. In fact, end-users can have a role in energy savings by shifting to local renewable energy sources (e.g., photovoltaic panels) or entering DR programs which allow energy providers to optimize energy generation and distribution, with positive effects on both the energy market and the household's bill. There is indeed a strong correlation between energy price/tariff and user savings [42–45]. To this end, Al Mamun et al. proposed applying energy incentives to low-energy users—who are those using less energy than their peers—as a motivation for curtailment, even if Wang et al. observed that savings are more affected by occupants' attitudes than the energy price itself [46]. Nevertheless, Deumling et al. suggested that low-energy users should be further studied to understand how they became "virtuous" and thus how they can act as pilot cases for others to follow [47].

Interestingly, a positive nexus can be frequently observed between energy efficiency investments and energy curtailment practices [48,49]: in general terms, this means that those who spend their money on energy retrofitting are more likely to implement conservation or saving behaviors/actions. This attitude can be reinforced by the environmental awareness of the household, but it is not always true. Gajdzik et al. argues that being a prosumer frequently leads to the adoption of pro-environmental behaviors [50]. Others note a controversial relationship between clean or energy-efficient homes and the adoption of energy use practices (i.e., the energy efficiency paradox) [51–53]: sometimes it appears that using clean energy or living in an energy-efficient home makes the occupant feel legitimized to care less about energy conservation measures.

4.2. How to Mobilize the Adoption of Energy Conservation Practices

Despite the consistent body of work in the literature and knowledge about the role of the end user in shaping the energy demand, and specifically in the range of saving practices that can be implemented, researchers all agree that there is a great discrepancy between willingness to change a certain energy consuming behavior and its actual uptake. This is usually called the knowledge–action, value–action, or intention–action gap and it represents one of the major challenges for this research field [54–57].

Therefore, many scholars have investigated how to overcome this gap and how energy-saving potentials related to users' behavioral change can be mobilized. Nonetheless, there is still great uncertainty about the leverage of this complex system (i.e., households' behavior). Determinants for occupants for the adoption of energy-saving practices can vary from values to knowledge, from personality traits to income, from subjective norms to age, up to dwelling ownership and features. Regardless of the determinants that are detected and examined in specific studies, all agree that several concurring factors may lead to a change in household behaviors [44,58–69].

Among these, social factors play a major role, which is why many authors have focused on the advantages or disadvantages of nudging (i.e., gently encouraging someone to do something) and behavioral interventions [70–74]. For instance, Sohre and Schubert have contended that top-down interventions, which are typical of these approaches, are ineffective in this field because they deal with users' preferences at home and their private behaviors, which can be challenging to control from the outside [26]. Even if nudging alone is not enough, it can be combined with other strategies and policy measures to overcome the traditional financial and social barriers of the energy transition, for which economic and financial means put in place in the last decade have been demonstrated as not being fully effective [70].

In addition, other scholars discuss how social comparison, competition among peers, and goal setting for energy conservation can become more effective means to mobilize the potential for households' energy transition at home [75–79]. Contrasting opinions emerge also on this point: while Schneider et al. illustrate the role of individual energy coaches for this purpose [80], Jorgensen et al. argue that group-targeted interventions are more successful than individualistic approaches [81].

Technology emerges as a powerful supporting tool for this purpose. Therefore, great effort has been put into developing digital and/or smart tools to support the optimization of energy demand in the last decades. Home energy management systems (HEMSs) have been extensively discussed in the literature, as well as the potential of utility apps and feedback in energy bills to enable or activate users' conservation practices [66,82–85]. However, up to date, most of the experts in the field agree that effective feedback from such devices is essential not only to activate one-shot changes but especially to engage users in the long run [31,61,86–99]. To this end, gamification is frequently mentioned as an effective means for the use of smart meters and other energy consumption feedback supports [100–103]. In connection with gamification, others suggest monetization of pro-environmental behaviors, as in the case of the study of Koasidis et al. [104].

Overall, it can be observed that a balanced mix of behavioral interventions, financial support, and technology-driven feedback can be good at motivating users' energy conservation, while none of them is enough by itself. Indeed, on the one hand, the effectiveness of smart devices for energy monitoring or management might be prevented if the building occupants are not properly engaged. On the other hand, a collective versus individual momentum towards energy transition at home is claimed, suggesting that behavioral interventions can be truly effective only if capable of engaging users as a community with a shared responsibility.

4.3. Contextual and User Differences in Energy Savings

Besides the attitudes and values of specific users, many authors agree that energy-saving behaviors can derive from cultural, geographical, or other contextual factors. Johansson et al., for example, discussed the different approaches and attitudes of users' energy conservation practices between developed and developing countries [105]. Similarly, Iwata et al. argued that behavioral interventions that are effective in the US might not be in Japan [106]. According to Long et al., the perception of each role in climate change—which is affected by the cultural background, among other factors—demonstrates diverse attitudes to energy savings in the US and Germany [107].

Several authors discuss the role of both users' background and dwelling features in shaping the energy demand [40,66,71,108–114]. In this regard, Spandagos et al., for example, have argued that socio-demographic factors impact less than value beliefs [109].

Some records then examine the effects of one or very few specific factors at a time. More specifically, the age of users seems to strongly affect energy saving practices: elders, for instance, need more energy for heating and cooling because they are more sensitive to thermal discomfort but use fewer digital devices so that this share of consumption is less than in the younger generation or adults working from home [115,116]. So, retirement may also imply a reduction in consumption [117].

In particular, the relationship between income and energy consumption in the residential sector is also investigated [77,118–122], with special concern for social housing occupants [123]. The income–energy use relationship is often examined along with other contextual factors, such as age and employment status, as is the case of the research of Godoy-Shimizu et al., who found that according to statistical analyses, high electricity use is significantly correlated with social class, large household size, unemployment, and middle age, while low electricity use is significantly correlated with single-person households, small homes, and retirement [124].

Genres also seem to affect energy consumption patterns at home [114,125]: generally speaking, it is found that women are more prone to save energy than men when dealing with air conditioning in bedrooms or other home activities. Some also discuss how ethnicity can impact energy consumption patterns [126], arguing that some groups tend to save more energy because of religious principles or cultural values.

Interestingly, the ownership status of the dwelling can also have an impact on whether to encourage the adoption of energy-saving actions or not. That is why both policymakers and energy providers are recommended to take this into account: Taneja and Mandys, for instance, found that in the UK, renters are more sensitive to changes in electricity and gas prices than homeowners [127], but less interested in energy-saving investments, as pointed out by Boudet at al. [128].

Based on these differences in user backgrounds, many researchers are attempting to define household profiles/archetypes related to energy use patterns to support policy targeting [54,129–136]: some try to profile users and associate energy use profiles with them; others discuss the barriers and drivers behind the adoption of energy saving practices in certain occupants' archetypes. However, it is also noted that determinants of users' behavioral change in the residential domain are so individual that generalization does not help lead to effective change.

4.4. Intersection with Environmental and Societal Issues

Energy consumption in the residential sector is strongly linked with some transversal topics which are mainstream in the literature, such as climate change (CC) and the effects of the COVID-19 pandemic on daily life.

The correlation between CC awareness and the adoption of energy-saving practices at home is indeed a hot topic in the literature. Despite the general increase in the first, some authors point out that there is limited knowledge of conservation behaviors and changes that can be implemented at home [68,137]. Others state that there is a positive link between increasing environmental awareness and energy saving at home (i.e., the climate–energy consumption nexus), but the intention to adopt certain practices is not always translated into real actions [121,138–142].

On the other hand, the increasing harshness of CC effects is leading to a rise in energy demand for cooling, especially in dense urban environments. A growing number of studies have been performed to this end: such as that of Du et al., who examined the combined effect of CC and COVID-19 on energy use especially for cooling [143], and Lam et al. who discussed the determinants of behavioral change in this regard [144]. So, even though energy saving for heating is still mainstream in the literature, the number of research projects addressing cooling-related energy conservation behaviors is increasing.

Regarding COVID-19, several authors have investigated how various occupancy patterns and the economic crisis brought on by the pandemic have either temporarily or permanently changed energy use at home [96,145–147], which was typically increased during the lockdown and reduced in the recovery phase to cope with resource shortages of low-income families.

4.5. Research Methods, Gaps, and Frontiers

Energy conservation in the domestic environment has been largely studied in its multiple facets, as proven by the available reviews and meta-studies [39,148–150]. It emerges that due to the complex nature of the topic, the survey is the most widely used research means [28,56,58,105,106,108,124,138,142,151].

As it is evident from both the co-occurrence visualization and the specific content of included articles, up to now, studies in this field have focused on the technological or the behavioral facets of the energy transition at home in a separate way.

Regarding the first, existing research is mostly focused on tools or methodologies to assess and encourage the energy-saving potential of users' behavior [152,153]. Overall, a controversial role for technology emerges in energy savings. Overall, ICT, smart meters, apps and home energy management systems (HEMSs) drive users to save energy either automatically or by encouraging them to change their behavior [154]. To a certain extent, sole access to the Internet can help save energy by allowing users to search for information about related practices and investments [155]. However, on the other hand, the growing use of ICT work or entertainment devices is accounted to drive an increase in the energy demand [156]. Additionally, it should be noted that a purely technological approach to managing home energy savings runs the risk of encouraging the energy efficiency paradox, which occurs when energy demand is reduced because of energy efficiency interventions, household income rises, and as a result, the residents tend to take fewer conservation measures [140]. Therefore, studies focused on digital tools alone appear outdated.

Regarding the social dimension, it can be observed that an increasing number of studies try to apply well-established behavioral theories to the topic to overcome existing research and action gaps. Among these, the theory of planned behavior (TPB) recurs the most, followed by the goal framing theory (GFT) [157–165]. It is frequently observed that intentions are the most direct contributors to actions and behaviors, and that they can be predicted by attitude, social norms, and especially perceived behavioral control (a feeling that one can act successfully). Without this perception, any top-down behavioral interventions would be ineffective because people will not change their behavior or spend money on something that does not directly affect major issues or give them a sense of empowerment. Thus, it is widely held that knowledge about energy-efficiency measures is insufficient to promote the transition (which explains, in large part, why smart devices or apps that only provide quantitative feedback are ineffective).

Overall, despite the expected potential, an integrated approach between technology and social sciences has been only occasionally implemented in this field. Therefore, in light of the discussed gaps, many authors recommend considering more seriously the interconnection between the different perspectives of the issue in further research [28,166–168].

5. Conclusions

This article reported the outcomes of a literature review about the role of end users in implementing energy saving practices at home. It emerged that studies in the field of energy transitions are progressively shifting from a technological-driven approach alone to a perspective including behavioral theories and tools, with particular implications in institutional and social-normative terms.

Traditionally, actions taken in this field have focused on and promoted energy-efficient technologies and infrastructure, which have proven to yield less energy savings than expected due to rebound effects or significant knowledge–action gaps. Most of the included records mention energy efficiency interventions as the first and most important measure to conserve energy in the long term (i.e., structural measures), even if the focus of this article is the behavioral change in households towards energy. It is indeed widely acknowledged that while the effects of the first type of measures are higher, and permanent once implemented, the second type are more unstable as they rely on the user's willingness to maintain virtuous practices in the long run. Notably, while data and figures related to the benefit of structural measures have led to wide recognition of their importance, there is still a large disagreement about the actual effect of user behavior on demand reduction. This is indeed recognized to contribute from less than 1% to over 50% of the whole energy demand of the house. Such a large span suggests that not only are the determinants of behavior unclear, but also terms and a shared understanding of energy-related user behavior are lacking. As a result, it is difficult to compare multiple studies across the world and from diverse disciplines, according to common and shared elements.

Nevertheless, the majority of authors of the included records agreed that along with designing new efficient buildings and retrofitting those that are already in use, the active role of end-users in shaping energy demand is becoming increasingly important to achieve sustainable development and carbon neutrality targets. However, the determinants of households' energy conservation behaviors are still being debated, making generalization difficult.

The analysis of both mainstream topics in discarded records and the detailed investigations of the selected ones suggests that the issue is highly multifaceted and there are no simple solutions. It can be deduced that more holistic and multi-scale approaches must be assumed to address it successfully.

Designing and targeting policies can be highly challenging, which is why many scholars suggest that additional research in the field is needed. The effectiveness (or not) of available policies and behavioral interventions to this end should also be considered, as well as the impact of pilot experiences, such as carbon-neutral districts, neighborhoods or communities. Institutional and community-based normative levels can indeed have a crucial role in shaping individual attitudes and behaviors, although these were out of the scope of the present review.

As for the building-user scale on which the review was focused, this article has outlined the most relevant studies that can be used for future research in the field to bridge the gap that is preventing both individual and collective efforts to achieve a rapid energy transition in the residential sector.

Overall, technical tools have largely proven to be ineffective alone. The same goes for the sole social (behavioral) sphere of individuals. This fact was claimed in more than 50 records included in the cluster about how to mobilize energy conservation practices and beyond. The most recent and promising studies then move in the direction of a combination of the two main domains. However, this does not ensure success and further studies should be performed to understand effective integration mechanisms, as suggested by at least four articles included in the cluster on research methods, gaps and frontiers.

In this regard, the few research projects that have tried to interlace several aspects still lack a wide implementation as they were mostly surveys taken on a case-by-case basis. This prevents academics and policymakers from making broader considerations: systematic or meta-studies could help knowledge advancement in this field.

In addition to methodological and general considerations, some interesting hints about specific aspects emerged.

Among the many considered factors and determinants, users' socioeconomic background seems to have a minor impact on the role of end-users in energy conservation. Rather, social norms and peers' comparison appear as strong drivers. This is understandable in the era of social media and networks, and both policies and energy app/device developers should consider it more carefully. In this regard, useful suggestions might come from the observation of "low energy" users, to better understand what, why and how they adopted virtuously to be potentially replicated by others. Instead, at present, very few articles focus on this.

Nonetheless, a consistent number of records (more than 14) indicate energy price as a strong determinant in shaping households' energy demand in the short term. When this is especially studied in relation to low-income families, it becomes an even more important driver. Unfortunately, when it comes to energy poverty—largely recurrent also in discarded records—social and subjective norms lose their positive relevance. Vulnerable consumers face an actual need to curtail their energy bills to face more vital expenses (e.g., food), at the expense of comfort and health. In these situations, public support for structural improvements to the dwelling energy performances is vital, but it also becomes crucial to increase awareness of the role that users' energy-saving actions can play. In other words, targeted feedback, and energy education (or literacy) can be more valuable than is deemed nowadays for such a specific share of the population.

Lastly, an interesting connection of the topic emerged with contextual factors that have strongly featured in the last five years, which are the COVID-19 pandemic and awareness raising toward climate change. Interestingly, both have had an impact on the way people use energy at home, in many cases positively affecting conservation attitudes. However, the effects related to the pandemic seem mostly to have disappeared with the end of the health crisis, but small seeds of change persist and deserve to be studied further. Environmental awareness and its complex socio-economical implications are also worth investigating more deeply, as, despite the general interest of citizens, the intention–action gap here is greater than elsewhere.

Author Contributions: Conceptualization, L.M. and J.G.; methodology, L.M. and J.G.; formal analysis, L.M.; investigation, L.M. and J.G.; writing—original draft preparation, L.M.; writing—review and editing, L.M. and J.G.; visualization, L.M.; supervision, J.G. All authors have read and agreed to the published version of the manuscript.

Funding: This research received no external funding.

Data Availability Statement: No new data were created or analyzed in this study. Data sharing is not applicable to this article.

Conflicts of Interest: The authors declare no conflict of interest.

References

1. United Nations. Transforming Our World: The 2030 Agenda for Sustainable Development (A/RES/70/1). Available online: https://www.un.org/en/development/desa/population/migration/generalassembly/docs/globalcompact/A_RES_70_1_E.pdf (accessed on 10 September 2023).
2. Lucon, O.; Ürge-Vorsatz, D.; Zain Ahmed, A.; Akbari, H.; Bertoldi, P.; Cabeza, L.F.; Eyre, N.; Gadgil, A.; Harvey, L.D.; Jiang, Y.; et al. Buildings. In *Climate Change 2014: Mitigation of Climate Change—Contribution of Working Group III to the Fifth Assessment Report of the Intergovernmental Panel on Climate Change*; Edenhofer, O., Pichs-Madruga, R., Sokona, Y., Farahani, E., Kadner, S., Seyboth, K., Adler, A., Baum, I., Brunner, S., Eickemeier, P., Eds.; Cambridge University Press: Cambridge, UK; New York, NY, USA, 2014; pp. 671–738.
3. Green Building: Improving the Lives of Billions by Helping to Achieve the UN Sustainable Development Goals. Available online: https://www.worldgbc.org/news-media/green-building-improving-lives-billions-helping-achieve-un-sustainable-development-goals (accessed on 4 July 2022).
4. Sustainable Development Goals. Available online: https://sdgs.un.org/goals (accessed on 14 July 2023).
5. Goubran, S. On the Role of Construction in Achieving the SDGs. *J. Sustain. Res.* **2019**, *1*, e190020. [CrossRef]
6. IEA. Buildings. Available online: https://www.iea.org/energy-system/buildings#overview (accessed on 4 March 2023).
7. BPIE. *Europe's Buildings under the Microscope. A Country-by-Country Review of the Energy Performance of Buildings*; Economidou, M., Ed.; Buildings Performance Institute Europe: Brussels, Belgium, 2011; ISBN 9789491143014.
8. IEA. Energy Efficiency 2022. Available online: https://www.iea.org/reports/energy-efficiency-2022 (accessed on 7 March 2023).
9. IEA. Tracking Clean Energy Progress 2023. Available online: https://www.iea.org/reports/tracking-clean-energy-progress-2023 (accessed on 7 March 2023).
10. BPIE (Buildings Performance Institute Europe). *Deep Renovation: Shifting from Exception to Standard Practice in EU Policy*; BPIE: Brussels, Belgium, 2021.
11. Levine, M.; Urge-Vorsatz, D.; Blok, K.; Geng, L.; Harvey, D.; Land, S.; Levermore, G.; Mongameli Mehlwana, A.; Mirasgedis, S.; Novikova, A.; et al. Residential and Commercial Buildings, Climate Change 2007: Mitigation, Contribution of Working Group III to the Fourth Assessment Report of the Intergovernmental Panel on Climate Change. In *Climate Change 2007 Mitigation of Climate Change*; Metz, B., Davidson, O.R., Bosch, P.R., Dave, R., Meyer, L.A., Eds.; Cambridge University Press: Cambridge, UK; New York, NY, USA, 2007; pp. 387–446, ISBN 9780511546013.
12. European Commission Energy Performance of Buildings Directive. Available online: https://ec.europa.eu/energy/topics/energy-efficiency/energy-efficient-buildings/energy-performance-buildings-directive_en (accessed on 7 April 2021).
13. GlobalABC/IEA/UNEP. *GlobalABC Roadmap for Buildings and Construction: Towards a Zero-Emission, Efficient and Resilient Buildings and Construction Sector*; GlobalABC/IEA/UNEP: Paris, France, 2020.
14. Artola, I.; Rademaekers, K.; Williams, R.; Yearwood, J. *Boosting Building Renovation: What Potential and Value for Europe?* European Union: Brussels, Belgium, 2016.
15. Clean Energy for All Europeans Package. Available online: https://energy.ec.europa.eu/topics/energy-strategy/clean-energy-all-europeans-package_en (accessed on 12 April 2023).
16. Medjdoub, B.; Larbi Chalal, M. Impact of Household Transitions on Domestic Energy Consumption and Its Applicability to Urban Energy Planning. *Front. Eng. Manag.* **2017**, *4*, 171. [CrossRef]
17. IEA. *Residential Behaviour Changes Lead to a Reduction in Heating and Cooling Energy Use by 2030*; IEA: Paris, France, 2022.
18. Energy Poverty in the EU. Available online: https://energy.ec.europa.eu/topics/markets-and-consumers/energy-consumer-rights/energy-poverty-eu_en (accessed on 10 May 2023).

19. BPIE (Buildings Performance Institute Europe). *How to Stay Warm and Save Energy: Insulation Opportunities in European Homes*; BPIE: Brussels, Belgium, 2023.
20. European Economic and Social Committee Energy Efficient Buildings Will Help Fight Energy Poverty. Available online: https://www.eesc.europa.eu/en/news-media/news/energy-efficient-buildings-will-help-fight-energy-poverty (accessed on 8 May 2023).
21. Fabbri, K. The Role of Building in the Reduction of Fuel Poverty. In *Urban Fuel Poverty*; Elsevier: Amsterdam, The Netherlands, 2019; pp. 63–103.
22. European Commission. *COM(2022) 230 Final—REPowerEU Plan*; European Commission: Brussels, Belgium, 2022.
23. European Commission. *COM(2022) 240 Final—EU "Save Energy"*; European Commission: Brussels, Belgium, 2022.
24. van den Brom, P.; Hansen, A.R.; Gram-Hanssen, K.; Meijer, A.; Visscher, H. Variances in Residential Heating Consumption—Importance of Building Characteristics and Occupants Analysed by Movers and Stayers. *Appl. Energy* **2019**, *250*, 713–728. [CrossRef]
25. Marchi, L.; Vodola, V.; Visconti, C.; Gaspari, J.; Antonini, E. Contribution of Individual Behavioural Change on Household Carbon Footprint. *E3S Web Conf.* **2021**, *263*, 05024. [CrossRef]
26. Sohre, A.; Schubert, I. The How and What of Bottom-up Governance to Change Household Energy Consumption Behaviour. *Energy Res. Soc. Sci.* **2022**, *89*, 102570. [CrossRef]
27. McKinsey & Company. Pathways to a Low-Carbon Economy. *Version 2 of the Global Greenhouse Gas Abatement Cost Curve*. Available online: https://www.mckinsey.com/~/media/mckinsey/dotcom/client_service/sustainability/cost%20curve%20pdfs/pathways_lowcarbon_economy_version2.ashx (accessed on 3 March 2021).
28. Gaspari, J.; Antonini, E.; Marchi, L.; Vodola, V. Energy Transition at Home: A Survey on the Data and Practices That Lead to a Change in Household Energy Behavior. *Sustainability* **2021**, *13*, 5268. [CrossRef]
29. Ehrhardt-Martinez, K.; Donnelly, K.A.; Laitner, J.A. *Advanced Metering Initiatives and Residential Feedback Programs: A Meta-Review for Household Electricity-Saving Opportunities*; ACEEE: Washington, DC, USA, 2010.
30. Joachain, H.; Klopfert, F. Smarter than Metering? Coupling Smart Meters and Complementary Currencies to Reinforce the Motivation of Households for Energy Savings. *Ecol. Econ.* **2014**, *105*, 89–96. [CrossRef]
31. Dalvi, S.D.; Chandrababu, D.; Kumar, V. Design and Development of Smart Electricity Meter—the Path towards Sustainable Globe. *Int. J. Ambient. Energy* **2016**, *37*, 348–353. [CrossRef]
32. Serrenho, T.; Zangheri, P.; Bertoldi, P. *Energy Feedback Systems: Evaluation of Meta-Studies on Energy Savings through Feedback*; JRC-IET: Ispra, Italy, 2015; ISBN 978-92-79-59778-7.
33. Martín-Martín, A.; Thelwall, M.; Orduna-Malea, E.; Delgado López-Cózar, E. Google Scholar, Microsoft Academic, Scopus, Dimensions, Web of Science, and OpenCitations' COCI: A Multidisciplinary Comparison of Coverage via Citations. *Scientometrics* **2021**, *126*, 871–906. [CrossRef]
34. Singh, V.K.; Singh, P.; Karmakar, M.; Leta, J.; Mayr, P. The Journal Coverage of Web of Science, Scopus and Dimensions: A Comparative Analysis. *Scientometrics* **2021**, *126*, 5113–5142. [CrossRef]
35. Pranckutė, R. Web of Science (WoS) and Scopus: The Titans of Bibliographic Information in Today's Academic World. *Publications* **2021**, *9*, 12. [CrossRef]
36. Rodgers, S.; Zhang, W. Evaluating Reliability of Google Scholar, Scopus, and Web of Science: A Study of Faculty in U.S. Advertising and Public Relations Programs. *J. Mass Commun. Educ.* **2022**, *77*, 292–307. [CrossRef]
37. Page, M.J.; McKenzie, J.E.; Bossuyt, P.M.; Boutron, I.; Hoffmann, T.C.; Mulrow, C.D.; Shamseer, L.; Tetzlaff, J.M.; Akl, E.A.; Brennan, S.E.; et al. The PRISMA 2020 Statement: An Updated Guideline for Reporting Systematic Reviews. *BMJ* **2021**, *372*, n71. [CrossRef] [PubMed]
38. VOSviewer. Available online: https://www.vosviewer.com/ (accessed on 2 August 2023).
39. Bouktif, S.; Ouni, A.; Lazarova-Molnar, S. Towards a Rigorous Consideration of Occupant Behaviours of Residential Households for Effective Electrical Energy Savings: An Overview. *Energies* **2022**, *15*, 1741. [CrossRef]
40. Karahan, E.E.; Göçer, Ö.; Göçer, K.; Boyacıoğlu, D. An Investigation of Occupant Energy-Saving Behavior in Vernacular Houses of Behramkale (Assos). *Sustainability* **2021**, *13*, 13476. [CrossRef]
41. Khani, S.M.R.; Haghighat, F.; Panchabikesan, K.; Ashouri, M. Extracting Energy-Related Knowledge from Mining Occupants' Behavioral Data in Residential Buildings. *J. Build. Eng.* **2021**, *39*, 102319. [CrossRef]
42. Nahiduzzaman, K.M.; Said Abdallah, A.; Moradzadeh, A.; Mohammadpour Shotorbani, A.; Hewage, K.; Sadiq, R. Impacts of Tariffs on Energy Conscious Behavior with Respect to Household Attributes in Saudi Arabia. *Energies* **2023**, *16*, 1458. [CrossRef]
43. Sloot, D.; Scheibehenne, B. Understanding the Financial Incentive Conundrum: A Meta-Analysis of the Effectiveness of Financial Incentive Interventions in Promoting Energy Conservation Behavior. *Renew. Sustain. Energy Rev.* **2022**, *168*, 112761. [CrossRef]
44. Al Mamun, A.; Hayat, N.; Mohiuddin, M.; Salameh, A.A.; Zainol, N.R. Modelling the Energy Conservation Behaviour among Chinese Households under the Premises of Value-Belief-Norm Theory. *Front. Energy Res.* **2022**, *10*, 1–14. [CrossRef]
45. Xu, Q.; Lu, Y.; Hwang, B.G.; Kua, H.W. Reducing Residential Energy Consumption through a Marketized Behavioral Intervention: The Approach of Household Energy Saving Option (HESO). *Energy Build.* **2021**, *232*, 110621. [CrossRef]
46. Wang, B.; Yuan, Z.; Liu, X.; Sun, Y.; Zhang, B.; Wang, Z. Electricity Price and Habits: Which Would Affect Household Electricity Consumption? *Energy Build.* **2021**, *240*, 110888. [CrossRef]

47. Deumling, R.; Poskanzer, D.; Meier, A. "Everyone Has a Peer in the Low User Tier": The Diversity of Low Residential Energy Users. *Energy Effic.* **2019**, *12*, 245–259. [CrossRef]
48. Matsumoto, S.; Sugeta, H. *Efficiency Investment and Curtailment Action*; Springer: Dordrecht, The Netherlands, 2022; Volume 83, ISBN 0123456789.
49. Never, B.; Kuhn, S.; Fuhrmann-Riebel, H.; Albert, J.R.; Gsell, S.; Jaramillo, M.; Sendaza, B. Energy Saving Behaviours of Middle Class Households in Ghana, Peru and the Philippines. *Energy Sustain. Dev.* **2022**, *68*, 170–181. [CrossRef]
50. Gajdzik, B.; Jaciow, M.; Wolniak, R.; Wolny, R.; Grebski, W.W. Energy Behaviors of Prosumers in Example of Polish Households. *Energies* **2023**, *16*, 3186. [CrossRef]
51. Wester, M. My Fault or Default—Household Behaviors Living in a Climate-Smart Building. *Buildings* **2022**, *12*, 245. [CrossRef]
52. Aydın, E.; Brounen, D.; Ergün, A. The Rebound Effect of Solar Panel Adoption: Evidence from Dutch Households. *Energy Econ.* **2023**, *120*, 106645. [CrossRef]
53. Moeller, S.; Bauer, A. Energy (in)Efficient Comfort Practices: How Building Retrofits Influence Energy Behaviours in Multi-Apartment Buildings. *Energy Policy* **2022**, *168*, 113123. [CrossRef]
54. Gaspar, R.; Antunes, D.; Faria, A.; Meiszner, A. Sufficiency before Efficiency: Consumers' Profiling and Barriers/Facilitators of Energy Efficient Behaviours. *J. Clean. Prod.* **2017**, *165*, 134–142. [CrossRef]
55. Yue, T.; Li, M.; Wang, Q.; Long, R.; Chen, H.; Liu, J.; Chen, F. Will Residents' Energy-Conservation Behavior Be Continued under Perception of Behavior Outcome? The Moderating Role of Attribution Style. *Resour. Conserv. Recycl.* **2023**, *189*, 106745. [CrossRef]
56. Zhang, Y.; Bai, X.; Mills, F.P.; Pezzey, J.C.V. Examining the Attitude-Behavior Gap in Residential Energy Use: Empirical Evidence from a Large-Scale Survey in Beijing, China. *J. Clean. Prod.* **2021**, *295*, 126510. [CrossRef]
57. Frederiks, E.R.; Stenner, K.; Hobman, E.V. Household Energy Use: Applying Behavioural Economics to Understand Consumer Decision-Making and Behaviour. *Renew. Sustain. Energy Rev.* **2015**, *41*, 1385–1394. [CrossRef]
58. Hori, S.; Kondo, K.; Nogata, D.; Ben, H. The Determinants of Household Energy-Saving Behavior: Survey and Comparison in Five Major Asian Cities. *Energy Policy* **2013**, *52*, 354–362. [CrossRef]
59. Belaïd, F.; Garcia, T. Understanding the Spectrum of Residential Energy-Saving Behaviours: French Evidence Using Disaggregated Data. *Energy Econ.* **2016**, *57*, 204–214. [CrossRef]
60. Stikvoort, B.; Juslin, P.; Bartusch, C. Good Things Come in Small Packages: Is There a Common Set of Motivators for Energy Behaviour? *Energy Effic.* **2018**, *11*, 1599–1615. [CrossRef]
61. Hagejärd, S.; Dokter, G.; Rahe, U.; Femenías, P. "It's Never Telling Me That I'm Good!" Household Experiences of Testing a Smart Home Energy Management System with a Personal Threshold on Energy Use in Sweden. *Energy Res. Soc. Sci.* **2023**, *98*, 103004. [CrossRef]
62. Jakučionytė-Skodienė, M.; Liobikienė, G. Changes in Energy Consumption and CO_2 Emissions in the Lithuanian Household Sector Caused by Environmental Awareness and Climate Change Policy. *Energy Policy* **2023**, *180*, 113687. [CrossRef]
63. Han, X.; Poblete-Cazenave, M.; Pelz, S.; Pachauri, S. Household Energy Service and Home Appliance Choices in Urban China. *Energy Sustain. Dev.* **2022**, *71*, 263–278. [CrossRef]
64. Owusu-Manu, D.G.; Quaigrain, R.A.; Edwards, D.J.; Hammond, M.; Hammond, M.; Roberts, C. Energy Conservation Literacy among Households in Sub-Sahara Africa. *Int. J. Energy Sect. Manag.* **2022**, *16*, 1130–1149. [CrossRef]
65. Duong, K.T.; Nguyen, T.T.; Do, T.A. The Relationship between the Factors Influencing Energy-Saving Behaviors in Households in Urban Vietnam. *Int. Energy J.* **2022**, *22*, 291–302.
66. Chen, C.-F.; Xu, X.; Adua, L.; Briggs, M.; Nelson, H. Exploring the Factors That Influence Energy Use Intensity across Low-, Middle-, and High-Income Households in the United States. *Energy Policy* **2022**, *168*, 113071. [CrossRef]
67. Xie, Y.; Noor, A.I.M. Factors Affecting Residential End-Use Energy: Multiple Regression Analysis Based on Buildings, Households, Lifestyles, and Equipment. *Buildings* **2022**, *12*, 538. [CrossRef]
68. Żywiołek, J.; Rosak-Szyrocka, J.; Mrowiec, M. Knowledge Management in Households about Energy Saving as Part of the Awareness of Sustainable Development. *Energies* **2021**, *14*, 8207. [CrossRef]
69. Wang, Q.C.; Chang, R.; Xu, Q.; Liu, X.; Jian, I.Y.; Ma, Y.T.; Wang, Y.X. The Impact of Personality Traits on Household Energy Conservation Behavioral Intentions—An Empirical Study Based on Theory of Planned Behavior in Xi'an. *Sustain. Energy Technol. Assess.* **2021**, *43*, 100949. [CrossRef]
70. Streimikiene, D. Use of Nudges for Promotion of Sustainable Energy Consumption in Households. *Contemp. Econ.* **2023**, *17*, 1–9. [CrossRef]
71. Sudarmaji, E.; Ambarwati, S.; Munira, M. Measurement of the Rebound Effect on Urban Household Energy Consumption Savings. *Int. J. Energy Econ. Policy* **2022**, *12*, 88–100. [CrossRef]
72. Williams, S.P.; Thondhlana, G.; Kua, H.W. Behavioural Interventions Yield Electricity Savings among High-Income Households in Johannesburg, South Africa. *Energy Environ.* **2022**, *33*, 1508–1523. [CrossRef]
73. Ruokamo, E.; Meriläinen, T.; Karhinen, S.; Räihä, J.; Suur-Uski, P.; Timonen, L.; Svento, R. The Effect of Information Nudges on Energy Saving: Observations from a Randomized Field Experiment in Finland. *Energy Policy* **2022**, *161*, 112731. [CrossRef]
74. Henry, M.L.; Ferraro, P.J.; Kontoleon, A. The Behavioural Effect of Electronic Home Energy Reports: Evidence from a Randomised Field Trial in the United States. *Energy Policy* **2019**, *132*, 1256–1261. [CrossRef]
75. Fraser, A. Success, Failure, and Information: How Households Respond to Energy Conservation Goals. *J. Assoc. Environ. Resour. Econ.* **2023**, *10*, 95–120. [CrossRef]

76. Bohdanowicz, Z.; Łopaciuk-Gonczaryk, B.; Kowalski, J.; Biele, C. Households' Electrical Energy Conservation and Management: An Ecological Break-through, or the Same Old Consumption-Growth Path? *Energies* **2021**, *14*, 6829. [CrossRef]
77. Vassileva, I.; Campillo, J. Increasing Energy Efficiency in Low-Income Households through Targeting Awareness and Behavioral Change. *Renew. Energy* **2014**, *67*, 59–63. [CrossRef]
78. Mukai, T.; Nishio, K.; Komatsu, H.; Sasaki, M. What Effect Does Feedback Have on Energy Conservation? Comparing Previous Household Usage, Neighbourhood Usage, and Social Norms in Japan. *Energy Res. Soc. Sci.* **2022**, *86*, 102430. [CrossRef]
79. Hafner, R.J.; Elmes, D.; Read, D. Promoting Behavioural Change to Reduce Thermal Energy Demand in Households: A Review. *Renew. Sustain. Energy Rev.* **2019**, *102*, 205–214. [CrossRef]
80. Schneider, P.T.; van de Rijt, A.; Boele, C.; Buskens, V. Are Visits of Dutch Energy Coach Volunteers Associated with a Reduction in Gas and Electricity Consumption? *Energy Effic.* **2023**, *16*, 42. [CrossRef]
81. Jorgensen, B.S.; Boulet, M.; Hoek, A.C. A Level-of-Analysis Issue in Resource Consumption and Environmental Behavior Research: A Theoretical and Empirical Contradiction. *J. Environ. Manag.* **2020**, *260*, 110154. [CrossRef]
82. Caldera, M.; Hussain, A.; Romano, S.; Re, V. Energy-Consumption Pattern-Detecting Technique for Household Appliances for Smart Home Platform. *Energies* **2023**, *16*, 824. [CrossRef]
83. Mostafa, N.A.; Grida, M.; Park, J.; Ramadan, H.S. A Sustainable User-Centered Application for Residential Energy Consumption Saving. *Sustain. Energy Technol. Assess.* **2022**, *53*, 102754. [CrossRef]
84. Hess, A.K.; Schubert, I.; Samuel, R.; Burger, P. Changing Routinized Household Energy Consumption Using the Example of Washing, Cooking, and Standby: A Randomized Controlled Field Experiment of Home Energy Advice. *Clean. Responsible Consum.* **2022**, *4*, 100052. [CrossRef]
85. Wood, G.; Day, R.; Creamer, E.; van der Horst, D.; Hussain, A.; Liu, S.; Shukla, A.; Iweka, O.; Gaterell, M.; Petridis, P.; et al. Sensors, Sense-Making and Sensitivities: UK Household Experiences with a Feedback Display on Energy Consumption and Indoor Environmental Conditions. *Energy Res. Soc. Sci.* **2019**, *55*, 93–105. [CrossRef]
86. LaMarche, J.; Cheney, K.; Akers, C.; Roth, K.; Sachs, O. Home Energy Displays: Consumer Adoption and Response. In *Home Energy Feedback Devices: Adoption and Analyses*; Nova Science Publishers: New York, NY, USA, 2014; pp. 1–41.
87. Pombeiro, H.; Santos, J.; Carreira, P.; Silva, C. Displaying Data Is Not Enough: Incorporating User Behavior Transformation in Domestic Reporting Systems. *Sustain. Cities Soc.* **2019**, *48*, 101451. [CrossRef]
88. Kendel, A.; Lazaric, N.; Maréchal, K. What Do People 'Learn by Looking' at Direct Feedback on Their Energy Consumption? Results of a Field Study in Southern France. *Energy Policy* **2017**, *108*, 593–605. [CrossRef]
89. Nilsson, A.; Wester, M.; Lazarevic, D.; Brandt, N. Smart Homes, Home Energy Management Systems and Real-Time Feedback: Lessons for Influencing Household Energy Consumption from a Swedish Field Study. *Energy Build.* **2018**, *179*, 15–25. [CrossRef]
90. Geelen, D.; Mugge, R.; Silvester, S.; Bulters, A. The Use of Apps to Promote Energy Saving: A Study of Smart Meter–Related Feedback in the Netherlands. *Energy Effic.* **2019**, *12*, 1635–1660. [CrossRef]
91. Straub, A.; Volmer, E. User's Perspective on Home Energy Management Systems. *Environments* **2018**, *5*, 126. [CrossRef]
92. Paneru, C.P.; Tarigan, A.K.M. Reviewing the Impacts of Smart Energy Applications on Energy Behaviours in Norwegian Households. *Renew. Sustain. Energy Rev.* **2023**, *183*, 113511. [CrossRef]
93. Garg, M.; Garg, V.; Srivastava, P.; Agarwal, R. Interface Design for Residential Energy Feedback, in the Indian Context. *Energy Inform.* **2023**, *6*, 12. [CrossRef]
94. Madsen, L.V.; Hansen, A.R.; Larsen, S.P.A.K. Embodied Competencies and Smart Home Technology in Energy Use: Three Ways Users Integrate Smart Heating Systems in Everyday Practices. *Energy Effic.* **2023**, *16*, 55. [CrossRef]
95. Kim, H.; Bilionis, I.; Karava, P.; Braun, J.E. Human Decision Making during Eco-Feedback Intervention in Smart and Connected Energy-Aware Communities. *Energy Build.* **2023**, *278*, 112627. [CrossRef]
96. Mataloto, B.; Ferreira, J.C.; Resende, R.P. Long Term Energy Savings Through User Behavior Modeling in Smart Homes. *IEEE Access* **2023**, *11*, 44544–44558. [CrossRef]
97. Al-Kababji, A.; Alsalemi, A.; Himeur, Y.; Fernandez, R.; Bensaali, F.; Amira, A.; Fetais, N. Interactive Visual Study for Residential Energy Consumption Data. *J. Clean. Prod.* **2022**, *366*, 132841. [CrossRef]
98. Chatzigeorgiou, I.M.; Andreou, G.T. A Systematic Review on Feedback Research for Residential Energy Behavior Change through Mobile and Web Interfaces. *Renew. Sustain. Energy Rev.* **2021**, *135*, 110187. [CrossRef]
99. Song, K.; Anderson, K.; Lee, S.H.; Raimi, K.T.; Sol Hart, P. Non-Invasive Behavioral Reference Group Categorization Considering Temporal Granularity and Aggregation Level of Energy Use Data. *Energies* **2020**, *13*, 3678. [CrossRef]
100. Csoknyai, T.; Legardeur, J.; Akle, A.A.; Horváth, M. Analysis of Energy Consumption Profiles in Residential Buildings and Impact Assessment of a Serious Game on Occupants' Behavior. *Energy Build.* **2019**, *196*, 1–20. [CrossRef]
101. Méndez, J.I.; Peffer, T.; Ponce, P.; Meier, A.; Molina, A. Empowering Saving Energy at Home through Serious Games on Thermostat Interfaces. *Energy Build.* **2022**, *263*, 21–26. [CrossRef]
102. Méndez, J.I.; Ponce, P.; Meier, A.; Peffer, T.; Mata, O.; Molina, A. Empower Saving Energy into Smart Communities Using Social Products with a Gamification Structure for Tailored Human–Machine Interfaces within Smart Homes. *Int. J. Interact. Des. Manuf.* **2023**, *17*, 1363–1387. [CrossRef]
103. Papineau, M.; Rivers, N. Experimental Evidence on Heat Loss Visualization and Personalized Information to Motivate Energy Savings. *J. Environ. Econ. Manag.* **2022**, *111*, 102558. [CrossRef]

104. Koasidis, K.; Marinakis, V.; Nikas, A.; Chira, K.; Flamos, A.; Doukas, H. Monetising Behavioural Change as a Policy Measure to Support Energy Management in the Residential Sector: A Case Study in Greece. *Energy Policy* **2022**, *161*, 112759. [CrossRef]
105. Johansson, T.; Pirouzfar, P. Sustainability Challenges in Energy Use Behaviour in Households: Comparative Review of Selected Survey-Based Publications from Developed and Developing Countries Zrównoważony Rozwój a Zużywanie Energii w Gospodarstwach Domowych: Przegląd Badań Odnoszących. *Probl. Ekorozwoju Probl. Sustain. Dev.* **2019**, *14*, 33–44.
106. Iwata, K.; Katayama, H.; Arimura, T.H. Do Households Misperceive the Benefits of Energy-Saving Actions? Evidence from a Japanese Household Survey. *Energy Sustain. Dev.* **2015**, *25*, 27–33. [CrossRef]
107. Long, C.; Mills, B.F.; Schleich, J. Characteristics or Culture? Determinants of Household Energy Use Behavior in Germany and the USA. *Energy Effic.* **2018**, *11*, 777–798. [CrossRef]
108. Zhao, X.; Cheng, H.; Zhao, H.; Jiang, L.; Xue, B. Survey on the Households' Energy-Saving Behaviors and Influencing Factors in the Rural Loess Hilly Region of China. *J. Clean. Prod.* **2019**, *230*, 547–556. [CrossRef]
109. Spandagos, C.; Yarime, M.; Baark, E.; Ng, T.L. "Triple Target" Policy Framework to Influence Household Energy Behavior: Satisfy, Strengthen, Include. *Appl. Energy* **2020**, *269*, 115117. [CrossRef]
110. Kumar, P.; Caggiano, H.; Shwom, R.; Felder, F.A.; Andrews, C.J. Saving from Home! How Income, Efficiency, and Curtailment Behaviors Shape Energy Consumption Dynamics in US Households? *Energy* **2023**, *271*, 126988. [CrossRef]
111. Lei, M.; Cai, W.; Liu, W.; Wang, C. The Heterogeneity in Energy Consumption Patterns and Home Appliance Purchasing Preferences across Urban Households in China. *Energy* **2022**, *253*, 124079. [CrossRef]
112. Sen, A.; Qiu, Y. Aggregate Household Behavior in Heating and Cooling Control Strategy and Energy-Efficient Appliance Adoption. *IEEE Trans. Eng. Manag.* **2022**, *69*, 682–696. [CrossRef]
113. Wang, X.; Fang, Y.; Cai, W.; Ding, C.; Xie, Y. Heating Demand with Heterogeneity in Residential Households in the Hot Summer and Cold Winter Climate Zone in China—A Quantile Regression Approach. *Energy* **2022**, *247*, 123462. [CrossRef]
114. Jareemit, D.; Limmeechokchai, B. Impact of Homeowner's Behaviours on Residential Energy Consumption in Bangkok, Thailand. *J. Build. Eng.* **2019**, *21*, 328–335. [CrossRef]
115. Pais-Magalhães, V.; Moutinho, V.; Robaina, M. Is an Ageing Population Impacting Energy Use in the European Union? Drivers, Lifestyles, and Consumption Patterns of Elderly Households. *Energy Res. Soc. Sci.* **2022**, *85*, 102443. [CrossRef]
116. Dai, M.; Chen, T. They Are Just Light Bulbs, Right? The Personality Antecedents of Household Energy-Saving Behavioral Intentions among Young Millennials and Gen z. *Int. J. Environ. Res. Public Health* **2021**, *18*, 13104. [CrossRef]
117. Lv, X.; Lin, K.; Chen, L.; Zhang, Y. Does Retirement Affect Household Energy Consumption Structure? Evidence from a Regression Discontinuity Design. *Sustainability* **2022**, *14*, 12347. [CrossRef]
118. Malama, A.; Makashini, L.; Abanda, H.; Ng'ombe, A.; Mudenda, P. A Comparative Analysis of Energy Usage and Energy Efficiency Behavior in Low-and High-Income Households: The Case of Kitwe, Zambia. *Resources* **2015**, *4*, 871–902. [CrossRef]
119. Podgornik, A.; Sucic, B.; Blazic, B. Effects of Customized Consumption Feedback on Energy Efficient Behaviour in Low-Income Households. *J. Clean. Prod.* **2016**, *130*, 25–34. [CrossRef]
120. Romero-Jordán, D.; del Río, P. Analysing the Drivers of the Efficiency of Households in Electricity Consumption. *Energy Policy* **2022**, *164*, 112828. [CrossRef]
121. Kaplowitz, S.A.; Boucher, J.L. Energy Conservation Behaviors, Climate Change Attitudes, Income, and Behavioral Plasticity. *Hum. Ecol.* **2022**, *50*, 937–952. [CrossRef]
122. Matthies, E.; Merten, M.J. High-Income Households—Damned to Consume or Free to Engage in High-Impact Energy-Saving Behaviours? *J. Environ. Psychol.* **2022**, *82*, 101829. [CrossRef]
123. Perez-Bezos, S.; Guerra-Santin, O.; Grijalba, O.; Hernandez-Minguillon, R.J. Occupants' Behavioural Diversity Regarding the Indoor Environment in Social Housing. Case Study in Northern Spain. *J. Build. Eng.* **2023**, *77*, 107290. [CrossRef]
124. Godoy-Shimizu, D.; Palmer, J.; Terry, N. What Can We Learn from the Household Electricity Survey? *Buildings* **2014**, *4*, 737–761. [CrossRef]
125. Shrestha, B.; Tiwari, S.R.; Bajracharya, S.B.; Keitsch, M.M.; Rijal, H.B. Review on the Importance of Gender Perspective in Household Energy-Saving Behavior and Energy Transition for Sustainability. *Energies* **2021**, *14*, 7571. [CrossRef]
126. Leslie, G.W.; Pourkhanali, A.; Roger, G. Electricity Consumption, Ethnic Origin and Religion. *Energy Econ.* **2022**, *114*, 106249. [CrossRef]
127. Taneja, S.; Mandys, F. Drivers of UK Household Energy Expenditure: Promoting Efficiency and Curbing Emissions. *Energy Policy* **2022**, *167*, 113042. [CrossRef]
128. Boudet, H.S.; Flora, J.A.; Armel, K.C. Clustering Household Energy-Saving Behaviours by Behavioural Attribute. *Energy Policy* **2016**, *92*, 444–454. [CrossRef]
129. Bedir, M.; Kara, E.C. Behavioral Patterns and Profiles of Electricity Consumption in Dutch Dwellings. *Energy Build.* **2017**, *150*, 339–352. [CrossRef]
130. Ben, H.; Steemers, K. Household Archetypes and Behavioural Patterns in UK Domestic Energy Use. *Energy Effic.* **2018**, *11*, 761–771. [CrossRef]
131. Ortiz, M.A.; Bluyssen, P.M. Developing Home Occupant Archetypes: First Results of Mixed-Methods Study to Understand Occupant Comfort Behaviours and Energy Use in Homes. *Build. Environ.* **2019**, *163*, 106331. [CrossRef]
132. Mi, L.; Xu, T.; Sun, Y.; Yang, H.; Wang, B.; Gan, X.; Qiao, L. Promoting Differentiated Energy Savings: Analysis of the Psychological Motivation of Households with Different Energy Consumption Levels. *Energy* **2021**, *218*, 119563. [CrossRef]

133. Akbari, S.; Haghighat, F. Occupancy and Occupant Activity Drivers of Energy Consumption in Residential Buildings. *Energy Build.* **2021**, *250*, 111303. [CrossRef]
134. Lu, Y.; Gao, W.; Kuroki, S.; Ge, J. Household Characteristics and Electricity End-Use under Dynamic Pricing in the Collective Housing Complex of a Japanese Smart Community. *J. Asian Archit. Build. Eng.* **2022**, *21*, 2564–2579. [CrossRef]
135. Heinrich, M.; Ruellan, M.; Oukhellou, L.; Samé, A.; Lévy, J.P. From Energy Behaviours to Lifestyles: Contribution of Behavioural Archetypes to the Description of Energy Consumption Patterns in the Residential Sector. *Energy Build.* **2022**, *269*, 112249. [CrossRef]
136. Chen, X.; Zanocco, C.; Flora, J.; Rajagopal, R. Constructing Dynamic Residential Energy Lifestyles Using Latent Dirichlet Allocation. *Appl. Energy* **2022**, *318*, 119109. [CrossRef]
137. Rosak-Szyrocka, J.; Żywiołek, J. Qualitative Analysis of Household Energy Awareness in Poland. *Energies* **2022**, *15*, 2279. [CrossRef]
138. Kopsakangas-Savolainen, M.; Juutinen, A. Energy Consumption and Savings: A Survey-Based Study of Finnish Households. *J. Environ. Econ. Policy* **2013**, *2*, 71–92. [CrossRef]
139. Sapci, O.; Considine, T. The Link between Environmental Attitudes and Energy Consumption Behavior. *J. Behav. Exp. Econ.* **2014**, *52*, 29–34. [CrossRef]
140. Zhang, J.; Ma, L.; Li, J. Why Low-Carbon Publicity Effect Limits? The Role of Heterogeneous Intention in Reducing Household Energy Consumption. *Energies* **2021**, *14*, 7634. [CrossRef]
141. Jaciow, M.; Rudawska, E.; Sagan, A.; Tkaczyk, J.; Wolny, R. The Influence of Environmental Awareness on Responsible Energy Consumption—The Case of Households in Poland. *Energies* **2022**, *15*, 5339. [CrossRef]
142. Chen, J.; Hassan, T.; Zhao, D. Does Awareness of Environmental Pollution Increase Electricity Consumption? A View from Household Survey of China. *Environ. Sci. Pollut. Res.* **2023**, *30*, 13532–13550. [CrossRef]
143. Du, J.; Pan, W. Cooling-Related Energy Uses and Adaptive Behaviors in High-Rise Residential Buildings in the Subtropical Climate: A Case Study in Hong Kong. *Build. Environ.* **2022**, *223*, 109456. [CrossRef]
144. Lam, T.W.L.; Tsui, Y.C.J.; Fok, L.; Cheung, L.T.O.; Tsang, E.P.K.; Lee, J.C.K. The Influences of Emotional Factors on Householders' Decarbonizing Cooling Behaviour in a Subtropical Metropolitan City: An Application of the Extended Theory of Planned Behaviour. *Sci. Total Environ.* **2022**, *807*, 150826. [CrossRef]
145. Ueno, T. Capturing Changes in Residential Occupant Behavior Due to Work from Home in Japan as a Consequence of the COVID-19 Pandemic. *Sustainability* **2022**, *14*, 2180. [CrossRef]
146. Khalil, M.A.; Fatmi, M.R. How Residential Energy Consumption Has Changed Due to COVID-19 Pandemic? An Agent-Based Model. *Sustain. Cities Soc.* **2022**, *81*, 103832. [CrossRef]
147. Balest, J.; Stawinoga, A.E. Social Practices and Energy Use at Home during the First Italian Lockdown Due to COVID-19. *Sustain. Cities Soc.* **2022**, *78*, 103536. [CrossRef]
148. Krishnan, D.; Kelly, S.; Kim, Y. A Meta-Analysis Review of Occupant Behaviour Models for Assessing Demand-Side Energy Consumption. *Energies* **2022**, *15*, 1219. [CrossRef]
149. Composto, J.W.; Weber, E.U. Effectiveness of Behavioural Interventions to Reduce Household Energy Demand: A Scoping Review. *Environ. Res. Lett.* **2022**, *17*, 063005. [CrossRef]
150. Hu, Z.; Wang, M.; Cheng, Z. Mapping the Knowledge Development and Trend of Household Energy Consumption. *Environ. Dev. Sustain.* **2022**, *24*, 6053–6071. [CrossRef]
151. Venkatesh, G.; Kjeang, A. Surveys as Means of Understanding User-Behaviour with Respect to Energy Usage in Households, Prior to Decision-Making to Promote Sustainability [Badania Ankietowe Jako Sposób Zrozumienia Zachowania Użytkowników w Odniesieniu Do Zużycia Energii w Gospodars. *Probl. Ekorozwoju* **2020**, *15*, 119–134. [CrossRef]
152. Kim, H.; Ham, S.; Promann, M.; Devarapalli, H.; Bihani, G.; Ringenberg, T.; Kwarteng, V.; Bilionis, I.; Braun, J.E.; Rayz, J.T.; et al. MySmartE—An Eco-Feedback and Gaming Platform to Promote Energy Conserving Thermostat-Adjustment Behaviors in Multi-Unit Residential Buildings. *Build. Environ.* **2022**, *221*, 109252. [CrossRef]
153. Andrade, S.H.M.S.; Contente, G.O.; Rodrigues, L.B.; Lima, L.X.; Vijaykumar, N.L.; Francês, C.R.L. Smart Home Tracking: A Smart Home Architecture for Smart Energy Consumption in a Residence with Multiple Users. *Wirel. Pers. Commun.* **2022**, *123*, 3241–3262. [CrossRef]
154. Bastida, L.; Cohen, J.J.; Kollmann, A.; Moya, A.; Reichl, J. Exploring the Role of ICT on Household Behavioural Energy Efficiency to Mitigate Global Warming. *Renew. Sustain. Energy Rev.* **2019**, *103*, 455–462. [CrossRef]
155. Qin, P.; Liu, M.; Su, L.; Fei, Y.; Tan-Soo, J.S. Electricity Consumption in the Digital Era: Micro Evidence from Chinese Households. *Resour. Conserv. Recycl.* **2022**, *182*, 106297. [CrossRef]
156. Pothitou, M.; Hanna, R.F.; Chalvatzis, K.J. ICT Entertainment Appliances' Impact on Domestic Electricity Consumption. *Renew. Sustain. Energy Rev.* **2017**, *69*, 843–853. [CrossRef]
157. Wang, S.; Lin, S.; Li, J. Exploring the Effects of Non-Cognitive and Emotional Factors on Household Electricity Saving Behavior. *Energy Policy* **2018**, *115*, 171–180. [CrossRef]
158. Xu, Q.; Hwang, B.G.; Lu, Y. Influencing Paths of the Behavior-Driven Household Energy-Saving Exploring the Influencing Paths of the Behavior-Driven Household Energy-Saving Intervention—Household Energy-Saving Option. *Sustain. Cities Soc.* **2021**, *71*, 102951. [CrossRef]

159. Yue, T.; Long, R.; Liu, J.; Liu, H.; Chen, H. Empirical Study on Households' Energy-Conservation Behavior of Jiangsu Province in China: The Role of Policies and Behavior Results. *Int. J. Environ. Res. Public Health* **2019**, *16*, 939. [CrossRef] [PubMed]
160. Qalati, S.A.; Qureshi, N.A.; Ostic, D.; Sulaiman, M.A.B.A. An Extension of the Theory of Planned Behavior to Understand Factors Influencing Pakistani Households' Energy-Saving Intentions and Behavior: A Mediated–Moderated Model. *Energy Effic.* **2022**, *15*, 40. [CrossRef] [PubMed]
161. Webb, D.; Soutar, G.N.; Mazzarol, T.; Saldaris, P. Self-Determination Theory and Consumer Behavioural Change: Evidence Fromahousehold Energy-Saving Behaviour Study. *J. Environ. Psychol.* **2013**, *35*, 59–66. [CrossRef]
162. Conradie, P.; Van Hove, S.; Pelka, S.; Karaliopoulos, M.; Anagnostopoulos, F.; Brugger, H.; Ponnet, K. Why Do People Turn down the Heat? Applying Behavioural Theories to Assess Reductions in Space Heating and Energy Consumption in Europe. *Energy Res. Soc. Sci.* **2023**, *100*, 103059. [CrossRef]
163. Le-Anh, T.; Nguyen, M.D.; Nguyen, T.T.; Duong, K.T. Energy Saving Intention and Behavior under Behavioral Reasoning Perspectives. *Energy Effic.* **2023**, *16*, 8. [CrossRef]
164. Nguyen, Q.N.; Hoang, T.H.L.; Mai, V.N. Applying the Theory of Planned Behavior to Analyze Household Energy-Saving Behavior. *Int. J. Energy Econ. Policy* **2022**, *12*, 287–293. [CrossRef]
165. Fatoki, O. Determinants of Household Energy Saving Behaviour: An Application of the Goal Framing Theory. *Int. J. Sustain. Dev. Plan.* **2022**, *17*, 1621–1628. [CrossRef]
166. Dietz, T.; Stern, P.C.; Weber, E.U. Reducing Carbon-Based Energy Consumption through Changes in Household Behavior. *Daedalus* **2013**, *142*, 78–89. [CrossRef]
167. Gram-Hanssen, K. Efficient Technologies or User Behaviour, Which Is the More Important When Reducing Households' Energy Consumption? *Energy Effic.* **2013**, *6*, 447–457. [CrossRef]
168. Savvidou, G.; Nykvist, B. Heat Demand in the Swedish Residential Building Stock—Pathways on Demand Reduction Potential Based on Socio-Technical Analysis. *Energy Policy* **2020**, *144*, 111679. [CrossRef]

energies

Review

Trends and Evolution of the GIS-Based Photovoltaic Potential Calculation

Sebastiano Anselmo [1] and Maria Ferrara [2,*]

[1] Interuniversity Department of Regional and Urban Studies and Planning, Politecnico di Torino, 10125 Turin, Italy; sebastiano.anselmo@polito.it
[2] Department of Energy, Politecnico di Torino, 10129 Turin, Italy
* Correspondence: maria.ferrara@polito.it; Tel.: +39-011-0904552

Abstract: In the current framework of energy transition, renewable energy production has gained a renewed relevance. A set of 75 papers was selected from the existing literature and critically analyzed to understand the main inputs and tools used to calculate solar energy and derive theoretical photovoltaic production based on geographic information systems (GISs). A heterogeneous scenario for solar energy estimation emerged from the analysis, with a prevalence of 2.5D tools—mainly ArcGIS and QGIS—whose calculation is refined chiefly by inputting weather data from databases. On the other hand, despite some minor changes, the formula for calculating the photovoltaic potential is widely acknowledged and includes solar energy, exploitable surface, performance ratio, and panel efficiency. While sectorial studies—targeting a specific component of the calculation—are sound, the comprehensive ones are generally problematic due to excessive simplification of some parts. Moreover, validation is often lacking or, when present, only partial. The research on the topic is in constant evolution, increasingly moving towards purely 3D models and refining the estimation to include the time component—both in terms of life cycle and variations between days and seasons.

Keywords: photovoltaic potential; solar radiation; GIS; renewable energy communities; zero-emission building; carbon-neutral cities; renewable energy directive

1. Introduction

In the current framework of energy transition, two concepts drafted at the European level have acquired relevance, i.e., Renewable Energy Communities and the Nearly Zero Energy Building. A common element of the two is the renewed attention given to renewable energy sources to meet—at least partially—energy needs.

The fit for 55 package [1]—aiming to bring the European Union in line with its decarbonization objectives by 2050—implied the revision of two key directives on energy matters, which are the Energy Performance of Buildings Directive and the Renewable Energy Directive. The proposal for the revision of the former [2] requires minimum performance requirements to be met in a twofold way. On the one hand, the building stock must be renovated according to the energy class, currently reported by the Energy Performance Certificates—carried out for a single building or building unit—but potentially estimated through alternative methodologies, as proposed in Deng et al. [3] or Anselmo et al. [4]. On the other hand, the share of energy produced from Renewable Energy Sources, which are the focus of the revised version of the Renewable Energy Directive [5], must be increased. Indeed, as claimed by Corgnati and Cattaneo [6], it is crucial that actions for the reduction of the energy demand are coupled with a coverage of the energy needs through renewable sources to meet the decarbonization goals set at the international level by both the European Union and the United Nations.

The Renewable Energy Directive also promoted the creation of Renewable Energy Communities, defined by Roberts as "collective cooperation[s] of an energy related activity around specific ownership, governance and a non-commercial purpose" [7]. These entities are meant to self-produce part of their energy needs and therefore require a dedicated study to assess their feasibility. In this framework, photovoltaic potential has generated specific interest, since PV panels are widely recognized as one of the most suitable renewable sources for urbanized areas thanks to the possibility of integrating them within roofs, as testified by worldwide data showing a rapid increase of installed capacity [8].

A particular commitment by the European Union to solar energy—especially integrated within roofs—is represented by the EU solar energy strategy [9], part of the REPowerEU plan. It targets an installed capacity of 320 GW of PV production by 2023 and 600 GW by 2030. It is articulated in three principal initiatives:

1. The European Solar Rooftops Initiative, promoting—and possibly setting requirements for—a growing installation of PV panels;
2. A large-scale skills partnership, developing an expert workforce in the sector;
3. An industry alliance, a forum of stakeholders that aims to maximize investment opportunities and diversify the supply chain.

Especially considering the first point, studies determining the most productive pitches are required. In this way, the path towards decarbonization would be sped up, ensuring that the potential is exploited as much as possible.

In this framework, Geographic Information Systems (GISs) are powerful tools for extensive modelling of solar potential, thanks to the possibility of integrating multiple geospatial data in complex analyses. GISs model entities—or objects—as made up of geometrical, alphanumerical, and relational components, allowing spatial analyses. These can be acquired for the scope of the research or gathered from public repositories and online geoportals or Volunteered Geographic Information–based websites such as OpenStreetMap. Archives are translated into GeoDataBases, storing data with the maximum accuracy possible, differently from traditional maps whose geographic information is subject to generalization and graphical errors [10]. The different data structures—further detailed in Section 4.1—can be processed to return multiple layers. The PV production layer, as specified in studies such as the ones cited in this review, can provide key information for Urban Energy Digital Twins, 3D platforms for the visualization and sharing of geodata. Digital Twins are virtual replicas of a physical object or system, allowing the simulation of the behavior of the object alone and in its environment for its whole lifecycle [11]. In this case, a Digital Twin makes it possible to identify the most productive pitches and simulate the productivity in different conditions.

2. Research Outline and Methodology

This study consists of a review of the use of three-dimensional GISs for the definition of the photovoltaic potential. The scientific literature to support the analysis was accessed through Scopus, an abstract and citation database curated by Elsevier, which lists papers from journals and conference proceedings from a wide set of disciplines. At first, the concepts "3D GIS" and "photovoltaic" were used as keywords. However, after having analyzed this first set of publications, it was chosen to also include the articles with "3D GIS" and "solar radiation" or "solar potential", because of the relevance this aspect—the one with the highest variability according to different simulations—has in the definition of the photovoltaic potential. The query—limiting the language to English only—returned a total of 125 papers. After the selection process, which excluded non-accessible and non-relevant papers, 75 of them were kept—equal to 60%—as shown in Table 1. The analysis was based on a set of parameters concerning the whole estimation process, from the required inputs to the scale of the analysis. Specific aspects—tackled only by a subset of the sample—were also deepened.

Table 1. List of papers considered in this study.

Reference	Author(s)	Year	Country	Scale
[12]	Achbab et al.	2022	Morocco	City
[13]	Agugiaro et al.	2011	Italy	District
[14]	Alam et al.	2012	Germany	District
[15]	Aleksandrowicz et al.	2020	Israel	District
[16]	Alomari et al.	2023	Jordan	Building
[17]	An et al.	2023	China	District
[18]	Beltran-Velamazan	2021	Spain	District
[19]	Bernabé et al.	2015	France	District
[20]	Biljecki et al.	2015	Monaco	District
[21]	Borfecchia et al.	2013	Italy	City
[22]	Borfecchia et al.	2014	Italy	City
[23]	Bremer et al.	2016	Austria	District
[24]	Carneiro et al.	2009	Switzerland	City
[25]	Catita et al.	2014	Portugal	Campus
[26]	Cheng et al.	2020	China	City
[27]	Chiabrando et al.	2017	Italy	Campus
[28]	Choi et al.	2011	USA	Campus
[29]	Chow et al.	2014	Canada	District
[30]	De Vries et al.	2020	Netherlands	District
[31]	Desthieux et al.	2018	Switzerland	Region
[32]	Dewanto et al.	2020	Taiwan	Campus
[33]	El-Bouzaidi et al.	2018	Morocco	District
[34]	Eldesoky et al.	2019	Italy	District
[35]	Esclapés et al.	2014	Spain	District
[36]	Fichera et al.	2018	Italy	District
[37]	Fijałkowska et al.	2022	Poland	Tram stop
[38]	Gawley et al.	2022	UK	District
[39]	Gergelova et al.	2020	Slovakia	District
[40]	Hafeez et al.	2015	Pakistan	District
[41]	Han et al.	2022	Taiwan	District
[42]	Harikesh et al.	2020	India	District
[43]	Helbich et al.	2013	Austria	District
[44]	Hippenstiel et al.	2012	USA	
[45]	Hofierka	2022	Slovakia	District
[46]	HosseiniHaghighi et al.	2022	Canada	City
[47]	Hubinský et al.	2023	Slovakia	District
[48]	Jakubiec et al.	2013	USA	City
[49]	Jately et al.	2021	Malta	Campus
[50]	Kazak et al.	2018	Poland	District
[51]	La Gennusa et al.	2011	Italy	City
[52]	Lagahit et al.	2019	Philippines	Building

Table 1. *Cont.*

Reference	Author(s)	Year	Country	Scale
[53]	Liang et al.	2014	USA	District
[54]	Liang et al.	2015	USA	City
[55]	Liang et al.	2017	China	District
[56]	Liang et al.	2020	China	City
[57]	Lindberg et al.	2015	Sweden	District
[58]	Liu et al.	2023	China	City
[59]	Lohani et al.	2018	India	Campus
[60]	Lu et al.	2022	Canada	City
[61]	Machete et al.	2018	Portugal	District
[62]	Mutani et al.	2018	Italy	City
[63]	Nakazato et al.	2021	Japan	District
[64]	Nakhaee et al.	2023	USA	District
[65]	Nex et al.	2013	Italy	City
[66]	Palliwal et al.	2021	Singapore	City
[67]	Pedrero et al.	2019	Spain	City
[68]	Peng et al.	2016	Hong Kong	Region
[69]	Prades-Gil et al.	2023	Spain	District
[70]	Prieto et al.	2019	Spain	City
[71]	Pružinec et al.	2022	Slovakia	District
[72]	Redweik et al.	2011	Portugal	Campus
[73]	Ren et al.	2022	Hong Kong	Bus stop
[74]	Ren et al.	2022	Hong Kong	Campus
[75]	Saadaoui et al.	2019	Morocco	City
[76]	Saran et al.	2015	India	Building
[77]	Singh et al.	2020	India	Campus
[78]	Soares et al.	2020	USA	District
[79]	Sun et al.	2019	China	Forest
[80]	Tara et al.	2021	Australia	District
[81]	Teofilo et al.	2021	Australia	Airport
[82]	Wate et al.	2015	India	Campus
[83]	Yan et al.	2023	China	City
[84]	Yoon et al.	2020	Korea	District
[85]	Zhang et al.	2019	China	Region
[86]	Zhu et al.	2022	Italy	District

Figure 1 reports the publication timeline, dividing the articles according to the keywords. With 3D GIS as the enabling technology—being therefore a keyword of all the research—the recurrence of the other keywords—namely photovoltaic (PV), solar radiation (SR), both (PV+SR), or solar potential (SP)—provides insights into the interest in the topic within the scientific community. It can be observed that the first articles were published in the early 2010s, a period characterized by different trends in Europe and worldwide. As described by IRENA [8], the installed solar capacity in Europe had a sharp increase from 2009 to 2012, followed by a stabilization that continued until 2019. On the other hand, photovoltaic production in Asia boomed from 2012 onward. Oceania and the Americas never had a boom in production values but only slight increases from 2011 onward. In this framework, there were several reasons for increased interest in the assessment of photovoltaic potential for both policymakers and industries. In Asia, it was necessary to conduct feasibility studies and analyses of the localization of photovoltaic panels to maximize the production, while in Europe similar research was required to rationalize the new installations. Indeed, several policies at international and national scales—such as the five *Conti Energia* (energy bills) in Italy, introduced in the Italian legislation with a Legislative Decree [87] following the European Directive [88]—were pushing the introduction of photovoltaic technologies across Europe, leading to alterations in the market. The subsidies favoring panel installations led to systems that did not receive enough solar energy to minimize the payback time and thus ensure profitability. A sharp increase in the publications per year was followed by a two-year plateau—2016 and 2017—and another increase in recent years, especially in 2020 and 2022, with the trend likely to be confirmed in 2023 as well.

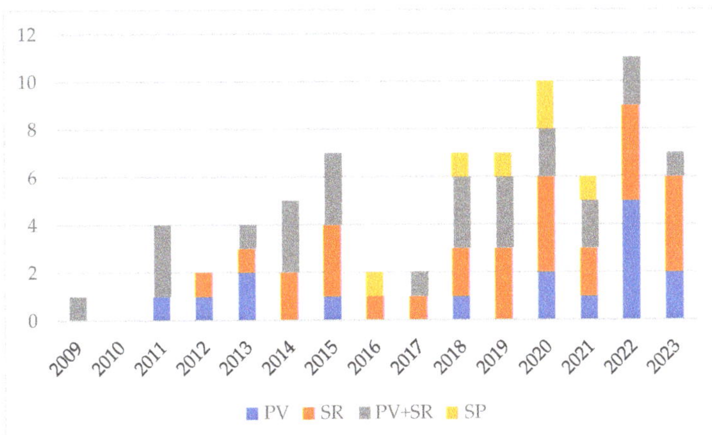

Figure 1. Publications per year, divided by topic.

Such widespread interest in PV is demonstrated by the countries in which the 69 studies were carried out. Indeed, Figure 2—plotting the spatial distribution of the analyzed papers—shows that despite the presence of several papers originating from Europe, the test studies are widespread across the world. Most of the case studies—35—are in Europe, followed by Asia, with 24. The two areas make up 85% of the total. However, considering countries instead of continents, the USA is the country with the third highest number of papers, after Italy and China. With further disaggregation, three of these studies (43%), namely [28,44,78], were carried out in Pennsylvania.

As for the type of paper, there is a clear prevalence of journal articles—52—over the conference proceedings, as shown in Figure 3. The conferences where most studies on the topic were presented belong to the International Society of Photogrammetry and Remote Sensing, oriented toward the advancements in Geomatics. On the contrary, there is a clear prevalence of energy-oriented journals compared to ones concerning remote sensing. In particular, seven papers [19,28,30,35,41,48,57] were published on Solar Energy and four [37,49,50,71] on Energies.

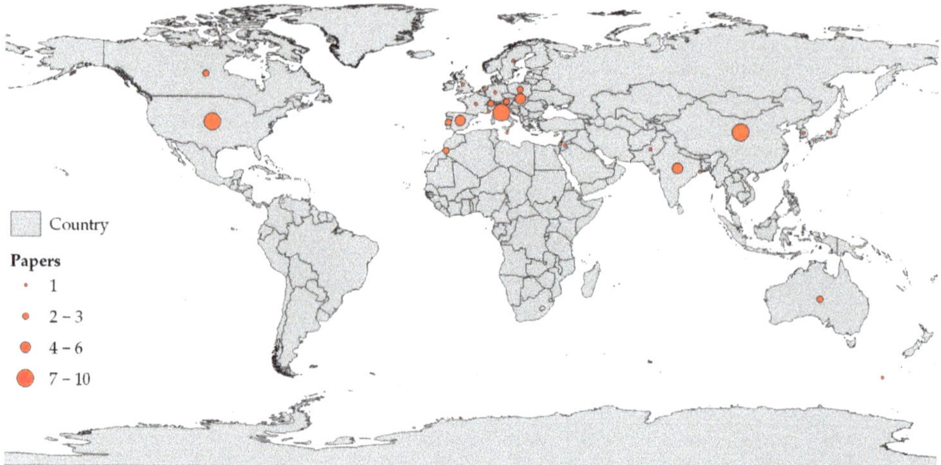

Figure 2. Spatial distribution of analyzed papers.

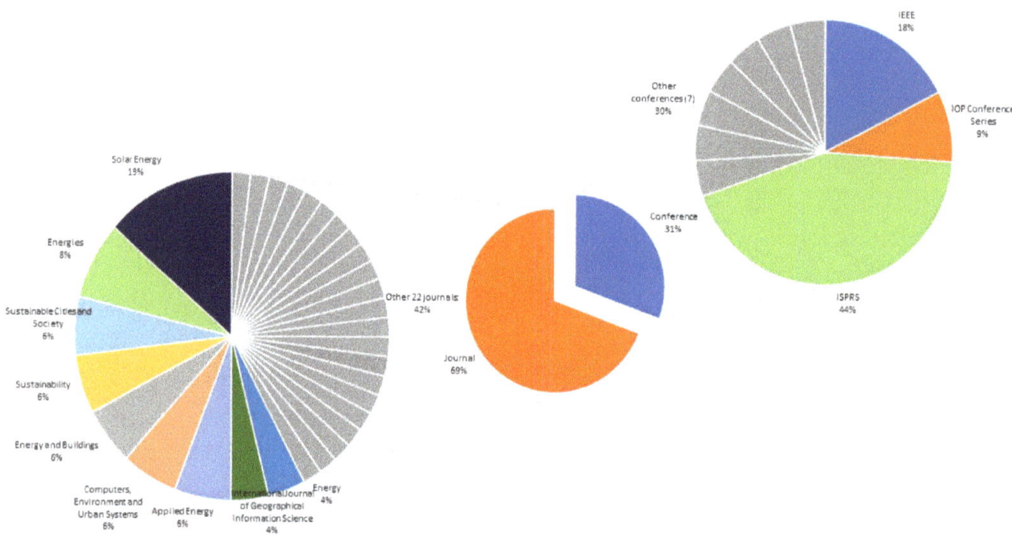

Figure 3. Conferences and journals of the selected papers.

A total of 257 authors contributed to the works included in this research, with 49 of them recurring at least twice. The most frequent author is Shuhua Zhang, currently at Xi'an University of Science and Technology (China). His name appears in seven studies, with one study citing him as the first author. Shuyi Li (Nanjing University) and Yongjun Sun (City University of Hong Kong) appear six and five times. They are most commonly last authors—generally recognized as project supervisors—three times each. Jianming Liang, from the Institute of Remote Sensing and Digital Earth of Beijing (China), is the first author—the one with the highest contribution to the research—in all of the four publications to which he contributed.

3. Results: Framework of the Studies

3.1. Scope

The first relevant aspect to be tackled is the scope of the work. The photovoltaic potential and the solar radiation are not only interesting per se but also considering other applications they could have when integrated with other datasets. Nevertheless, based on the specificity of the query, it is not surprising that most papers are focused on photovoltaic potential—22—and solar radiation—31. The other three thematic groups of the papers—according to which this section is structured—are shown in Figure 4.

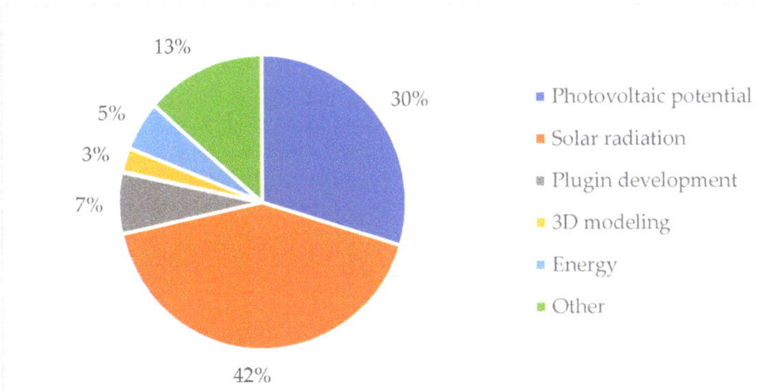

Figure 4. Classification of papers based on the scope.

The photovoltaic potential is mostly analyzed in terms of produced energy, with one case in which Mutani et al. [73] related the PV production to its costs, thus assessing the economic feasibility of a wide-scale installation of panels on bus stops and optimizing the system accordingly. Similar research was described in [39], which gives specific attention to public incentives to quantify the financial requirements for economically sustainable PV production.

However, an element to be highlighted emerged in the contrast between the two queries. Some articles, such as [21,35], claim in the title and abstract to have photovoltaic energy as the research objective but stop at the calculation of solar radiation. In this case, PV is the framework to contextualize the research, making the relevance of the topic explicit, but it may lead to misunderstandings. The same applies to other studies in the opposite situation, e.g., [17,26,74,81]. These papers limit their application to solar energy—in two cases adding the term "potential"—in the title and abstract, despite the fact that they also include the calculation of the photovoltaic potential. For the proper dissemination of the research, this emerges as an extremely critical element, requiring scholars to widen the research in order to collect all the pertinent information.

Given the interest generated about the topic, some authors proposed the development of tools to calculate solar radiation or coded new plugins for GISs. Liang et al. [56] extended r.sun—a processing tool of GRASS GIS integrated into QGIS (for a better understanding, see Section 4.3)—to three-dimensional models, implementing a tool to be used to evaluate the solar energy received not only by the roofs, as with traditional methods, but also the facades. SURFSUN3D, whose implementation is reported in [54], was realized by the same authors 5 years later, perfecting the previous study. Also, SOLIS, developed by Kazak and Świąder [50], is an integration of existing tools—in this case, the ones included in ArcGIS—that returns the radiation divided by roof. Another study [64] implemented a deep learning algorithm to estimate solar radiation starting from imagery and an Energy Plus Weather dataset. Finally, Pružinec and Ďuračiová [71] describe the realization of a point cloud solar radiation tool, which uses the ESRA model to calculate solar radiation by aggregating points into a voxel. From these tools, it is evident that there is a multiplicity of inputs that can be used to calculate solar energy, an aspect that is further addressed in the following paragraphs.

The previous paragraphs highlight the introduction of the vertical component in the estimation of solar radiation with GIS. For this, it is crucial to have a reliable 3D model. Hippenstiel et al. [44] studied the accuracy of a 3D model realized from LiDAR data, while [27] concerns 3D modeling starting from a DSM and structure using motion. In both cases, solar radiation estimation is used to evaluate the model, given the complexity it entails. In particular, the second study compares the outcomes of the solar radiation when using a validation DSM and the 3D model created during the research: the maximum discrepancy is 0.32%, with most cases not above 0.1%. Similarly, Beltran-Velamazan et al. [18] studied how to produce an accurate 3D model—to be reconverted in a DSM for solar processing—and the discrepancies between alternative techniques and datasets.

A group of papers is more related to energetic aspects, specifying some of the aspects that contribute to a comprehensive study of photovoltaic potential. Fijałkowska et al. [37] consider the impacts of shadows on the panel producibility, focusing on energy losses, while de Vries et al. [30] assess the energy yield by comparing alternative algorithms and defining a coefficient to increase the accuracy of high-precision evaluations. Energy aspects are also tackled when comparing the producible energy to the consumption, as in [69]. Thermal demand and retrofitting benefits are assessed together with the PV potential in [4], thus targeting the definition of a comprehensive energy model able to orient future energy policies. An energy model was produced also in [46,76], which have a specific focus on the capability of the CityGML data structure to integrate energy aspects. The former simply returns the energy balance between production and consumption, while the latter aims to define a comprehensive repository of energy-related data, grouped in geometric and non-geometric parameters.

In contrast, there is an extremely specific paper [51] tackling a single aspect of the calculation of PV producibility. By addressing the disturbances, it quantifies the exploitable surface to be used in the equations. It considers the shadowing factor, the tilt, and the type of covering to define correction factors to move from the whole roof surface to the actual area that can be considered for the installation of photovoltaic panels.

Finally, some works include the calculation of the solar energy or the PV potential as instrumental for research not related to energy production. The following are examples of this:

- Helbich et al. [43], looking for a correlation between the potential energy production and the housing market through the hedonic prices method;
- Hofierka [45], using solar energy predictions to correct land surface temperature estimations;
- Sun et al. [79], adopting solar radiation as one of the factors influencing the choice of adequate points to install a wireless sensor network;
- Palliwal et al. [66], requiring the quantification of solar energy to define the feasibility of urban farming in Singapore;
- Peng et al. [68], analyzing Urban Heat Islands and thus requiring solar energy as one of the parameters to predict the incidence of this phenomenon.

3.2. Scale of Analysis

The assessment of the photovoltaic potential can be conducted on different scales, according to the final purpose of the study. Figure 5 plots a classification of the selected papers based on the analyzed scale. A GIS is generally used for larger-scale analyses, making it different from Building Information Modeling, which is specifically designed for building scale. Moreover, raster datasets enable better visualization of extremely variable phenomena, thanks to the continuity of the data and the homogeneous level of detail, with the Ground Sampling Distance reaching values of a few centimeters.

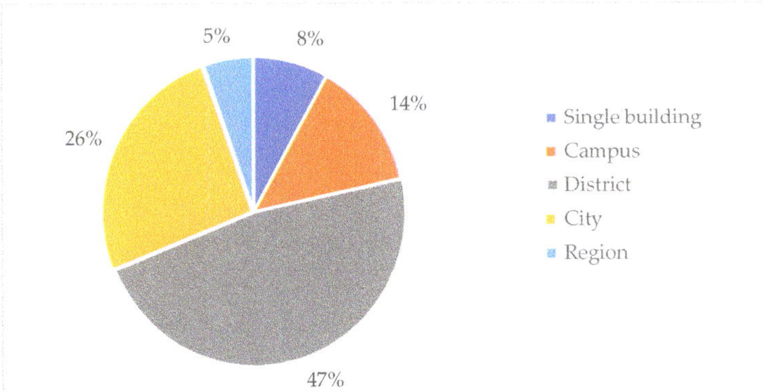

Figure 5. Classification of papers based on the analyzed scale.

Three of the analyzed studies [16,52,76] focused on single buildings, with two others [37,73] assessing public transport stops. While the attention was focused on single elements, the surrounding context was crucial. Indeed, when considering bus stops—at comparatively low height levels in an urban setting—particular attention was paid to the shadowing effect of buildings nearby, resulting in relevant constraints for PV production.

As mentioned previously, one of the possible uses of these studies is to identify the most suitable pitches and slopes for the installation of photovoltaic panels. This is especially relevant when looking at groups of buildings; the impact that each building has on the others is taken into account—at this scale, other elements of disturbance, such as trees, can be included in the model—together with slope and orientation in order to define, among a restricted set of alternatives, the most profitable. In this group, all the papers considering this scale [25,27,28,32,49,59,72,74,77,82] focused on a university campus. However, the number of buildings in this category of papers ranges from two and three in [49,72], respectively, to tens in [32]. This results in heterogeneous processing methods, with some studies being more similar to single-building studies and others closer to district-scale analyses. In all cases, the possibility to work on a campus makes it possible to have a controlled environment—without requiring the support of external bodies such as public institutions—and potentially to validate the results with devices whose installation does not require additional authorizations.

The most prominent group of papers—47%—belongs to studies conducted on a district scale. In this study, districts are defined as aggregations of buildings covering more than two blocks but not identifiable as a whole city nor controlled by the same body (such as university campuses). Still, it is to be recognized that a district can be very small or can be comparable to a small city. The former is the case of [17], whose square study areas have borders of 500 m only; in contrast, [53] takes into account a large part of Boston consisting of more than 22,000 buildings.

Another set of articles concerned studies conducted on a city scale. This is the most relevant for policymakers, who use this kind of research to identify the most suitable areas for the installation of photovoltaic panels on public land and buildings. Moreover, based on the current attention towards Positive Energy Districts and Renewable Energy Communities—whose creation is supported at the European level—it is crucial to define the most productive areas in order to identify where self-production would be sufficient to meet the entire energy demand. Despite the definition of city being agreed upon, in this context, some differences may emerge too, with Shanghai—whose population is more than 26 million—and Turin—with less than 1 million inhabitants—as case studies of [62,83], respectively.

A multi-scale approach is used by Aleksandrowicz et al. [15], starting their analyses from a pilot study area and extending it to the whole city of Tel Aviv. This explores the possibility to assess the methodology on a relatively small sample—to validate it more easily—and scale it up immediately to display its applicability.

Finally, a limited number of studies focused on larger scales. This is the case of [79], in which solar radiation is used to determine the most prominent nodes for installing wireless sensors in a whole forest. In three papers, namely [31,68,85], a whole region is taken into account. In this case, the heterogeneity is given not only by the different surfaces but also by the land use. Indeed, while districts and cities have an intrinsic urban character, despite minor differences in terms of form and morphology, regions can differ based on urban or rural characteristics. In particular, the three papers considered the Canton of Geneva (Switzerland), consisting of 300 km^2 with a central urban area—Geneva city—and rural surroundings; Kowloon Peninsula (Hong Kong), a 160 km^2 fully urbanized area with one of the highest population densities in the world; and the Tizinafu watershed (China), an arid mountain region with a negligible population.

3.3. Relationships between Scope and Scale

By comparing the different findings of this chapter—as shown in Table 2—it is possible to highlight some major elements. First, there is not a category for which all scales were covered, nor a scale on which the studies of all categories were conducted.

Table 2. Papers divided according to scale and scope.

	Single Building	Campus	District	City	Region
3D modeling	0	1	0	0	0
Energy	2	0	2	0	0
Plugin development	0	0	4	1	0
Other	1	0	4	3	2
Photovoltaic potential	2	3	9	8	0
Solar radiation	1	5	16	7	2

On average, there are 2.43 papers per cell, or 4.06 when excluding the zero values. As described in previous chapters, most papers—55% of the total—are conducted on a district or city scale and concern solar radiation or photovoltaic potential. In particular, solar radiation studies on a district scale are the most numerous—16.

It is then possible to read Table 2 by rows and columns, which are categories and scales, respectively. Looking at how studies of the same category are distributed on different scales, it emerges that 80% of the studies on plugin developments are carried out at a district scale. From this emerges the need to test the proposed methodology on a sufficiently wide and heterogeneous scale while at the same time keeping the process as compact as possible for efficiency. Once the developed tool passes the research phase, it would be possible to scale up the prototype. For both photovoltaic potential and solar radiation, around 20% of studies are conducted on single buildings or campuses, while district-scale analyses are the most recurrent (41% and 52%, respectively). From this, it emerges that, despite the use of validated methodologies, the need for high-consuming technologies—in terms of both computing power and required time—results in the scholars orienting their research toward medium scales.

Observing instead the incidence of the different scopes inside the same scale of analysis, it emerges that for all scales except for single buildings, photovoltaic potential and solar radiation are the most recurring topics, totaling, on average, 75% of studies. As mentioned, an exception is represented by the building scale, for which energy-related papers account for 33% of the total. Energy studies are especially relevant on this scale—with half of the research concerning single buildings—because of the need to deepen technical parameters, reducing the use of archetypes for estimating the aspects under investigation.

4. Results: Solar Energy Calculation

As mentioned, one of the most prominent aspects to be considered when calculating the photovoltaic potential is solar radiation. GIS-based calculations for the estimation of the incident solar energy are multiple and can be classified according to two main parameters, i.e., the required geographic dataset to be used as input—according to which they can be divided into purely three-dimensional or 2.5D methods—and the tool used for the estimation of solar radiation. Moreover, the latter requires in most cases the setting of specific parameters involving the average weather conditions, thus representing a further aspect to be explored.

4.1. Geographical Input

A GIS-based solar radiation estimation tool requires the input of a geographical dataset to establish a direct connection between the geographical location and the calculated solar energy received. In GIS software, two data structures can be used, i.e., vector and raster. The former is object-based, using the geometrical primitives to describe the spatial component and a table to report the known characteristics of each entity; the latter is field-based and represents surfaces as divided into regular tiles. A specific kind of raster is the Digital Surface Model (DSM), which describes the morphology of the observed area by reporting the height of the surfaces of both terrain and emerging objects (buildings, trees, etc.). DSMs are referred to as 2.5D because they include the vertical component—thus allowing the three-dimensional reconstruction—in a bidimensional representation. 3D models—known as city models—are vector representations in which polygons are not only planar but variously oriented in the space, potentially representing a solid when topological relationships between them allow the identification of a closed boundary.

Fifty-seven of the analyzed papers are based either on 3D models or DSMs. In particular, 27 used 3D models as input, and 30 used DSM. From this division—plotted in Figure 6—an aspect of the original query emerges. In the field, 2.5D analyses are widely recognized as included in the three-dimensional category, not considering only vectorial representations.

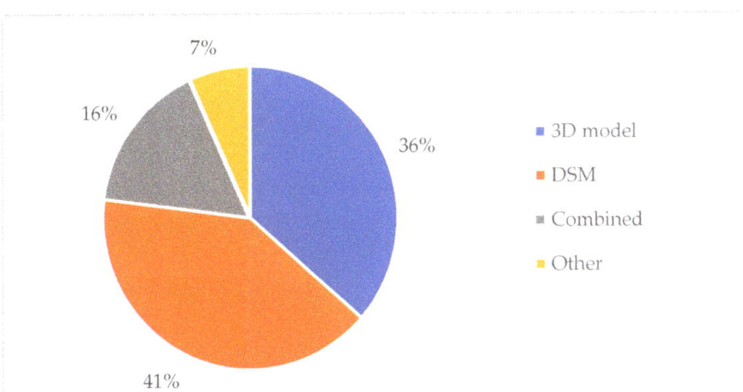

Figure 6. Classification of papers based on the geographical input.

In most cases, it is not mentioned how the city model was realized. Nevertheless, in three cases it is made explicit. Yan et al. [83] modelled the urban environment with a deep learning network that correlates high-resolution pictures for the segmentation and three-dimensional reconstruction of roofs where PV panels could be installed. The picture correlation process, used to create the city model, collimates homologous points—observable in two or more pictures—to create a point cloud, with which the actual model is created. LiDAR point clouds remove a step of the process—the collimation—thus simplifying the model production. Point clouds of this kind were used in [23,63], which keep the process hidden and do not make the workflow for the model production explicit. Nevertheless, two main methods can be used, i.e., feature extraction and machine learning. The former is based on radiometric information, and the latter takes advantage of recent evolutions in IT.

In twelve cases, LiDAR point clouds were also used for the creation of the DSM. However, it is particularly interesting to note that in [57] there is a combined use of LiDAR-derived DSM and fully 3D objects; the DSM is first used to calculate the shadows, with the output further enriched by the processing of the three-dimensional objects. In contrast, Bremer et al. [23] divided the shadowing objects into buildings and trees, using the DSM for the former and a 3D model for the latter. Other combinations of LiDAR-based DSMs and 3D models are reported in [37,72]. The former—like the previously mentioned study—cast the shadows through the city model, while the latter used the 3D model values to calibrate the algorithm to be later implemented.

Another combined application of 2.5D and 3D, with the source of the first not being explicated, is reported in [56], where the DSM is used to produce the solar radiation map, later draped on the 3D model through correction algorithms. In [41], the 3D model is used to derive a DSM for the solar energy analysis. Liang et al. [54] presented a merger between the two methodologies: the plugin developed in this study converts the 3D model into a DSM to be used for calculating solar radiation, then moves back to the three-dimensional representation by draping the results on the original model.

Just as in city models, DSMs can be produced from imagery through collimation. This is the case in [40], whose DSM was realized by LMKR (a Pakistan company) from stereo paired satellite imagery, and [42], whose pictures were acquired with a UAV. Both Teofilo et al. [81] and Achbab et al. [12] mention their use of stereo imagery for the realization of the products for the calculations, but they do not describe whether the dataset was acquired using a plane or from a satellite. The former explains that images were taken from world imagery in ArcGIS Pro, whose support describes the dataset as composed of mixed acquisitions. Other DSMs were produced by extruding bidimensional vectors, as in Eldesoky et al. [34].

Some papers considered alternative inputs for their calculation, especially when developing procedures alternative to the standard GIS tools—mostly depending on DSMs. This is the case of DeepRadiation [64], which processes 360° panoramic pictures for the punctual estimation of solar radiation. This implies the possibility of working with both acquired pictures and Google Street View imagery. The study in [59] is specifically focused on mobile-acquired pictures. A RedMi 3S prime camera was calibrated through MATLAB, and its pictures are processed by segmenting the content into sky and non-sky zones, thus creating masks to compute the shadowing.

Inputs can be evaluated in terms of spatial resolution. About half of the studies—32—made the resolution explicit, with approximately one-third having a value of 0.5 m, as in [29, 80]. While in five cases the resolution is higher than 0.5 m, nine other studies used models with a 1 m resolution. It is interesting to note that five studies use multi-resolution inputs, thus enabling an assessment of the proper value to use and the variations with different datasets. Finally, it is not surprising that two of the three studies mentioned in Section 3.2 as having small scales are the ones with lower spatial resolution: 30 m for [68] and 1 km for [85]. The difference between the two derives from the fact that—even on small scales—urban areas need to be more detailed compared to a natural region in which the principal disturbing element is the terrain.

As for city models, the Open Geospatial Consortium codified the CityGML data structure, which uses five Levels of Detail (LoDs). The concept of LoD involves spatial resolution, requiring increasing precision when moving up in the scale. Twelve papers describe the LoD of the used dataset: [53,66] are limited to LoD1, [45,46,82] to LoD2, and [47,76,84] to LoD3. Moreover, it is interesting to note that in three cases, multiple Levels of Detail are considered in the same study, making the study multi-resolution. In particular, [55] ranges from LoD1 to LoD4, showing the improvements of their ray-casting method while increasing the Level of Detail.

4.2. Weather Parameter Sources

Thirty-four papers specify the source of the additional information used to calibrate the solar radiation calculation. Indeed, in this elaboration, along with the latitude—which determines the angle of incidence of the sun rays—it is relevant to input additional parameters describing the average weather conditions. The cloudiness and the potential presence of other shading elements—such as high humidity or smog—determine the share of solar radiation actually striking the Earth's surface. According to this source, it is possible to classify the papers into six main groups based on database, observed data, online tools, algorithms, and combined sources. These are described in Figure 7.

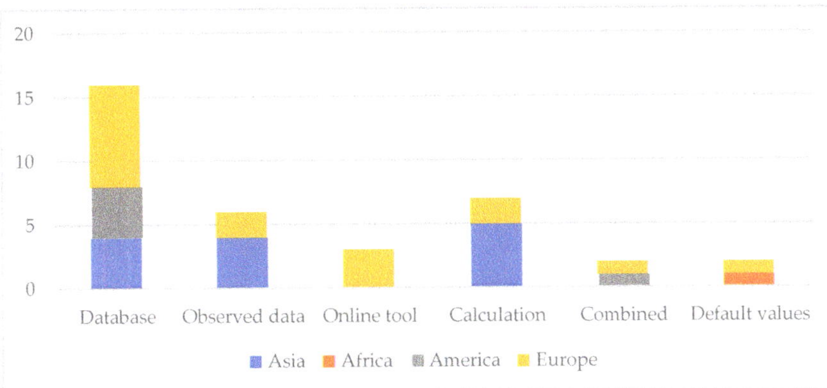

Figure 7. Classification of papers based on the weather input.

First, 16 papers accessed various databases to retrieve the necessary data, half of them in Europe. In three cases [24,26,31], Meteonorm was accessed. This is a software that includes historical time series of irradiation, temperature, humidity, precipitation, and wind, thus allowing not only a precise definition of the parameters required for the calculation but also the validation of the elaborated data. It reports the information gathered by five satellites and 8320 weather stations worldwide [89]; their even distribution makes it possible to have reliable data for both Switzerland—where [24,31] calculated solar radiation—and China—where the case study of An et al. [17] is located. Meteonorm also includes TMY3 (Typical Meteorological Year 3), data collection based on 1020 USA locations' data from a 1976–2005 time series. TMY3 was used in [28,41], whose test areas are in the USA and Taiwan, respectively. Therefore, it can be assumed that the accuracy of the latter is slightly lower. The NASA climatic dataset—accessed by default by ArcGIS—was used for [78], while the EPW (Energy Plus Weather) format—reporting climatic information calculated through Energy Plus, software specifically designed for energy calculations—is inputted in the tools used in [15,60,64]. Three other studies—namely [13,37,71]—calibrated their calculation based on SoDa (the Solar radiation Database), specifically devoted to solar energy. This database can be accessed on different time scales—from 1 minute to 1 month. It is based on Meteosat-11 images with a time series starting from February 2004 [90]. Mutani et al. [62] used the database of the regional agency for environmental protection (ARPA), taking advantage of the possibility of a site-specific time series, while the Japanese Meteorological Agency provided the cloud clover data (from 2015 to 2020) for [63]. National data were also provided for [26], on a 30-year time series. In another case [20], the historical irradiance data measured by a single station was downloaded from the International Weather for Energy dataset (published by the American Society of Heating, Refrigerating and Air-Conditioning Engineers), while cloud cover data from World Weather Online [91] were used in Zhu et al. [86], combined with an online tool.

Five studies used observed data gathered for the purpose of the research. In particular, Lindberg et al. [57] measured data concerning solar radiation and its components—direct and diffuse—while only the monthly average daily total radiation at the site was measured for [82]. Gergelova et al. [39] gathered data over 11 years (2009–2019) about the weather conditions (sunny, partly cloudy, cloudy, or rainy) of each month in a specific Slovakian municipality.

When it is not possible to access an existing database or directly measure the phenomenon, online tools that interpolate existing data can be used. The European Commission developed PVGIS (Photovoltaic Geographic Information System), which returns weather and irradiation data with different aggregations—monthly, daily, and hourly. Borfecchia et al. [21] used PVGIS values of monthly direct and diffuse irradiance, while Borfecchia et al. [22] used this information to compute the Linke turbidity factor. This online tool also allows the user to compute the Typical Meteorological Year data, aggregating values of the PVGIS-SARAH 2 database, ranging from 2005 to 2020. Hubinský et al. [47] calculated annual irradiance values in this way, thus reducing the possibility of having the results biased by extraordinary events. PVGIS is not the only energy-related online tool. Zhu et al. [86] used Sun Earth Tools to calculate the position of the sun. This website [92] includes several widgets to compute information related to solar energy and photovoltaic production, such as the payback time and the position of the sun, as well as basic geographic tools for calculating distances and visualizing GPS tracks. It can be observed that all three studies using online tools have European case studies; the reliability of institutional widgets for calculation—such as PVGIS—led multiple scholars to consider those values as usable, despite the spatial resolution not always being the best available.

Another option is to use algorithms to compute the necessary inputs. An example is provided by Helbich et al. [43], who used the SOLPOS code, developed by the National Renewable Energy Laboratory, to calculate the position of the sun. Similarly, PySolar is a Python library that can be used for calculations related to solar energy. It is able to compute the PV production, but it can also return intermediate values. In this case, it was used to calculate the sun's location. Another example of code is MODTRAN, whose 4.0 version was used in [16] to estimate atmospheric transmissivity and radiances.

Some authors combined various sources to calibrate their models, thus solving problems of data availability. This is the case of [85], which used MODIS satellite data for the cloudiness factor and algorithms from the literature e.g., [93,94], for the atmospheric transmittance in clear and cloudy sky conditions. In another case [48], some assumptions—on the implication of geometrical simplifications—are used together with an institutional database that improves the calculation of reflected energy.

Nevertheless, two studies [12,36] adopted the default values proposed by the calculation tool to compute solar radiation.

4.3. Tools for Solar Radiation Estimation

As previously mentioned, GIS 3D models are quite new compared to 2.5D datasets. Therefore, it is not surprising that out of the 67 papers that report the software used to compute solar radiation, nearly half of them—23—are based on widespread tools requiring a DSM for the principal geographical input, namely ArcGIS Area Solar Radiation and QGIS r.sun. Despite these being the most common ones, they are part of a wider set of tools, which are included in a complete review by Freitas et al. [95]. The subdivision of papers according to the software they used to calculate the incoming solar energy is plotted in Figure 8.

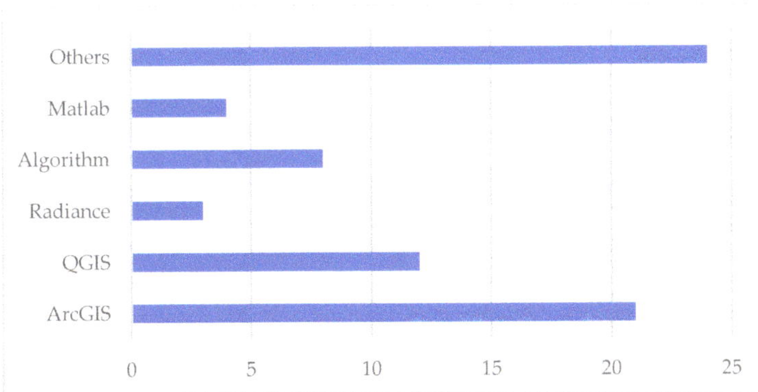

Figure 8. Classification of papers based on the tool used to compute solar radiation.

Fourteen papers used the ArcGIS Area Solar Radiation tool, based on the viewshed algorithm. It is part of the Spatial Analyst, calculating solar energy based on a DSM. Its parameters are grouped into two principal clusters:

- Topographic parameters: the "z factor" corrects biases deriving from the use of different scales in planar and vertical units; the "slope and aspect input type" establishes whether the elaboration assumes a planar surface or requires the calculation of orientation and inclination of the receiving surfaces; "calculation directions" refers to the viewshed calculation.
- Radiation parameters: zenith and azimuth divisions are instrumental for the definition of the sky map; the correlation between diffuse radiation and zenith angle is defined according to the "diffuse model type"; "diffuse ratio" quantifies the share of diffuse radiation over the global radiation; "transmissivity" is the fraction of radiation passing through the atmosphere.

It is possible to run the simulation in different time configurations, from single hours to whole years. ESRI, the developer of ArcGIS, has already developed two tools for the estimation of solar radiation on three-dimensional elements. Points Solar Radiation works on point clouds using the same inputs as Area Solar Radiation. The difference lies in the possibility of having as output not only a raster dataset—on which 3D facades are hidden—with solar radiation values but a cloud of 3D points enriched with solar energy information. Calculate Solar Radiation is a configuration of ArcGIS Pro that can be used by local governments to calculate solar energy on 3D city models; however, there is no documentation available. Two studies mention they used ArcGIS without specifying the tool. However, from the inputs they use, it can be assumed that both Hippenstiel et al. [44] and Nakazato et al. [63] used the three-dimensional configuration. Therefore, in total, thirteen studies used Area Solar Radiation, four used Calculate Solar Radiation, and two used Points Solar Radiation. Moreover, three studies can be added to this group, because they used ArcGIS in combination with other tools. Fijałkowska et al. [37] targeted a 3D extension of Area Solar Radiation output by casting shadows through the Sun Shadow Volume tool. The authors created a script through the ArcPy library by which they automatized the process, merging different data sources in a single output. Machete et al. [61] compared Area Solar Radiation with ECOTECT—a three-dimensional algorithm integrated in the Revit product family—concluding by highlighting the easier calibration of the latter, which reads Energy Plus Weather files to apply local weather conditions to the elaboration. Finally, as already mentioned in Section 3.1, one study [28] aimed at developing a 3D extension—alternative to Calculate Solar Radiation—for ArcGIS, named PV Analyst.

Ten works were carried out using r.sun, a GRASS GIS model that is informed by ESRA (European Solar Radiation Atlas) model. Its reliability is confirmed by its institutional use, being the basis of the PVGIS database [96]. It has been integrated into QGIS, and—differently from ArcGIS—it is also able to compute the reflected solar energy. r.sun works in two modes, calculating solar incidence and solar irradiance for a specific time or summing daily values of solar radiation. However, in the analyzed studies, there is a prevalence of uses in the first mode. Its parameters are more disaggregated compared to ArcGIS Area Solar Radiation—with which it shares the DSM as principal input—thus requiring more complete knowledge about the territory under analysis. It does not compute every time slope and aspect, whose maps can be calculated ex ante and inputted directly into the model. As for the Linke turbidity factor, the user can choose between inserting a single value or a raster reporting the variations in turbidity inside the study area. The same applies to the albedo. Further parameters quantify the beam and diffuse radiation, the horizon values for the shadowing calculation, the days of the year for which the radiation is calculated, the time step for the computation of daily radiation sums (in the second mode), and the solar constant. The multiple options it gives in terms of producible outputs lead to different usages in the analyzed studies. For instance, Agugiaro et al. [13] chose to focus on the direct component only because of the interest the authors had in shadowing, while Liang et al. [54] considered global radiation as a whole and Nex et al. [67] disaggregated it into basic components. Similarly to [28], Ref. [55] also proposed an extension of the r.sun calculation to 3D, though an add-in for city models—named v.sun—had been previously developed. This was used in [45] for the city of Košice (Slovakia), where a LoD2 city model was available.

Another QGIS plugin is UMEP (Urban Multi-scale Environmental Predictor). It also uses the DSM as the principal input, together with a climatic file in the Energy Plus Weather format, outputting a global radiation map. It was used in two cases in Spain, namely for the city of Irun in [67] and the city of Vitoria-Gasteiz in [70]. An UMEP plugin called SEBE—Solar Energy on Building Envelopes—which can also calculate irradiance from a DSM on vertical elements, was used by Aleksandrowicz et al. [15]. Nevertheless, its main limitation is the high computation time, resulting in the need to adopt ArcGIS Area Solar Radiation to calculate solar energy for the whole city of Tel Aviv.

The ESRA model was used not only for the r.sun implementation but also in a newly developed tool reported in [71]. Indeed, it proved to be sound in the calculation of direct and diffuse components of solar radiation and therefore dependable for the development of new instruments.

Five other studies are based on Radiance, a validated ray-tracing tool. In two cases [58,60], it was integrated into an environmental analysis program, Honeybee, while [17,48] rely on derived software, Daysim. The principal characteristic of Radiance is the possibility to customize the optical surface properties, such as reflectance and absorptivity. In contrast to the tools analyzed previously, it supports 3D models without the need for extensions. Therefore, it was integrated into 3D modelling software such as Rhinoceros (with the Ladybug tool) and Grasshopper (which takes advantage of a Honeybee plugin). Also, Blender (open-source software for 3D modelling)—in particular its add-on package VI-Suite—incorporates Radiance in the modules to calculate the sun path, shadows, and lighting analysis. Palliwal et al. [66] calculated irradiance values through the Climate-Based Daylight Modeling, starting from the lighting values.

Similarly, the RayMan model, used by Ren et al. [74], can process three-dimensional data by considering the sunshine duration and shadowing. It requires the input of basic meteorological data, like temperature and humidity, as well as the characterization of solid surfaces, mainly in terms of emissivity.

A wide set of papers concerned the definition and application of algorithms. While in five cases [24,25,57,72,73] it is specified that MATLAB was used to implement the algorithm, and in another case [85] a raster calculator was chosen, eight other studies [19,26,30,31,41,59,69,77] based on algorithms specify the formulas but not how they were applied to the test study from a software perspective. The development of specific algorithms allowed scholars to include or exclude distinct factors, thus improving the quality of the results or simplifying the calculation by limiting the inputs. In particular, in [19] the authors claim they aim to develop a simplified version of the solar energy calculation.

Finally, there are multiple studies that used other tools for estimating the solar energy striking the analyzed case studies:

- The paper by Nakhaee and Paydar [64] was specifically devoted to the elaboration of a tool to compute solar radiation through artificial intelligence. The authors used data from QGIS and Ladybug for training but then ran elaborations on their own.
- Esclapés et al. [35] used the open-source software gvSIG, designing with Java a specific code to be run in its environment.
- PySolar, which was mentioned in the previous paragraph because of the intermediate outputs it can provide, was used by Lagahit and Blanco [52]. It is based on data from the USA, and it can compute direct clear-sly solar irradiation.
- Saran et al. [76] used the SunCast application in Virtual Environment Software to calculate the solar intensity on wall surfaces, using a CityGML LoD2 model as geometrical input.

From this brief overview of the main processing tools, it is evident that scholars have widely recognized the limits of the most widespread processing tools—mainly the oversimplification of ArcGIS Area Solar Radiation compared to the need for extensive data to be inputted in GRASS GIS r.sun—thus exploring a wide set of possibilities for the calculation of the solar energy.

5. Results: Photovoltaic Potential Calculation

After the gathering of the necessary parameters, the actual photovoltaic potential can be calculated. Before addressing the equations used for the calculation, the following section is devoted to the understanding of the conditions that make the roofs exploitable for the calculation of the photovoltaic potential according to the scientific community. It is to be mentioned that data reported in this section pertain to a subsample of the analyzed papers, including only the buildings that concern the photovoltaic potential calculation, approximately one-third of the whole sample.

5.1. Pre-Filtering

In thirteen studies, the calculation of the photovoltaic potential was preceded by a selection of suitable roofs for the installation of the panels. A total of seven parameters—shown in Figure 9—was considered, with most studies defining the suitability based on at least two parameters.

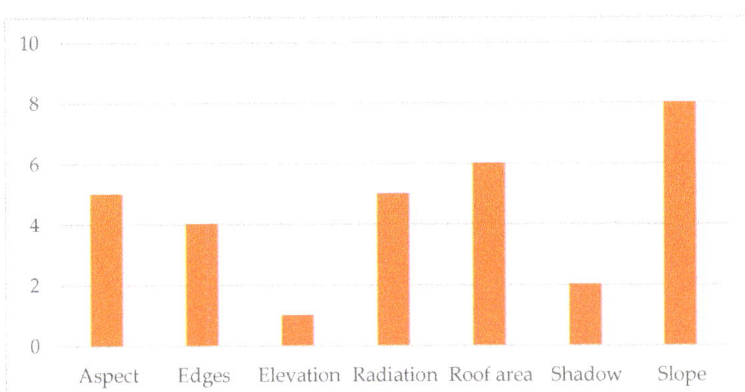

Figure 9. Classification of papers based on the filtering parameter.

In one case only [40], the elevation is taken into account, setting a threshold at 3 m. Despite not being justified in the paper, it can be assumed that this is meant to remove the areas below the first floors, thus the most shadowed ones.

Shadowing is another parameter for the pre-filtering. In Hafeez et al. [40], only the shadows deriving from the vegetation are considered. They are calculated through Ecotect software, inputting a 3D model realized with Google Sketchup according to an unsupervised classification of aerial pictures, for 10:30 AM and 3:30 PM of the solstices. By contrast, the authors of [12] calculated the shadows with the Hillshade tool in ArcGIS for every hour of the year 2020, considering as non-suitable all the areas that were shadowed in at least one elaboration.

When installing a photovoltaic panel, it is relevant to consider the surface. From this, it derives the dimensioning of the system and therefore the producibility; however, some constraints must be taken into account. First, installation and maintenance require the definition of buffer areas located on the edges, which should be removed from the theoretically useful area. It is generally accepted that this buffer is equal to 1 m, a value considered in [12,75,78]. In one study [75], this constraint is combined with the requirement of a minimum surface for panel installation, equal to 5 m^2. Indeed, this and five other papers assumed that it is necessary to have a minimum available surface to grant sufficient revenues for justifying the installation costs, therefore being able to grant a minimum production. Moreover, there is a minimum dimensioning of the panels. The value to be set as the threshold is not shared: another study [47] takes 5 m^2 as the reference value, one [50] requires an even smaller surface—2 m^2—and a third one [46] requires at least 40 m^2. For this purpose, it is interesting to note that Gawley et al. [38] set not only a minimum requirement but also a maximum, according to the values included in the British Energy Saving Trust sizing guide [97]. These return a producible output from 1 kWp—generally used in European studies—to 4 kWp—representing an average UK configuration.

Another aspect directly affecting the producibility is the radiation striking the surface. In five cases, this was used as a limiting element, with different thresholds, ranging from 700 kWh/m^2 [38] to 1355 kWh/m^2 [40]. This depends on the case study, with some places receiving less solar energy compared to others, e.g., 700 kWh/m^2 is set for Northern Ireland, while 1355 kWh/m^2 for Pakistan. Moreover, considering the direct correlation between the radiation striking the panel and the photovoltaic potential, it is also possible to assume that, when focusing on the economical aspect, places where the electricity price is lower will require a bigger surface to ensure a good payback time.

Finally, two elements correlated to solar radiation were tackled when defining the parameters for filtering, i.e., aspect and slope. Out of the five times that aspect is considered, slope is also taken into account. In the case of [12], there is no clear indication of the orientations and inclinations to exclude because of the use of a formula to calculate the losses—with 20% loss set as threshold.

As for the aspect, this filter aims to orient the panels towards the portion of the sky where the sun remains for longer, thus maximizing the production. South-facing panels are the most productive, but a degree of tolerance is generally accepted. This depends on the author and—as for solar radiation—on the geographical position of the case study. Two studies exclude only 45° from their calculations, namely [38,39]. The former aims to avoid the panels oriented from 315° to 0°, while the latter tends to exclude north-facing panels—oriented from 337.5° to 22.5°. These two studies pertain to Switzerland and Northern Ireland, respectively, thus needing a broader usable surface to receive the necessary amount of solar radiation. On the other hand, [40,75], studying areas in Morocco and Pakistan, can be more selective, excluding 270°. The Pakistani research is shifted westwards compared to the Moroccan research, with the two selecting portions from 135° to 225° and from 115° to 205°, respectively.

A similar distribution can be observed for the thresholds defined for the slope. All authors except for Gawley and McKenzie [38] define only an upper threshold. Gawley and McKenzie [38] include a range of permissible slopes from 30° to 60°, defining as optimal inclination the one between 39° and 40°. Three other studies [39,48,78] set as the maximal slope 60°, excluding vertical elements, while [47]—dividing roof planes in two inclination categories—does not exclude vertical walls, setting the threshold at 90°. Nevertheless, there are more restrictive studies, which limited the maximum slope at 5° [56] and 35° [40]. The former justifies this extreme selection based on the claim of loss of geometrical information caused by the rasterization of three-dimensional surfaces, thus using the DSM only for quantifying heights and not inclination.

5.2. Equation Used

Once having selected the buildings suitable for assessing the photovoltaic potential and gathered the necessary inputs, the actual calculation can be conducted. Several methods have been explored in the analyzed sample, thus requiring an overview of the different equations that can be used.

The most used equation is as follows:

$$E = I \times \eta \times PR \times A, \quad (1)$$

where E is the electricity output, I is the annual solar radiation expressed in kWh/m^2, η is the conversion efficiency of the panel—the ratio of solar energy converted into produced electricity, PR is the performance ratio—returning the efficiency of the supporting system, and A is the area for the installation of PV panels.

It emerges that the impact of the performance ratio and the cell efficiency is relevant. However, the values adopted for both are generally similar. PR is generally about 80%, a percentage considered in three studies [12,38,83]. Similar values, in the range of ±5%, are adopted in [46,58,67]. Pedrero et al. [67] calculated the PV potential based on two different PR values, 79% and 83%, much closer to each other compared to the values used by El-Bouzaidi et al. [33], 66% and 85%. This 19% difference would lead to considerable discrepancies between production values in comparable contexts, thus highlighting the need for a reliable supporting system. Low values, with a reduced producibility—resulting in higher payback times—were adopted also by Nakazato et al. [63].

As for the conversion efficiency, this strongly depends on the technology considered. For example, [12] is the only study that took into account the possibility of installing thin film modules, with an efficiency equal to 8%. By contrast, [47] considered technologies that reach an efficiency of 39%. In general, values range from 15% to 22%, standard values for polycrystalline and monocrystalline silicon cells, respectively. This is the case of [26,46,58,62,78,83]. However, solar cell efficiency tables, such as specified in [98], can be used for a better quantification of this parameter.

Despite Equation (1) being widely recognized as a simple yet reliable method to calculate the PV potential according to four macro-parameters, in some cases it is modified. Sometimes, the performance ratio is not taken into account, thus limiting the analysis at the level of the panel, not considering the actual electricity that can be injected into the grid. Soares et al. [78] calculate the photovoltaic potential together with its potential revenues, but the energy part can be isolated by removing the economic indicators. A formula emerges that considers a 15% conversion efficiency but no performance ratio, thus overestimating the theoretical benefits (≅+20%). Similarly, Cheng et al. [26] quantify the power production as the product of efficiency and solar irradiation—estimated in absolute values, not referred to as square meters, and corrected with sunshine duration. On the contrary, Liu et al. [58] elaborated on Equation (1) to include additional elements, thus providing a more stable assessment, using Equation (2).

$$E = \frac{I \times \eta \times PR \times A \times \sum_{i=1}^{N} (1-R)^{i-1}}{N} \quad (2)$$

where N is the life cycle of the panel, 25 years; and R is the yearly attenuation rate of the module, 0.7%.

Therefore, Equation (2)—differently from Equation (1)—takes into account not only the productivity at the moment of the installation but also the trends throughout the life cycle. This is particularly relevant, considering that in this study the decrease is equal to 0.7% and the life cycle is 25 years.

Another study [38] started from the assumption of panels with a known peak production. The concept of peak production involves the efficiency of the panel and its area, so it was sufficient to multiply this value by the solar energy and the performance ratio.

Nakazato et al. [63] focus on daily rather than on yearly production. They compute the productivity based on Equation (3).

$$E = \frac{I \times PR \times P}{D} b \tag{3}$$

where P quantifies the system capacity, 0.208 kW/m^2; and D is the solar radiation intensity under standard conditions.

Therefore, Equation (3) can also be considered as an extension of Equation (1) because of the inclusion of the efficiency and the area in the calculation of the system capacity. However, the radiation intensity is added.

Another set of papers used correction factors on solar radiation to compute the photovoltaic potential. Nex et al. [65] take into account an alpine area, where the impact of temperature on the module's efficiency is highly relevant. Therefore, the authors adopted correction factors based on temperature to estimate the losses derived from unfavorable conditions. By contrast, Ren et al. [74] consider a densely urbanized area, where shadows have a high impact. Therefore, they reduced the incoming solar radiation according to shading, considering the result as electricity that is potentially producible. Two other papers [35,48] correct the incoming solar energy according to geometric parameters, slope and orientation, whose relevance was cited in the previous paragraph, with the latter including shading too.

Hafeez et al. [40] took advantage of an automatized procedure embedded in PVsyst software, not explicating the equation used. Similarly, Choi et al. [28] integrated TRNSYS into ArcGIS, using the first to compute the photovoltaic potential by applying its own four- and five-parameter PV array performance models. Both include reference values at standard test conditions, the electron charge constant, the Boltzmann constant, voltage, and the module series resistance, with the latter adding the module shunt resistance. All these parameters, in contrast to basic technical information about the panel and the supporting systems, require an extensive analysis that is much more complex compared to the one required for Equation (1).

6. Discussion

6.1. Validation of Results

Twenty-six papers with different focuses were validated, with fourteen related to solar radiation and eight assessing the photovoltaic potential. They can be divided into two main groups: validation using observed data (11) and an elaboration on an alternative tool or software (12).

When observing the validation process of the observed data, the first element that emerges is that in many cases only a component of the final calculation is validated, for both PV potential and solar radiation analysis. While [48,65] used real PV production data—despite considering two validation sets vastly different in terms of dimension, with respectively 2 and 252 panels—other studies validated partial results only. For example, in [58] only the global radiation was measured. However, this could be considered sufficient when having reliable data on the technical parameters of the panel and the system. Further problems arise in [74], which validated the shadowing analysis only by comparing the simulation with acquired pictures. The same applies to [59,77], whose focus is on solar radiation.

In other studies, the results are validated by running the simulation on a platform or software and comparing the outcomes. In one case [13], the outcome was compared to PVGIS, with the problem of strong discrepancies in the resolution. Indeed, the study worked on a DSM with a resolution of up to 25 cm, while PVGIS calculates based on 1 km^2

cells. This is also the case in [38], which compared its 40 cm accurate results with PVGIS and two other online tools—PVWatts by NREL and a tool by the Energy Saving Trust—whose accuracy depends on data availability (with some results being interpolated). NREL data were also accessed for validation by Harikesh et al. [42]. Different scholars confronted their results with solar radiation tools that were mentioned previously—r.sun, Area Solar Radiation, and Ladybug—while some others evaluated the reliability of their models on a subsample. Prades-Gil et al. [69] realized the model of six test buildings in Design Builder—a 3D modelling software that uses EnergyPlus as a calculation engine—while Liang et al. [54] coupled the single-building validation—conducted with Ecotect—to a comprehensive test using SURFSUN3D. Another example of validation carried out in a twofold way is [31], which used an approach developed by the University of Geneva for validating the solar energy striking horizontal surfaces and PVsyst software for the analysis of vertical walls.

The three papers not included in the previous two groups validated their outcomes in the following ways:

- Hafeez and Atif [40] used the Marksim model.
- Zhang et al. [85] compared the results to previous studies on the same case study.
- Biljecki et al. [20] calculated the RMSE with a model whose reliability was assessed.

6.2. Limitations and Future Perspectives

Before the advent of GIS, different calculation methods for solar energy had already been validated. This is the case of the algorithm proposed by Zhang et al. [99], whose reliability is attested to by the more than 800 citations. This method starts from the estimation of the daily irradiation and—by calculating the hourly diffuse and beam radiation—makes it possible to estimate the cumulative daily or monthly radiation as well as how radiation changes throughout the day. However, this approach has several limitations. First, its level of detail is quite poor, requiring a multitude of acquisitions for refining the results in wider areas; in the case in which a single acquisition is performed, all buildings will have the same irradiation value. Another constraint is given by the impossibility of considering aspect, inclination, and shadowing, including the hours of light but not the sun's altitude or the sun's path. This makes this approach unsuitable to quantify the solar potential, not being able to identify the surfaces struck by the most by sun rays—i.e., vertical elements at high latitudes and roofs at low latitudes.

GIS technologies make it possible to include the third dimension—height—in the estimation of solar potential. First, it is possible to interpolate data, in order to model local changes with a lower number of measurements compared to traditional approaches. As explained in Section 4.2, this enabled the use of databases with different spatial resolutions. Second, accurate DSMs allow modelling of urban contexts to create a 3D reconstruction from which it is possible to cast shadows. These are relevant especially in contemporary cities, with the emergence of tall buildings shadowing wide neighboring areas. Nevertheless, these models, despite their accuracy, are simplified versions of complex three-dimensional objects, e.g., trees are returned as cylinders, not differentiating between trunk and crown. It is possible to perfect the calculation by using 3D models, as was done in some of the studies presented herein. According to the Level of Detail—in particular from LoD2, it is possible to better observe the variations in the objects rather than returning simple blocks. However, it was mentioned that pure three-dimensional tools are intensive in terms of both time and computational capacity.

It was previously mentioned that some studies over-simplify the calculation by using default or inaccurate values as input for the estimation. Some of the analyzed studies use highly accurate geographical input while lacking accuracy in the weather parameters. This makes it difficult to ensure satisfactory results, especially in areas characterized by extreme variability. For example, big cities with relevant urban heat island phenomena can show different cloudiness factors from district to district due to air moves caused by this phenomenon. Another simplification that leads to discrepancies is that—despite the diffuse ratio and transmissivity being computed by all tools—traditional GIS approaches compute the solar radiation in clear sky conditions, not being able to predict the variations caused by additional shadows derived from compact clouds. At the same time, it is likely that—for areas with a high average diffuse ratio—the software returns lower irradiation values compared to the real ones in clear sky days.

7. Conclusions

This research highlighted that progress is being made in the field of three-dimensional calculation of photovoltaic potential and its components—such as the solar radiation or the evaluation of suitable areas. With growing interest in the field following the issue of the EU Solar Strategy, this topic is likely to be further explored, increasing the body of studies and refining the outcomes.

In summary, it emerged that 3D GIS is an evolving field, thus highlighting the importance of intermediate technological progress—such as 2.5D models—in the calculation of the incident solar energy. Nevertheless, a wide variety of inputs, from three-dimensional models to point clouds, have been used by the authors of the sampled papers, resulting in a heterogeneous bunch of tools used for highly detailed solar radiation estimation, either directly embedded into widespread GIS software or running on dedicated programs or energy-oriented modelling tools. Moreover, some authors developed their own algorithms, implemented through programming code or working on pixels in GIS. In some cases, the calculation of photovoltaic potential is preceded not only by data gathering but also by a filtering phase in which suitable roofs are selected considering a wide set of parameters—the most relevant are slope and available roof area. The actual evaluation of photovoltaic potential takes place with different formulas, whose principal parameters are the cell efficiency—which varies according to the used technology—the solar radiation, and the available surface. Some authors added other elements, first the performance ratio of the supporting system, while others kept the necessary values only. Finally, validation is often conducted partially, not being able to prove the reliability of the model in all its components.

Therefore, from this review, a complex framework emerges. On the one hand, extensive studies have been conducted on the topic, but on the other, some critical elements often recur. For example, often the input parameters for the calculation of solar radiation are standard, kept by default or partially improved with information from low-precision databases. This is the case of the Linke turbidity factor, with values aggregated in big areas or estimated based on the urban or rural setting of the case study. Once having computed the solar radiation, there are not widely accepted criteria for defining suitable roofs, and the photovoltaic potential is then calculated based on assumed parameters, with few studies providing technical evidence on the reasons to adopt a specific conversion efficiency value; this does not take into account the degradation of the panel throughout its life cycle. Finally, the differences in productivity during the day are rarely considered, limiting the studies on the possible necessity of storage systems.

As for the geomatic aspects, data availability is often a discriminating factor, with spatial resolutions ranging from a few centimeters to one kilometer. Similarly, three-dimensional modelling must be improved, using at least LoD 2, to establish the slope and orientation of the pitches. Moreover, considering that not all energy experts are knowledgeable in the programming and GIS fields, it will be necessary to publish validated tools for solar energy calculations on 3D models.

A final aspect that was little considered is the difference in the output of alternative tools on a heterogeneous set of buildings. It is likely that different GIS tools will return similar values in some contexts and very different outcomes in others. Therefore, it is necessary to perform comparative analyses on the basis of a validated set of inputs, as discussed in Section 6.2.

In conclusion, future works should focus on the assessment of the differences between alternative tools in the same case study. A complete assessment of the reliability of the inputs should precede the work, making the outputs of tools and formulas comparable. Moreover, it is necessary to consider real values—of both solar radiation and photovoltaic production—for the validation of the results. Finally, the time component should be appropriately tackled, thus defining the sizing of the system and the requirements in terms of storage capacity and efficiency.

Author Contributions: Conceptualization, M.F.; Methodology, M.F. and S.A., investigation, S.A.; writing—original draft preparation, S.A.; writing—review and editing, M.F.; visualization, S.A.; supervision, M.F. All authors have read and agreed to the published version of the manuscript.

Funding: This paper reports part of the work developed within the project NODES, which has received funding from the MUR - M4C2 1.5 of PNRR with grant agreement no. ECS00000036. Maria Ferrara's activity was funded by Italian MUR within the PON "Ricerca e Innovazione" 2014–2020, Asse IV "Istruzione e ricerca per il recupero"—Azione IV.4—"Dottorati e contratti di ricerca su tematiche dell'innovazione" e Azione IV.6—"Contratti di ricerca su tematiche Green. Further, this work was developed in accordance with the framework agreement between the City of Turin and the Polytechnic of Turin, signed on 9th February 2023, for the realisation of pilot projects towards the implementation of a Digital Twin.

Data Availability Statement: No new data were created or analyzed in this study. Data sharing is not applicable to this article.

Conflicts of Interest: The authors declare no conflict of interest.

References

1. Council of the European Union Fit for 55. Available online: https://www.Consilium.Europa.Eu/En/Policies/Green-Deal/Fit-for-55-the-Eu-Plan-for-a-Green-Transition/ (accessed on 10 October 2023).
2. European Commission. Proposal for a Directive of the European Parliament and of the Council on the Energy Performance of Buildings. Recast. 2021.
3. Deng, Y.; Dewil, R.; Baeyens, J.; Ansart, R.; Zhang, H. The "Screening Index" to Select Building-Scale Heating Systems. *IOP Conf. Ser. Earth Environ. Sci.* **2020**, *586*, 012004. [CrossRef]
4. Anselmo, S.; Ferrara, M.; Corgnati, S.; Boccardo, P. Aerial Urban Observation to Enhance Energy Assessment and Planning towards Climate-Neutrality: A Pilot Application to the City of Turin. *Sustain. Cities Soc.* **2023**, *99*, 104938. [CrossRef]
5. Directive (EU) 2018/2001 of the European Parliament and of the Council of 11 December 2018 on the Promotion of the Use of Energy from Renewable Sources. Recast. 2018.
6. Corgnati, S.P.; Cattaneo, S. Efficientamento Energetico Degli Edifici e Soluzioni Innovative Di Decarbonizzazione. In *Le città a Impatto Climatico Zero: Strategie e Politiche*; Ministero delle Infrastrutture e della Mobilità Sostenibili: Rome, Italy, 2022; pp. 37–56.
7. Roberts, J.; Frieden, D.; d'Herbemont, S. Energy Community Definitions. 2019. Available online: https://main.compile-project.eu/wp-content/uploads/Explanatory-note-on-energy-community-definitions.pdf (accessed on 10 October 2023).
8. IRENA. IRENASTAT Online Data Query Tool. Available online: https://www.irena.org/Data/Downloads/IRENASTAT (accessed on 28 September 2023).
9. European Commission. Communication from the Commission to the European Parliament, the Council, the European Economic and Social Committee and the Committee of the Regions—EU Solar Energy Strategy 2022. Available online: https://eur-lex.europa.eu/legal-content/EN/TXT/?uri=COM%253A2022%253A221%253AFIN (accessed on 10 October 2023).
10. Amadio, G. *Introduzione Alla Geomatica*; Dario Flaccovio Editore: Palermo, Italy, 2012.
11. Minerva, R.; Lee, G.M.; Crespi, N. Digital Twin in the IoT Context: A Survey on Technical Features, Scenarios, and Architectural Models. *Proc. IEEE* **2020**, *108*, 1785–1824. [CrossRef]
12. Achbab, E.; Lambarki, R.; Rhinane, H.; Saifaoui, D. Estimation of Photovoltaic Potential at the Urban Level from 3d City Model (Solar Cadaster): Case of Casablanca City, Morocco. *Int. Arch. Photogramm. Remote Sens. Spat. Inf. Sci.* **2022**, *46*, 9–16. [CrossRef]
13. Agugiaro, G.; Remondino, F.; Stevanato, G.; De Filippi, R.; Furlanello, C. Estimation of Solar Radiation on Building Roofs in Mountainous Areas. *Int. Arch. Photogramm. Remote Sens. Spat. Inf. Sci.* **2013**, *38*, 155–160. [CrossRef]
14. Alam, N.; Coors, V.; Zlatanova, S.; Oosterom, P.J.M. Shadow Effect on Photovoltaic Potentiality Analysis Using 3D City Models. *Int. Arch. Photogramm. Remote Sens. Spat. Inf. Sci.* **2012**, *39*, 209–214. [CrossRef]
15. Aleksandrowicz, O.; Zur, S.; Lebendiger, Y.; Lerman, Y. Shade Maps for Prioritizing Municipal Microclimatic Action in Hot Climates: Learning from Tel Aviv-Yafo. *Sustain. Cities Soc.* **2020**, *53*, 101931. [CrossRef]
16. Alomari, K.; AlTal, B.; Ayadi, O. The Efficiency of Using of Solar Cells on the Multistory Residential Buildings in Jordan (Housing Building as a Case Study). *Int. J. Sustain. Dev. Plan.* **2023**, *18*, 1833–1839. [CrossRef]
17. An, Y.; Chen, T.; Shi, L.; Heng, C.K.; Fan, J. Solar Energy Potential Using GIS-Based Urban Residential Environmental Data: A Case Study of Shenzhen, China. *Sustain. Cities Soc.* **2023**, *93*, 104547. [CrossRef]

18. Beltran-Velamazan, C.; Monzón-Chavarrías, M.; López-Mesa, B. A Method for the Automated Construction of 3d Models of Cities and Neighborhoods from Official Cadaster Data for Solar Analysis. *Sustainability* **2021**, *13*, 6028. [CrossRef]
19. Bernabé, A.; Musy, M.; Andrieu, H.; Calmet, I. Radiative Properties of the Urban Fabric Derived from Surface Form Analysis: A Simplified Solar Balance Model. *Sol. Energy* **2015**, *122*, 156–168. [CrossRef]
20. Biljecki, F.; Heuvelink, G.B.M.; Ledoux, H.; Stoter, J. Propagation of Positional Error in 3D GIS: Estimation of the Solar Irradiation of Building Roofs. *Int. J. Geogr. Inf. Sci.* **2015**, *29*, 2269–2294. [CrossRef]
21. Borfecchia, F.; Pollino, M.; De Cecco, L.; Martini, S.; La Porta, L.; Marucci, A.; Caiaffa, E. *Integrated GIS and Remote Sensing Techniques to Support. PV Potential. Assessment of Roofs in Urban. Areas*; Springer: Berlin, Germany, 2013; Volume 7973, LNCS; ISBN 9783642396458.
22. Borfecchia, F.; Caiaffa, E.; Pollino, M.; De Cecco, L.; Martini, S.; La Porta, L.; Marucci, A. Remote Sensing and GIS in Planning Photovoltaic Potential of Urban Areas. *Eur. J. Remote Sens.* **2014**, *47*, 195–216. [CrossRef]
23. Bremer, M.; Mayr, A.; Wichmann, V.; Schmidtner, K.; Rutzinger, M. A New Multi-Scale 3D-GIS-Approach for the Assessment and Dissemination of Solar Income of Digital City Models. *Comput. Environ. Urban. Syst.* **2016**, *57*, 144–154. [CrossRef]
24. Carneiro, C.; Morello, E.; Desthieux, G. *Assessment of Solar Irradiance on the Urban Fabric for the Production of Renewable Energy Using Lidar Data and Image Processing Techniques*; Springer: Berlin, Germany, 2009; ISBN 9783642003172.
25. Catita, C.; Redweik, P.; Pereira, J.; Brito, M.C. Extending Solar Potential Analysis in Buildings to Vertical Facades. *Comput. Geosci.* **2014**, *66*, 1–12. [CrossRef]
26. Cheng, L.; Zhang, F.; Li, S.; Mao, J.; Xu, H.; Ju, W.; Liu, X.; Wu, J.; Min, K.; Zhang, X.; et al. Solar Energy Potential of Urban Buildings in 10 Cities of China. *Energy* **2020**, *196*, 117038. [CrossRef]
27. Chiabrando, F.; Danna, C.; Lingua, A.; Noardo, F.; Osello, A. 3D Roof Model Generation and Analysis Supporting Solar System Positioning. *Geomatica* **2017**, *71*, 137–153. [CrossRef]
28. Choi, Y.; Rayl, J.; Tammineedi, C.; Brownson, J.R.S. PV Analyst: Coupling ArcGIS with TRNSYS to Assess Distributed Photovoltaic Potential in Urban Areas. *Sol. Energy* **2011**, *85*, 2924–2939. [CrossRef]
29. Chow, A.; Fung, A.S.; Li, S. GIS Modeling of Solar Neighborhood Potential at a Fine Spatiotemporal Resolution. *Buildings* **2014**, *4*, 195–206. [CrossRef]
30. de Vries, T.N.C.; Bronkhorst, J.; Vermeer, M.; Donker, J.C.B.; Briels, S.A.; Ziar, H.; Zeman, M.; Isabella, O. A Quick-Scan Method to Assess Photovoltaic Rooftop Potential Based on Aerial Imagery and LiDAR. *Sol. Energy* **2020**, *209*, 96–107. [CrossRef]
31. Desthieux, G.; Carneiro, C.; Camponovo, R.; Ineichen, P.; Morello, E.; Boulmier, A.; Abdennadher, N.; Dervey, S.; Ellert, C. Solar Energy Potential Assessment on Rooftops and Facades in Large Built Environments Based on LiDAR Data, Image Processing, and Cloud Computing. Methodological Background, Application, and Validation in Geneva (Solar Cadaster). *Front. Built Environ.* **2018**, *4*, 14. [CrossRef]
32. Dewanto, B.G.; Novitasari, D.; Tan, Y.C.; Puruhito, D.D.; Fikriyadi, Z.A.; Aliyah, F. Application of Web 3D GIS to Display Urban Model and Solar Energy Analysis Using the Unmanned Aerial Vehicle (UAV) Data (Case Study: National Cheng Kung University Buildings). In *Proceedings of the IOP Conference Series: Earth and Environmental Science*; IOP Publishing: Bristol, UK, 2020; Volume 520.
33. El-Bouzaidi, R.D.; Rhinane, H.; Hilali, A.; El Hassan, E.A.; Maanan, M.; Saddiqi, O. Technical and Economical Photovoltaic Potential Assessment on Flat Roofs in Urban Area Case Study: Casablanca, Morocco. In Proceedings of the 2018 4th International Conference on Renewable Energies for Developing Countries (REDEC), Beirut, Lebanon, 1–2 November 2018.
34. Eldesoky, A.H.; Colaninno, N.; Morello, E. An Integrated, Agile Approach for Estimating Solar Radiation on Building Facades in Complex Urban Environments. *J. Phys. Conf. Ser.* **2019**, *1343*, 012015. [CrossRef]
35. Esclapés, J.; Ferreiro, I.; Piera, J.; Teller, J. A Method to Evaluate the Adaptability of Photovoltaic Energy on Urban Façades. *Sol. Energy* **2014**, *105*, 414–427. [CrossRef]
36. Fichera, A.; Gagliano, A.; Nocera, F.; Pagano, A.; Volpe, R.; Bisegna, F. Application of a Geographical Information System to Plan Energy Policy at a Neighborhood Scale. In Proceedings of the 2018 IEEE International Conference on Environment and Electrical Engineering and 2018 IEEE Industrial and Commercial Power Systems Europe (EEEIC/I and CPS Europe), Palermo, Italy, 12–15 June 2018.
37. Fijałkowska, A.; Waksmundzka, K.; Chmiel, J. Assessment of the Effectiveness of Photovoltaic Panels at Public Transport Stops: 3D Spatial Analysis as a Tool to Strengthen Decision Making. *Energies* **2022**, *15*, 1230. [CrossRef]
38. Gawley, D.; McKenzie, P. Investigating the Suitability of GIS and Remotely-Sensed Datasets for Photovoltaic Modelling on Building Rooftops. *Energy Build.* **2022**, *265*, 112083. [CrossRef]
39. Gergelova, M.B.; Kuzevicova, Z.; Labant, S.; Kuzevic, S.; Bobikova, D.; Mizak, J. Roof's Potential and Suitability for PV Systems Based on LiDAR: A Case Study of Komárno, Slovakia. *Sustainability* **2020**, *12*, 10018. [CrossRef]
40. Hafeez, S.; Atif, S. 3D Rooftop Photovoltaic Potential Calculation Using GIS Techniques: A Case Study of F-11 Sector Islamabad. In Proceedings of the 12th International Conference on Frontiers of Information Technology, Islamabad, Pakistan, 17–19 December 2014; pp. 187–192.
41. Han, J.-Y.; Chen, Y.-C.; Li, S.-Y. Utilising High-Fidelity 3D Building Model for Analysing the Rooftop Solar Photovoltaic Potential in Urban Areas. *Sol. Energy* **2022**, *235*, 187–199. [CrossRef]
42. Harikesh; Singh, S.; Shrivastava, V.; Sharma, V. *CityGML Based 3D Modeling of Urban Area Using Uav Dataset for Estimation of Solar Potential*; Springer: Berlin, Germany, 2020; Volume 51.

43. Helbich, M.; Jochem, A.; Mücke, W.; Höfle, B. Boosting the Predictive Accuracy of Urban Hedonic House Price Models through Airborne Laser Scanning. *Comput. Environ. Urban. Syst.* **2013**, *39*, 81–92. [CrossRef]
44. Hippenstiel, R.; Brownson, J.R.S. Computing Solar Energy Potential of Urban Areas Using Airborne Lidar and Orthoimagery. In Proceedings of the World Renewable Energy Forum, WREF 2012, Including World Renewable Energy Congress XII and Colorado Renewable Energy Society (CRES) Annual Conference, Denver, Colorado, 13–17 May 2012; Volume 3, pp. 2004–2008.
45. Hofierka, J. Assessing Land Surface Temperature in Urban Areas Using Open-Source Geospatial Tools. *Int. Arch. Photogramm. Remote Sens. Spat. Inf. Sci.* **2022**, *48*, 195–200. [CrossRef]
46. HosseiniHaghighi, S.; de Uribarri, P.M.Á.; Padsala, R.; Eicker, U. Characterizing and Structuring Urban GIS Data for Housing Stock Energy Modelling and Retrofitting. *Energy Build.* **2022**, *256*, 111706. [CrossRef]
47. Hubinský, T.; Hajtmanek, R.; Šeligová, A.; Legény, J.; Špaček, R. Potentials and Limits of Photovoltaic Systems Integration in Historic Urban Structures: The Case Study of Monument Reserve in Bratislava, Slovakia. *Sustainability* **2023**, *15*, 2299. [CrossRef]
48. Jakubiec, J.A.; Reinhart, C.F. A Method for Predicting City-Wide Electricity Gains from Photovoltaic Panels Based on LiDAR and GIS Data Combined with Hourly Daysim Simulations. *Sol. Energy* **2013**, *93*, 127–143. [CrossRef]
49. Jately, V.; Balaji, V.V.; Azzopardi, S.; Azzopardi, B. Design and Performance Investigation of a Pilot Micro-Grid in the Mediterranean: Mcast Case Study. *Energies* **2021**, *14*, 6846. [CrossRef]
50. Kazak, J.K.; Świąder, M. SOLIS—A Novel Decision Support Tool for the Assessment of Solar Radiation in ArcGIS. *Energies* **2018**, *11*, 2105. [CrossRef]
51. La Gennusa, M.; Lascari, G.; Rizzo, G.; Scaccianoce, G.; Sorrentino, G. A Model for Predicting the Potential Diffusion of Solar Energy Systems in Complex Urban Environments. *Energy Policy* **2011**, *39*, 5335–5343. [CrossRef]
52. Lagahit, M.L.R.; Blanco, A.C. Using Openly Sourced 3D Geographic Information Systems (GIS) in Determining the Photovoltaic Potential of Quezon City Hall in Terms of Received Direct Solar Radiation. *Int. Arch. Photogramm. Remote Sens. Spat. Inf. Sci.* **2019**, *42*, 263–270. [CrossRef]
53. Liang, J.; Gong, J.; Li, W.; Ibrahim, A.N. A Visualization-Oriented 3D Method for Efficient Computation of Urban Solar Radiation Based on 3D-2D Surface Mapping. *Int. J. Geogr. Inf. Sci.* **2014**, *28*, 780–798. [CrossRef]
54. Liang, J.; Gong, J.; Zhou, J.; Ibrahim, A.N.; Li, M. An Open-Source 3D Solar Radiation Model Integrated with a 3D Geographic Information System. *Environ. Model. Softw.* **2015**, *64*, 94–101. [CrossRef]
55. Liang, J.; Gong, J. A Sparse Voxel Octree-Based Framework for Computing Solar Radiation Using 3d City Models. *ISPRS Int. J. Geo-Inf.* **2017**, *6*, 106. [CrossRef]
56. Liang, J.; Gong, J.; Xie, X.; Sun, J. Solar3D: An Open-Source Tool for Estimating Solar Radiation in Urban Environments. *ISPRS Int. J. Geo-Inf.* **2020**, *9*, 524. [CrossRef]
57. Lindberg, F.; Jonsson, P.; Honjo, T.; Wästberg, D. Solar Energy on Building Envelopes—3D Modelling in a 2D Environment. *Sol. Energy* **2015**, *115*, 369–378. [CrossRef]
58. Liu, J.; Wu, Q.; Lin, Z.; Shi, H.; Wen, S.; Wu, Q.; Zhang, J.; Peng, C. A Novel Approach for Assessing Rooftop-and-Facade Solar Photovoltaic Potential in Rural Areas Using Three-Dimensional (3D) Building Models Constructed with GIS. *Energy* **2023**, *282*, 128920. [CrossRef]
59. Lohani, B.; Kumar Singh, S.; Choudhary, D.; Nagarajan, B. A New Approach to Determine Solar Potential Using Terrestrial Images. *Remote Sens. Lett.* **2018**, *9*, 636–645. [CrossRef]
60. Lu, Y.; McCarty, J.; Sezto, J.; Cheng, Z.; Martino, N.; Girling, C.; Rysanek, A.; Barron, S.; Matasci, G. Modeling the Shading Effect of Vancouver's Urban Tree Canopy in Relation to Neighborhood Variations. *Arboric. Urban For.* **2022**, *48*, 95–112. [CrossRef]
61. Machete, R.; Falcão, A.P.; Gomes, M.G.; Moret Rodrigues, A. The Use of 3D GIS to Analyse the Influence of Urban Context on Buildings' Solar Energy Potential. *Energy Build.* **2018**, *177*, 290–302. [CrossRef]
62. Mutani, G.; Casalengo, M.; Ramassotto, M.A. The Effect of Roof-Integrated Solar Technologies on the Energy Performance of Public Buildings: Tthe Case Study of the City of Turin (IT). In Proceedings of the INTELEC, International Telecommunications Energy Conference (Proceedings), Turino, Italy, 7–11 October 2018; Volume 2018.
63. Nakazato, R.; Yokogawa, S.; Ichikawa, H.; Ushirokawa, T.; Takeda, T. Compact Model for Estimating Area-Level Photovoltaic Power Generation on Facade Surface Using 3D City Model and Solar Radiation Simulation. In Proceedings of the 2021 IEEE PES Innovative Smart Grid Technologies—Asia (ISGT Asia), Brisbane, Australia, 5–8 December 2021.
64. Nakhaee, A.; Paydar, A. DeepRadiation: An Intelligent Augmented Reality Platform for Predicting Urban Energy Performance Just through 360 Panoramic Streetscape Images Utilizing Various Deep Learning Models. *Build. Simul.* **2023**, *16*, 499–510. [CrossRef]
65. Nex, F.; Remondino, F.; Agugiaro, G.; De Filippi, R.; Poletti, M.; Furlanello, C.; Menegon, S.; Dallago, G.; Fontanari, S. 3D SolarWeb: A Solar Cadaster in the Italian Alpine Landscape. *Int. Arch. Photogramm. Remote Sens. Spat. Inf. Sci.* **2013**, *40*, 173–178. [CrossRef]
66. Palliwal, A.; Song, S.; Tan, H.T.W.; Biljecki, F. 3D City Models for Urban Farming Site Identification in Buildings. *Comput. Environ. Urban. Syst.* **2021**, *86*, 101584. [CrossRef]
67. Pedrero, J.; Hermoso, N.; Hernández, P.; Munoz, I.; Arrizabalaga, E.; Mabe, L.; Prieto, I.; Izkara, J.L. Assessment of Urban-Scale Potential for Solar PV Generation and Consumption. In *Proceedings of the IOP Conference Series: Earth and Environmental Science*; IOP Publishing: Bristol, UK, 2019; Volume 323.

68. Peng, F.; Wong, M.S.; Nichol, J.E.; Chan, P.W. Historical GIS Data and Changes in Urban Morphological Parameters for the Analysis of Urban Heat Islands in Hong Kong. *Int. Arch. Photogramm. Remote Sens. Spat. Inf. Sci.* **2016**, *41*, 55–62. [CrossRef]
69. Prades-Gil, C.; Viana-Fons, J.D.; Masip, X.; Cazorla-Marín, A.; Gómez-Navarro, T. An Agile Heating and Cooling Energy Demand Model for Residential Buildings. Case Study in a Mediterranean City Residential Sector. *Renew. Sustain. Energy Rev.* **2023**, *175*, 113166. [CrossRef]
70. Prieto, I.; Izkara, J.L.; Usobiaga, E. The Application of Lidar Data for the Solar Potential Analysis Based on Urban 3D Model. *Remote Sens.* **2019**, *11*, 2348. [CrossRef]
71. Pružinec, F.; Ďuračiová, R. A Point-Cloud Solar Radiation Tool. *Energies* **2022**, *15*, 7018. [CrossRef]
72. Redweik, P.M.; Catita, C.; Brito, M.C. 3D Local Scale Solar Radiation Model Based on Urban LiDAR Data. *Int. Arch. Photogramm. Remote Sens. Spat. Inf. Sci.* **2012**, *38*, 265–269. [CrossRef]
73. Ren, H.; Ma, Z.; Ming Lun Fong, A.; Sun, Y. Optimal Deployment of Distributed Rooftop Photovoltaic Systems and Batteries for Achieving Net-Zero Energy of Electric Bus Transportation in High-Density Cities. *Appl. Energy* **2022**, *319*, 119274. [CrossRef]
74. Ren, H.; Xu, C.; Ma, Z.; Sun, Y. A Novel 3D-Geographic Information System and Deep Learning Integrated Approach for High-Accuracy Building Rooftop Solar Energy Potential Characterization of High-Density Cities. *Appl. Energy* **2022**, *306*, 117985. [CrossRef]
75. Saadaoui, H.; Ghennioui, A.; Ikken, B.; Rhinane, H.; Maanan, M. Using GIS and Photogrammetry for Assessing Solar Photovoltaic Potential on Flat Roofs in Urban Area Case of the City of Ben Guerir/Morocco. *Int. Arch. Photogramm. Remote Sens. Spat. Inf. Sci.* **2019**, *42*, 155–166. [CrossRef]
76. Saran, S.; Wate, P.; Srivastav, S.K.; Krishna Murthy, Y.V.N. CityGML at Semantic Level for Urban Energy Conservation Strategies. *Ann. GIS* **2015**, *21*, 27–41. [CrossRef]
77. Singh, S.K.; Lohani, B.; Arora, L.; Choudhary, D.; Nagarajan, B. A Visual-Inertial System to Determine Accurate Solar Insolation and Optimal PV Panel Orientation at a Point and over an Area. *Renew. Energy* **2020**, *154*, 223–238. [CrossRef]
78. Soares, P.; Bayrakci-Boz, M.; Brownson, J.R.S. GIS Information for Solar PV Energy Siting: A Case Study in the Borough of State College, PA, USA. In Proceedings of the Conference Record of the IEEE Photovoltaic Specialists Conference, Calgary, AB, Canada, 15 June–21 August 2020; Volume 2020, pp. 1749–1753.
79. Sun, D.; Tan, Y.; Zhang, S. Research and Application of GIS in Wisdom Forestry Wireless Sensor Networks Node Location Selection. In Proceedings of the 2019 IEEE 4th Advanced Information Technology, Electronic and Automation Control Conference (IAEAC), Chengdu, China, 20–22 December 2019; pp. 232–235.
80. Tara, A.; Patuano, A.; Lawson, G. Between 2d and 3d: Studying Structural Complexity of Urban Fabric Using Voxels and Lidar-Derived Dsms. *Fractal Fract.* **2021**, *5*, 227. [CrossRef]
81. Teofilo, A.; Sun, Q.C.; Radosevic, N.; Tao, Y.; Iringan, J.; Liu, C. Investigating Potential Rooftop Solar Energy Generated by Leased Federal Airports in Australia: Framework and Implications. *J. Build. Eng.* **2021**, *41*, 102390. [CrossRef]
82. Wate, P.; Saran, S. Implementation of CityGML Energy Application Domain Extension (ADE) for Integration of Urban Solar Potential Indicators Using Object-Oriented Modelling Approach. *Geocarto Int.* **2015**, *30*, 1144–1162. [CrossRef]
83. Yan, L.; Zhu, R.; Kwan, M.-P.; Luo, W.; Wang, D.; Zhang, S.; Wong, M.S.; You, L.; Yang, B.; Chen, B.; et al. Estimation of Urban-Scale Photovoltaic Potential: A Deep Learning-Based Approach for Constructing Three-Dimensional Building Models from Optical Remote Sensing Imagery Imagery. *Sustain. Cities Soc.* **2023**, *93*, 104515. [CrossRef]
84. Yoon, D.H.; Song, J.H.; Koh, J.H. Estimation of Solar Radiation Potential in the Urban Buildings Using CIE Sky Model and Ray-Tracing. *J. Korean Soc. Surv. Geod. Photogramm. Cartogr.* **2020**, *38*, 141–151. [CrossRef]
85. Zhang, S.; Li, X.; She, J.; Peng, X. Assimilating Remote Sensing Data into GIS-Based All Sky Solar Radiation Modeling for Mountain Terrain. *Remote Sens. Environ.* **2019**, *231*, 111239. [CrossRef]
86. Zhu, R.; Cheng, C.; Santi, P.; Chen, M.; Zhang, X.; Mazzarello, M.; Wong, M.S.; Ratti, C. Optimization of Photovoltaic Provision in a Three-Dimensional City Using Real-Time Electricity Demand. *Appl. Energy* **2022**, *316*, 119042. [CrossRef]
87. Decreto Legislativo 387/2003 2003. Available online: https://www.parlamento.it/parlam/leggi/deleghe/03387dl.htm (accessed on 10 October 2023).
88. Directive 2001/77/CE of the European Parliament and of the Council of 27 September 2001 on the Promotion of Electricity Produced from Renewable Energy Sources in the Internal Electricity Market 2001.
89. Meteonorm Intro. Available online: https://Meteonorm.Com/En/ (accessed on 11 October 2023).
90. SoDa. Available online: https://Www.Soda-pro.Com/ (accessed on 11 October 2023).
91. World Weather Online. Available online: https://Www.Worldweatheronline.Com/ (accessed on 11 October 2023).
92. Sun Earth Tools. Available online: https://Www.Sunearthtools.Com/ (accessed on 11 October 2023).
93. Iqbal, M. *An Introduction to Solar Radiation*; Elsevier: Amsterdam, The Netherlands, 1983; ISBN 9780123737502.
94. Xie, Y.; Sengupta, M.; Dudhia, J. A Fast All-Sky Radiation Model for Solar Applications (FARMS): Algorithm and Performance Evaluation. *Sol. Energy* **2016**, *135*, 435–445. [CrossRef]
95. Freitas, S.; Catita, C.; Redweik, P.; Brito, M.C. Modelling Solar Potential in the Urban Environment: State-of-the-Art Review. *Renew. Sustain. Energy Rev.* **2015**, *41*, 915–931. [CrossRef]
96. Šúri, M.; Huld, T.A.; Dunlop, E.D. PV-GIS: A Web-Based Solar Radiation Database for the Calculation of PV Potential in Europe. *Int. J. Sustain. Energy* **2005**, *24*, 55–67. [CrossRef]

97. Energy Saving Trust Solar Energy Calculator Sizing Guide. Available online: https://www.pvfitcalculator.energysavingtrust.org.uk/Documents/150224_SolarEnergy_Calculator_Sizing_Guide_v1.pdf (accessed on 10 October 2023).
98. Green, M.A.; Dunlop, E.D.; Yoshita, M.; Kopidakis, N.; Bothe, K.; Siefer, G.; Hao, X. Solar Cell Efficiency Tables (Version 62). *Prog. Photovolt. Res. Appl.* **2023**, *31*, 651–663. [CrossRef]
99. Zhang, H.L.; Baeyens, J.; Degrève, J.; Cacères, G. Concentrated Solar Power Plants: Review and Design Methodology. *Renew. Sustain. Energy Rev.* **2013**, *22*, 466–481. [CrossRef]

energies

MDPI

Article

Environmental Design on Site-Specific Energy Solidary Communities around Public High Schools in the Metropolitan Area of Naples (Italy)

Roberto Bosco [1,*], Savino Giacobbe [1], Salvatore Losco [1], Louise Anna Mozingo [2] and Renata Valente [1]

[1] Department of Engineering, Università degli Studi della Campania Luigi Vanvitelli, 81031 Aversa, Italy; savino.giacobbe@unicampania.it (S.G.); salvatore.losco@unicampania.it (S.L.); renata.valente@unicampania.it (R.V.)

[2] Department of Landscape Architecture and Environmental Planning, University of California at Berkeley, Berkeley, CA 94720, USA; lmozingo@berkeley.edu

* Correspondence: roberto.bosco@unicampania.it; Tel.: +39-3475437047

Abstract: Renewable energy communities (RECs) around photovoltaic systems on public buildings are optimal solutions to counter energy poverty, ensuring all stakeholders access to cheap, reliable, sustainable, and modern energy systems. As the neighborhood is the minimum suitable unit for the implementation of highly sustainable settlements, this article discusses the potential and criticality of RECs at this scale in southern Italy. Starting with the concept of RECs, this study presents a methodology to size sustainable urban communities around school buildings. It integrates practical energy indicators with those defining performance in vegetation and water management. The impact of these factors is analyzed to identify the ideal community size in terms of energy efficiency, economic value, and social cohesion. An interactive scorecard ranks high school sites suitable for transformation into community hubs, taking into consideration the scale of substation distribution. The findings provide empirically validated operational guidelines and best practices to support the transition to smart, efficient, and socially inclusive communities. At the urban scale, the analysis evaluates different urban morphologies, microclimates, characteristics and density of buildings, and population around each assumed community hub. The study provides valuable guidance to local designers, planners, and administrators for the implementation of sustainable technologies by preparing a map of potential RECs.

Keywords: construction; sustainability; renewable energies; process model; urban equipment regeneration

1. Introduction

The European Union set targets for reducing CO_2 emissions through renewable energy use to achieve carbon neutrality [1]. Cities are key places to implement transformations toward the development and welfare of the population, consistent with the planet's capacity. Contemporary transitions (technological–digital, energy–environmental, and demographic–social transitions) lead to redefining citizens' social responsibilities, evolving the current urban development pattern, promoting social cohesion, and adopting circular economy models [2].

Renewable energy communities (RECs) are solutions for pursuing environmental, economic, and social sustainability [3]. The EU's Clean Energy Package for 2019 includes provisions to help local societies in the transition through energy communities. These are associations of citizens, companies, businesses, and territorial and local authorities coming together to produce and share electricity from renewable sources in a decentralized grid [4]. In general, renewable energy communities and energy communities involve multiple disciplines and many scientific studies examine them from one or more factors, such as technical, economic, and ecological [5]. Studies also [6,7] examine techno-economic aspects, defining the best technical frameworks in relation to individual or community economic return. Other studies identify combinations between forms of prosumership and renewable energy models to foster payback on initial investments [8,9].

Several studies [10–12] outline RECs in terms of architectural and urban design by structuring the elements into physical and symbolic networks. These allow a progressive strengthening of community education, representation, and function by crossing technologies and cultures. Decarbonization issues, the importance of renewable energy, and the social repercussions of reactivating community dynamics and resource sharing should be explained to all categories of users. The duty of public agencies is to plan and promote cultural and social rearrangements at the necessary scale for the expected change.

The international literature suggests different organizational interpretive models of RECs, which can be traced back to those driven by public governance (top-down approach); pluralists, which apply horizontal models of community (bottom-up); and hybrids, as a result of the intermediation between local projects and individual consumers [13]. Typical distinctions relate to stakeholders, types of generated benefits, and triggered participatory processes.

In the wide range of interpretations of the meaning of "community" within the RECs, broad agreement exists regarding the importance of maintaining a certain level of both local involvement and shared benefits to maximize acceptance and support, as well as broader catalytic effects [14]. To this end, it is critical to prevent RECs from becoming solely a business strategy tool for energy companies or mere profit vehicles for corporations.

With their provocative 1957 Solar Do-Nothing Machine, Charles and Ray Eames challenged the traditional conception of utility and productivity in favor of a more sustainable approach in the modern era. This solar-powered kinetic sculpture was criticized by some as an elitist and niche exercise, a superfluous toy for intellectuals detached from the real needs of society (Figure 1). A similar perception of elitism could threaten renewable energy communities, as promoted today. Although they propose a virtuous model of clean-source energy production and consumption that is accessible and participatory, there is a risk that they will be seen as niches reserved for the privileged few. To avoid this, it is essential to clearly define the preferred legal form and actively promote the process of establishing communities [15].

Figure 1. The Solar Do-Nothing Machine (1957) www.eamesoffice.com (accessed on 25 March 2024).

In practice, a general lack of overall strategy and uneven development of RECs both at the spatial level and in association models are noted, resulting in uncertainties about the potential for scalability. There is a need for a scientific assessment of the actual utility and universal applicability of the model, which has limitations in terms of affordability, technical feasibility, and environmental sustainability in certain contexts. This analysis is crucial to avoid improper generalizations of the paradigm and ensure its balanced and truly effective diffusion. Energy communities, whether conceived from spontaneous bottom-up actions or from non-profit administrative or private entities, encompass a transformation of the socio-technical systems that form the basis of citizens' daily lives. This necessitates understanding the cultural, technical, and economic value of energy transformed communities, currently challenged by an individualistic, poorer, inequitable, and fearful society and emphasizes the collective to the collaborative [16].

This research, which is part of the PRIN 2022 "FASTECH" research project (a model for rapid technological building retroFit to cut energy demAnd and GHG emiSsions of housing Toward rEnewable energy self-Consumption and tHe smart energy communities; Principal Investigator: F. Tucci; Research Lead at the University of Campania 'Luigi Vanvitelli': R. Valente with the collaboration of Louise A. Mozingo, director of the Center for Resource Efficient Communities (CREC) at the College of Environmental Design, University of California at Berkeley, CA, USA.), tests a process for defining sustainable urban communities, capable of sharing different natural, material, and immaterial resources, starting with the broad theme of renewable energy communities.

The phases of this research are articulated in a nonlinear way to define scientifically based solutions that complement the actions currently in place (Figure 2). The resulting reflections aim to harmonize and expand the beneficial effects of community initiatives. The creation of a repository of existing and planned RECSs in Italy, selected from those with more pronounced characteristics of social commitment and public involvement, is preparatory to the identification of economic and social indicators describing their potential, strengths, and weaknesses. To fully understand the organizational, administrative, and energy aspects of RECS, a multi-methodological survey is underway.

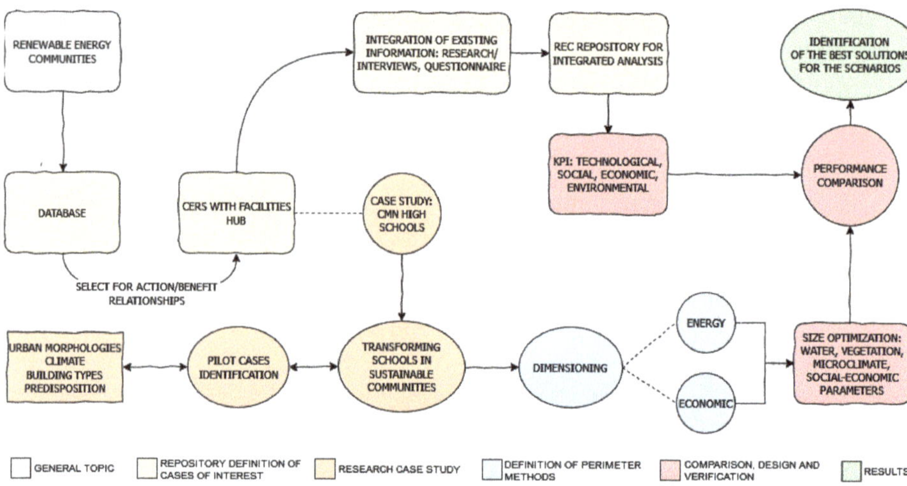

Figure 2. The research process (source: authors).

The study conducted an initial, comprehensive review of relevant documentation. Then, semi-structured interviews with key stakeholders and the administration of targeted questionnaires (see Appendix A) capture the contextual peculiarities, the internal organizational dynamics, and the operational challenges faced by these entities, as well as collecting quantitative data.

An integrated database records the collected information, meant as a knowledge base for developing multidimensional optimization models. These allow assessment of the replicability and potential impact of similar initiatives on lots around school buildings in the Metropolitan City of Naples, selected based on similar conditions regarding housing context, climatic characteristics, and socio-economic condition.

This contribution demonstrates the methodology to size constituent communities around school buildings, integrating energy indicators with those defining performance in vegetation and water management. We also study their impact by identifying the ideal size of communities in terms of energy efficiency, economic value, and strengthened social cohesion. We look for empirically validated operational guidelines and best practices to support the transition to smart, efficient, and socially inclusive urban communities.

This research studies the development of additional integrated factors that give greater meaning, social value, and environmental effectiveness to the concept of an energy community in southern Italy. Since 2021, our research group from the Department of Engineering at the University of Campania, "L. Vanvitelli", has been working with the Metropolitan City of Naples School Area (Campania Region in southern Italy), which manages 292 school buildings for higher education in an area of 1171 sqkm that includes 92 municipalities.

The Register of School Buildings (ARES Anagrafe Regionale dell'Edilizia Scolastica) database is a tool adopted at the national level in 2021 to survey the school building stock managed by local administrations. It was analyzed—) in order to build a sustainable integrated process model for the redevelopment of higher education school buildings, establishing associations ated according to the clear logic of influence radii and catchment areas. This develops an assessment of a high sustainability neighborhood, delving into aspects of energy consumption and the possibilities and critical issues related to the implementation of a Renewable and Solidary Energy Community (RESC). The aim is to bridge the gap between the knowledge acquired by the scientific community and the implementation and dissemination of best practices. We describe a tool for assessing the suitability of public buildings to serve as energy hubs, in reference to to the size and scope of the community served. This provides the basis for a model integrating morphological-urban assessments with the socio-territorial characteristics of the schools' sites.

The purpose is to develop a procedure for the design of sustainable urban communities, considering collective facilities as centers of energy, sustainability, and citizen health management and conceiving them as interconnected spatial networks capable of providing community services and generating positive spatial impacts. To this end, we also consider the crucial role of green infrastructure in the urban environment for integrated natural resource management. They can perform a range of environmental, economic, and social functions, such as regulating the microclimate and water cycle, conserving biodiversity, and ensuring the well-being of citizens [17]. The new tool is directed at local governments to define and foster communities that can share and manage multiple natural resources (energy, water, and vegetation), optimizing operational energy consumption and greenhouse gas emissions. Around these poles of physical, cultural, energy, and social aggregation, the perimeters of the connected community also consider green areas and water runoff subbasins. These factors contribute to identifying communities of citizens that can share and appropriately manage multiple natural resources and processes of physical transformation in places.

GIS solutions are already being used to organize and visualize spatial data when performing energy calculations, considering construction data, activity types, energy consumption, load profiles, fuel rates and costs, and climate data. They identify potential sites and analyze the distribution of customer characteristics [18,19]. Organically processing the data, enriched with information from environmental contexts, identifies strategies for enhancing renewable energy communities, referring to the latest regulatory provisions to design feasible application solutions. At the urban scale, different spatial areas are analyzed, depending on the building density around each assumed energy community hub. In addition, an interactive worksheet classifies high schools to transform into REC hubs. The end result of the PRIN research will be a digital tool allowing local governments to derive potential community perimeters based on savings parameters and by renewable energy plant type on public sites.

2. Energy Transition and Electrical Network Infrastructure at the Neighbourhood Scale

Buildings are the largest consumers of energy, accounting for 40 percent of total consumption in the EU and being the main sources of CO_2. In Italy, the first regulations on energy efficiency date back to the mid-1970s, and 65% of buildings in our country were constructed before these important normative provisions [20].

Neighborhoods should be redesigned as effective powerhouses of energy production. To reduce energy consumption, construction intervention on individual buildings has objective limitations and precludes the fulfillment of system (neighborhood–city–territory) results that would enable significant improvements in sustainability with an inter-scalar and systemic logic. It is therefore essential to refer to the aggregations of buildings and their interactions in terms of material–energy inputs and outputs. The intermediate scale of the neighborhood represents the minimum compliant unit suitable for the implementation of the characteristics that highly sustainable settlements should possess.

According to the international scientific literature [21–24], the four macro categories of sustainable neighborhoods are: morphological and functional organization, emission control, waste cycle, and energy consumption. To reduce the latter, it is necessary to intervene both in buildings as well as optimizing energy and bioclimatic factors of the neighborhood and city, understanding the interrelationships with the other three macro categories. Spatial planning tools, both general and sectoral, lend themselves to directing and encouraging transformation processes with high environmental sustainability in existing and new settlements. To achieve this aim neccesitates addressing the fundamental problem of the relationship between the distribution network suitable for this new type of spatial organization, such as primary urbanization and infrastructure, and the morphology of neighborhoods. A further issue is the storage of non-programmable renewable energy to compensate for the variability of energy sources and operate an effective match between availability and demand.

Decarbonization and electrification can only happen with a modern, widespread, and robust distribution network consisting of points and grids as in all primary urbanization [25]. This was designed on the assumption that electricity would be produced at a few points in the territory, transported along large national backbones, and finally distributed to end users. The tree-like logic resulted in a territorial configuration and management of the grid that presently constrains the energy transition in the current production, transportation, and distribution model.

Instead, a semi-lattice logic is needed [26], at least as a theoretical reference, almost undifferentiated and not very hierarchical, covering the service territory. Although neglected in the last two decades of debate on the energy transition, the importance of this infrastructure is supported by the International Energy Agency (IEA), which declared only two years ago that it must invest conspicuously in the adaptation of the electricity grid according to the requirements and performance targets that the infrastructure will need.

Today, the typical size of a power-generating installation worldwide is generally 1000 MW, while in Europe it is 560 MW. In 2050, the average size in Europe will have to drop to 32 MW (source: Enel Grid (https://www.enel.com/it/azienda/storie/articles/2021/05/nuovo-modello-rete-distribuzione-energia accessed on 25 March 2024), which will entail a much higher number of points of energy input to the grid distribution networks not designed for this function. Within two decades, this condition will have to change radically, as on the one hand the energy model will be subverted and on the other hand the consumption of electricity will triple to meet new needs from utilities that today use gas, fuel oil, and gasoline. Consumption of about 20% of electricity by 2045/2050 will have to reach 50% to meet the emission level deemed safe according to the Paris roadmap [27].

Electrification will take place first through consumer choices to decarbonize consumption with the widespread use of photovoltaic panels, micro-eolic heat pumps, electric cars, and induction cookers. This change in the energy model and increase in consumption will involve the low-voltage distribution network, which is the most widespread and dispersed but also the most antiquated and fragile. This transformation involves a change in the role of the distributor, who becomes a system operator, an organizer of energy flows, and collects the excess self-consumption and production connected to the distribution network. The operator then harmonizes flows primarily for local consumption, possibly using primary electrical substations to transport them elsewhere. These substations, however, were designed and built to transform energy from high voltage to low voltage, while in this new energy model they will need to work in the reverse direction as well. This change of role will involve technological and regulatory transformations, as well as the upgrading of the grid, which must connect many more plants. It should have hosting capacity, i.e., the ability to accept power from distributed generation and handle new discontinuous consumption flows.

Digital technology provides an important contribution, providing the ability for networks to self-regulate and monitor consumption and failures. By changing configurations in milliseconds and intervening in devices deployed along the route in the event of a failure, computer-aided systems allow changes in distribution systems to leave as few consumers as possible without supply. This ability to self-correct is possible only through digital tools connected to advanced telecommunications networks. These are inevitably conditioned and/or benefited by the actual characteristics of existing urbanization, which in turn are consequential both to the morphology and the logical structure that generated them over time. Distinguishing between the dynamics determining settlements formed spontaneously over many years and those of planned interventions will determine a lesser or greater susceptibility to the transformation of networked urbanization and especially of the electrical infrastructure distribution.

Alongside these spatial characteristics, socio-economic ones must also be considered when assessing the greater or lesser susceptibility of a context to reorganize itself as an energy community to improve sustainability. An assessment based exclusively on spatial characteristics also conflicts with the definition of the neighborhood unit [28], which constitutes a first community level within the anthropized territory and has represented the theoretical reference model in the planning, design, and realization (also spontaneous) of Modern Movement neighborhoods. Spatial equity, which proper spatial planning must strive for, increasingly involves energy issues; a free and unplanned process of energy transition risks accentuating energy poverty and injustice. Physical transformations of the territory and changes in the socio-economic conditions of communities and societies mutually influence each other. The analysis of the state of spatial layout must also involve the analysis of socio-economic conditions, and the integrated understanding of both, favored by governance techniques, must plan their correct interaction to structure a harmonious, livable, and more sustainable neighborhood, including the identification and organization of renewable and supportive energy communities.

3. Legislative Aspects

From the point of view of EU policies, the Covenant of Mayors for Climate and Energy requires signatories to draw up a Climate and Sustainable Energy Action Plan with the aim of reducing CO_2 emissions by at least 40% by 2030 and increasing resilience to climate change. Following the Paris Agreement on reducing greenhouse gas emissions, the EU issued Directive No. 2001 on 11/12/2018, called "Renewable Energy Directive Recast", which requires that at least 32% of the energy produced in Europe come from renewable sources by 2030. It regulates rights recognized by European citizens so that they can establish a REC to create a sustainable energy market [29].

Italy transposes this directive with Legislative Decree No. 199/2021, which defines RECs as legal entities whose associates maintain their customer rights. In Mission 2 of the PNRR (Piano Nazionale di Ripresa e Resilienza—National Recovery and Resilience Plan), approximately 2.2 billion euros were allocated for the construction of RECs. Regarding the sizing of the REC, its connection, and the age of the usable PV plants, resolution 727/2022/R/eel of 27 December 2022, published by ARERA (Autorità di Regolazione per Energia Reti e Ambiente—Regulatory Authority for Energy Networks and Environment), establishes that the plants may have a total power of up to 1 MW and must be connected to the electrical grid through the same primary substation (Italy's electricity distribution and metering company, E-Distribuzione, has brought online the interactive map to check the primary substation and the conventional geographical area of belonging). DL 162/2019 provides for incentive and payback mechanisms by valuing all energy fed into the grid at the market price, fostering the economic return on investment for REC participants [30]. Public and private entities provide various online tools to estimate the bill savings achievable by indicating the number of housing units that will comprise the REC. ENEA (Agenzia nazionale per le nuove tecnologie, l'energia e lo sviluppo economico sostenibile) offers the Recon tool (a web application developed by ENEA to support preliminary energy, economic, and financial assessments for the establishment of renewable energy, which is currently being updated in communities https://recon.smartenergycommunity.enea.it, accessed on 20 March 2024), while Energy Service Companies (ESCo) have developed simplified models useful for calculating reimbursement of system installation and maintenance costs, as well as payback times. However, such calculation models are linked to individual buildings, which is insufficient to define a holistic neighborhood-scale approach.

Italian energy planning is based on Law No. 10/1991, which prescribes the regions to draw up PER (Piani Energetici Regionali—Regional Energy Plans). They must indicate the actions necessary for the rational use of energy, its saving, and the development of renewable sources. Among the provisions of art. 5, the PERs must contain indications of energy balances, the identification of territorial energy basins (areas and energy islands suitable for the most significant and profitable interventions), and procedures for the identification and location of energy production plants up to 10 MW. The same art. 5, paragraph 5, establishes that the regulatory plans of municipalities with a population of over 50,000 inhabitants must provide for a specific sector plan concerning the rational use of energy and the use of renewable sources, i.e., the municipal energy plan.

There are two possible configurations for electricity sharing under Italian law: the renewable energy community when users are not located in the same building and the collective autoconsumption groups when users reside in the same apartment block. The latest Italian decree (Decreto Pichetto Fratin—9 December 2023, No. 181) [31] directs that the power of individual plants connected to the power grid through the same primary substation is limited to preserve the social character and prevent them from turning into energy-producing enterprises. It is also reiterated that participation in the REC is open, voluntary, and with no obligation of duration. Overall, member entities access the incentives provided for the energy transition by depositing any surplus in storage systems or feeding it into the grid for sale. Revenue sharing among community members is freely determined through a private agreement.

The first energy community activated in Italy resides in the municipality of Magliano Alpi, dating to 2020; later, examples such as the Community of East Naples or Ferla (Syracuse, Sicily) inspired the birth of the Network of Renewable and Solidary Energy Communities, an informal entity that aims to fight energy poverty by aspiring to greater environmental and social justice through bottom-up participatory processes. This involvement encourages civic participation and reinforces a sense of belonging and environmental responsibility by characterizing RESCs (Renewable and Solidary Energy Communities) as potential catalysts for initiatives that, while starting with energy issues, can influence local development at all levels in an integrated way.

In Italy, the establishment of RECs around photovoltaic plants built on public buildings represents an optimal solution to fight energy poverty, ensuring the actors involved access to economic, reliable, sustainable, and modern energy systems [32]. To date, according to GSE (Gestore dei Servizi Energetici—Energy Services Manager) data, 67 CACER (Configurazioni di Autoconsumo per la Condivisione dell'Energia Rinnovabile—renewable energy shared energy self-consumption configurations) operate in our country, including renewable energy communities (30%) and collective self-consumption configurations (70%), with a total of 501 connected end customers [33].

The latest regulatory directives provide for the association of public and private entities supported by the capital invested by the GSE, which is responsible for the installation and maintenance of the plant until the financing is extinguished [34]. The main recipients of funding for the promotion of renewable energy communities are local governments, which are considered the most appropriate actors to mobilize and coordinate other local actors.

American Legislations

There is no specific national regulatory framework for energy communities in the United States as there is in Europe, but there are several similar initiatives and models emerging locally. Legislation about shared renewable energies has been enacted in 24 states [35]. California, along with the Hawaiian Islands, was the first state to legislatively promote incentive tariffs for renewable energy production [36].

Community solar, sometimes referred to as a "solar garden", is currently the most popular form of shared renewables, with about 1600 projects nationwide [37]. Other models of renewable energy sharing and consumption can be found in Community Choice Aggregation (CCA) and community microgrids. In the former, local communities can purchase power from alternative suppliers to the traditional utility by increasing the share of renewables in the energy mix; in the latter, distributed renewable generation, storage, and connected utilities are integrated at the neighborhood or community level, often in an infrastructure isolated from the main grid. It is also interesting to note that blockchain-based energy trading platforms are being experimented with to enable peer-to-peer exchange of renewable energy between producers and consumers within an "energy network".

Although the U.S. regulatory environment is different, these initiatives pursue the same goals of democratization, decarbonization, and increased energy self-sufficiency in local communities through renewables and participatory models.

4. Methodology

The process proposed by the research group involved cyclic repetition of the analysis, design, and verification phases to identify the best multi-purpose solutions. Predimensioning is carried out by using a balance equation between the self-consumption of the school building and the perimeter of the community, considering the installed power. The proposed model integrates morphological-urban evaluations through thematic maps and the interactive worksheet.

Based on ARES data, we initially identified 48 secondary school buildings distributed across 24 municipalities, defining their microclimatic, urban, and energy characteristics. For each school, urban densities were categorized as low (involving 22 schools), medium (involving 14 schools), and high density (involving 12 schools). Among these identified institutions, five (A, B, C, D, and E) were selected, i.e., two with low density, two with medium density, and one with high density. The selected institutions underwent analysis regarding roofing type, orientation of the photovoltaic system, energy production capacity, and sky-view factor [38,39] (Table 1).

Table 1. Interactive worksheet of sample school buildings. Description of the urban and technological characteristics of the considered high schools (source: authors).

School	A	B	C	D	E
Building code	######	######	######	######	######
Municipality	Melito di Napoli	Giugliano in Campania	Giugliano in Campania	Giugliano in Campania	Giugliano in Campania
Latitude	######	######	######	######	######
Longitude	######	######	######	######	######
Primary electrical substation code	AC001E00225	AC001E00226	AC001E00226	AC001E00226	AC001E00226
Building density	Low	Medium	Medium	High	Low
Technology of bearing structure	Reinforced Concrete	Reinforced Concrete	Reinforced Concrete	Reinforced Concrete	Reinforced Concrete
Roofing surface [sqm]	1890	4407	4657	6977	4622
Roofing tipology	Plain	Plain	Plain	Plain	Plain
PV orientation	South	South	South	South-East	South
Sky-view Factor	1	1	1	1	1
Albedo of roof	0.13	0.13	0.13	0.13	0.13
PV surface [sqm]	1062.88	1860.56	2463.88	3518.22	2070
Energy produced annually [kWh]	252,429.21	446,639.55	601,391.08	830,052.1	467,062.39

building code and geographical data are not reported due to school identification concerns.

The study of urban, energy, and water spatial characteristics highlights an interesting focus for Institutes D and E, both of which are located within the same municipality. The overlapping of the spatial boundaries, primary electrical substations, and drainage micro-basins shows, for case D, a special condition in which the interventions cannot be planned and implemented by a unique institution as they fall within two different administrative boundaries. On the other hand, for case E, the coincidence of the perimeters of the various analysis factors defines an ideal community in which the different perimeters (territory, energy, and water) characterize an ideal sustainable community scenario.

In the analysis, the energy consumption of the schools and the amount of energy produced on the roof by the PV system were first drawn using the solar field sizing software "Blumatica Impianti Solari" (version 2.0.1.0). The energy consumption of each single POD (points of delivery) was estimated using the withdrawal profiles related to domestic utilities [40].

Once the RECs were sized as in Bosco et al. [41], the indicators useful for identifying the most energy-efficient solutions were selected within the most cost-effective ones. The study compares the following two perimeters for each institution: the first identifies a community in which all participants cut the cost of the power bill to zero, while the second considers halving costs for users.

To map the membership of users in the identified communities, the number of *PODs* in a building was estimated using ISTAT data through the following relationship:

$$POD = P \times A, \tag{1}$$

where P is the resident population and A is the number of housing units per resident (housing/residents) (the number of dwellings for residents can be found in the datasets "DICA_ALLOGGI" and "DICA_EDIFICIRES" provided by ISTAT, the Italian National Institute of Statistics).

Simultaneous microclimatic, vegetation, and hydraulic analyses enable design interventions to improve thermal comfort by impacting the energy performance of buildings. Interventions include tree planting, the replacement of road pavements to reduce albedo, and the installation of Green Stormwater Infrastructures (GSI). In the verification phase, microclimatic, vegetation, and water management performance indicators allow comparison of the energy and environmental performance of the assumed technological solutions and the selection of the most suitable ones. Results are reported in Files S1 and S2 in the Supplementary Materials.

4.1. Pre-Sizing and Checking of RECs

Starting from the studies identified in the literature [42–46], we propose for the first time a new model for sizing energy communities, divided into the following two stages: pre-dimensioning, arising from the identification of cost-benefit ratios related to each individual building, and verification, through appropriate indicators, of the solution that provides the best energy balance. The theoretical experimentation considered two time frames, i.e., one encompassing the span of a day (T); the other is the energy production interval of the PV system (T_h). In the first stage of the procedure, for greater technical and economic simplification, we give the case of a REC with a hub affected by self-consumption from a renewable source plant (PV) with no storage elements.

The theoretical assumptions for the development of the two phases derive from considerations of the magnitudes of hourly energy flows occurring within the assumed configuration [10]. The entities are as follows:

- The electricity that can be produced by PV over time T_h is as follows:

$$E_{prod} = \sum_j E^j_{PV}(T_h), \tag{2}$$

where $E^j_{PV}(t)$ represents the power that the plant produces over time T_h.

- The self-consumed electricity is as follows:

$$E_{sc} = \sum_j E^j_{sc}(T_h), \tag{3}$$

where $E^j_{sc}(T_h)$ represents the self-consumed power of the hub over time T_h.

- The energy fed into the grid over time T_h is as follows:

$$E_{inj} = \sum_j \left[E^j_{PV}(T_h) - E^j_{sc}(T_h)\right]; \tag{4}$$

- The energy withdrawn from the grid over time T_h is as follows:

$$E_{with} = \sum_j E^j_{grid,out}(T_h), \tag{5}$$

where $E^j_{grid,out}(T_h)$ represents the hourly withdrawal of PODs belonging to the REC, only during the time when the plants produce energy.

- The energy shared within the REC over time T_h, as per regulations [47], is as follows:

$$E_{sh} = \min(E_{inj};\ E_{with}); \tag{6}$$

- The energy fed into the grid and not shared in the REC over time T_h is as follows:

$$E_{to,grid} = E_{inj} - E_{sh}; \tag{7}$$

- The energy consumed by all REC electric utilities is as follows:

$$E_{cons} = E_{with} + E_{sc}; \tag{8}$$

- The REC users' energy demand over time T is as follows:

$$U_e = \sum_j U_e^j(T); \tag{9}$$

Here C_{HUB} is the average annual power consumption in the Hub (kWh/year); C_{POD} is the average annual power consumption of a *POD* (kWh/year); c is the unit cost of power (EUR/kWh) (net of transport and management costs placed at EUR 0.12; data were collected from ARERA on 8 March 2024). Assuming that the produced energy totally or partially cancels out the costs in the bill over time T_h, it was estimated here:

$$R_{Tot} = \text{c}\cdot E_{prod}, \tag{10}$$

$$R_{\text{HUB}} = \text{c}\cdot C_{\text{HUB}}(T_h), \tag{11}$$

$$R_{POD}(T_h) = \frac{R_{Tot} - R_{\text{HUB}}(T_h)}{n}, \tag{12}$$

where (10) defines the total bill savings (EUR); (11) defines the Hub's annual bill savings (EUR); (12) defines the annual bill savings in euros of a *POD*; n is the REC's *POD* number; T_h is the time in which energy is produced and shared with the community.

Given the annual bill savings of a *POD* and its annual power consumption, the annual power cost was derived as follows:

$$C_{\text{n}} = c\cdot C_{\text{POD}}(T_h) - R_{POD}. \tag{13}$$

By substituting (9), (10), and (11) into (12), we obtain the following:

$$\text{n}(C) = \frac{\text{c}\cdot E_{prod} - \text{c}\cdot C_{\text{HUB}}(T_h)}{c\cdot C_{\text{POD}}(T_h) - C_{\text{n}}}. \tag{14}$$

where (14) is true for $C_{\text{n}} \neq c\cdot C_{\text{POD}}$. As C_{POD} varies, it is possible to derive the number of PODs of the REC. Similarly, it is possible to derive n as a function of R_{POD} by substituting in (12), (10), and (11):

$$\text{n}(R) = \frac{\text{c}\cdot E_{prod} - \text{c}\cdot C_{\text{HUB}}(T_h)}{R_{POD}}. \tag{15}$$

The graphical representation of Equations (14) and (15), in which the x-axis shows the values of costs C_{POD} and revenues R_{POD} and the y-axis shows the numbers of PODs, defines a parabolic symmetrical diagram (Figure 3). The $f(C_{\text{POD}}; \text{n})$ represents an increasing parabolic function with a vertical asymptote, which defines the number of PODs as the cost increases. On the other hand, $f(R_{POD}; \text{n})$ represents a decreasing parabolic function that defines the number of PODs as the economic savings in the bill increase.

Once the number of PODs has been derived as a function of bill savings or costs, the second step is to verify the community size as a function of energy efficiency, using the indicators of Cielo et al. [42]:

- Self-Consumption Index (*SCI*)

$$SCI(\%) = \frac{E_{cons}}{E_{prod}}, \tag{16}$$

which expresses the amount of energy that can be shared within the community versus the amount of energy that can be produced.

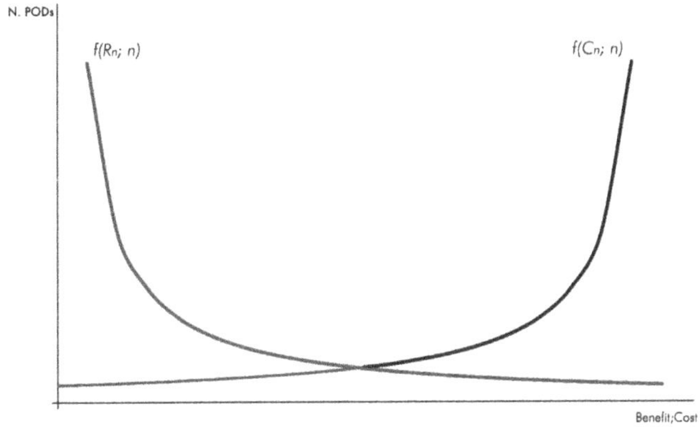

Figure 3. Graphical representation of functions (14) and (15) (source: S. Giacobbe).

- Self-Sufficiency Index (*SSI*)

$$SSI(\%) = \frac{E_{cons}}{U_e}, \tag{17}$$

which expresses how much renewable energy is consumed by the community compared to the users' needs.

- The CO_2 index, which expresses the amount of CO_2 avoided in one year by establishing a REC, is equal to the following:

$$\Delta CO_2(\%) = \frac{\varepsilon_o - \varepsilon_{CER}}{\varepsilon_o}, \tag{18}$$

where:

$$\varepsilon_o = \varepsilon_{rete} \cdot E_{cons}, \tag{19}$$

$$\varepsilon_{CER} = \varepsilon_{rete} \cdot [E_{cons} - (E_{sc} + E_{sh})] + \varepsilon_{PV} \cdot E_{prod}, \tag{20}$$

with:

- ε_o carbon dioxide emissions in the absence of a REC;
- ε_{CER} carbon dioxide emissions with the REC;
- ε_{rete} CO_2 emission coefficient, which is equal to 0.276 kg of CO_2/kWh, derived from the literature;
- ε_{PV} CO_2 emission coefficient related to the production and disposal of the photovoltaic system, obtained from the literature, is equal to 0.05 kg of CO_2/kWh [36].

4.2. Assessing the Effects of Vegetation on REC

The potential effects of environmental design on microclimate are evaluated through the strategic use of vegetation to optimize community energy consumption. Vegetation can supply a variety of ecosystem services, including energy services through its thermoregulatory function as well as the absorption of air pollutants and stormwater [48]. Planting trees at less than 18 m (60 ft) from buildings can decrease the amount of energy required for indoor heating and cooling by reducing temperatures and wind and providing shade [49]. Using the i-Tree Eco simulation software (rel. 6), the energy and economic benefits provided by trees were determined, allowing for community implementation. Indeed, by subtracting the cost of electricity saved in (11) and (13), we obtain the following:

$$R_{\mathrm{HUB}} = c \cdot C_{\mathrm{HUB}}(T) - R_{\mathrm{HUB.Tree}}(T), \tag{21}$$

$$C_{\mathrm{n}} = c \cdot C_{\mathrm{POD}} - R_{POD} - R_{\mathrm{POD.Tree}}(T). \tag{22}$$

where $C_{\text{HUB.Tree}}$ is the energy savings of the HUB due to the effect of trees; $C_{\text{POD.Tree}}$ is the POD energy savings generated by the trees. Then, (13) and (14) become the following:

$$\mathrm{n}(C) = \frac{\mathrm{c}\cdot E_{prod} - \mathrm{c}\cdot C_{\text{HUB}}(T) - R_{\text{HUB.Tree}}(T)}{c\cdot C_{\text{POD}}(T) - C_{\text{c}} - R_{\text{POD.Tree}}(T)}. \tag{23}$$

$$\mathrm{n}(R) = \frac{\mathrm{c}\cdot E_{prod} - \mathrm{c}\cdot C_{\text{HUB}}(T) - R_{\text{HUB.Tree}}(T)}{R_{POD}}. \tag{24}$$

With (23) and (24) and the savings provided by trees, it is possible to gain more PODs than with (14) and (15) (Figure 4).

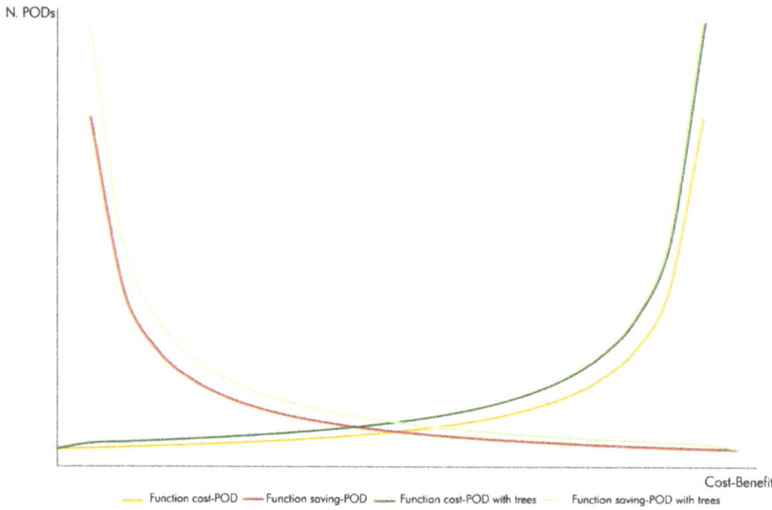

Figure 4. Comparison of functions (14) and (15) with (23) and (24) (source: S. Giacobbe).

4.3. *Evaluation of the Effects of Microclimate and Water*

Although the best-known urban climate phenomenon observed at the surface is the urban heat island (UHI), the processes controlling it act across a range of spatial and temporal scales that span the depth of the urban boundary layer (UBL) [50]. The sparse presence of bodies of water, lawns, and trees results in significant temperature increases within this layer of the atmosphere. In summer, the heat emitted by the HVAC units' engines further increases the temperature values in urban areas. Due to virtues such as shading and evapotranspiration, planting trees provides a particularly effective solution to reducing air temperatures. For this purpose, using the fluid-dynamic simulation software for calculating microclimatic conditions ENVI-met (rel. 6.1), we identified critical areas where the most unfavorable conditions occur, particularly in the summer. The integrated design of rainwater harvesting systems (GSI) and vegetation allows for improved microclimatic conditions [51], indirectly impacting the energy performance of buildings [52]. The reduction of outdoor temperatures results in a further reduction of energy consumption from air conditioning systems, resulting in further energy retrofits.

5. Results and Discussion

The first part of the analysis phase on the two pilot case studies as a comparison sample (D and E) enables an estimation of the number of PODs served by the REC through (14) and (15), according to an energy–cost balance. Subsequent calculation of indicators (16), (17), and (18) allows verification of PODs suitable to meet the economic and energy needs of the renewable community. The study focuses on the minimum and maximum eligible perimeters. The former defines a REC capable of feeding surplus energy to the power grid as well as ideally meeting the total renewable energy needs of the community. The latter harnesses almost all the energy produced for the community while feeding a negligible portion of it into the grid (Figures 5 and 6). Different REC configurations were bordered through the GIS application, starting from the primary substations to which the hubs belong.

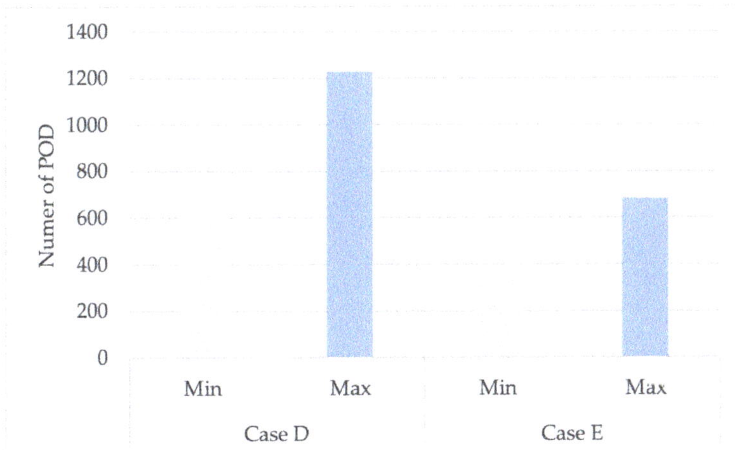

Figure 5. Maximum and minimum expansion values of the RECs (source: authors).

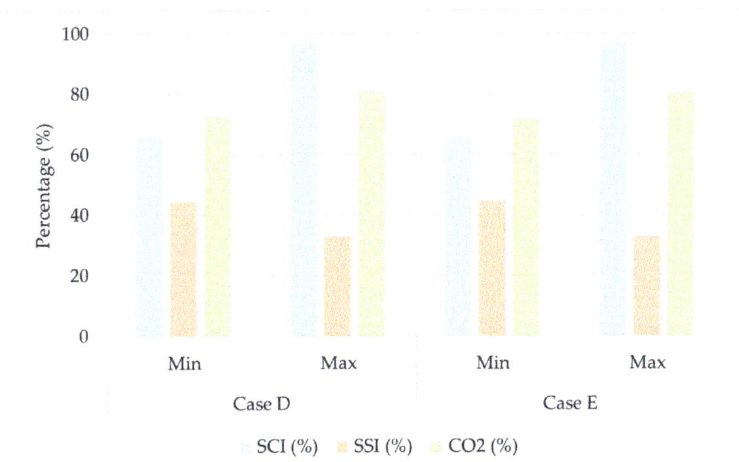

Figure 6. Index value for maximum expansion of the RECs (source: authors).

The subsequent identification of the runoff basins according to the trends of the outflow lines defines two different configurations. In case E, the local catchment area, given by the sum of the sub-basins, coincides with the perimeter of the REC, defining an ideal condition. In case D, the results are only partially coincident with the community.

This case study was chosen as an example of foreshadowing a particular condition in which the hydraulic intervention area includes both a portion of the primary electrical substation and a municipal border (Figure 7). The GSI project allows up to 28% of rainwater to be harvested for case E and up to 32% for case D (rainwater volume calculated for 30 min with a 20-year return time). Encouraging infiltration and absorption reduces the risks associated with extreme weather events while avoiding costly upgrades to the urban sewer system (Figures 8 and 9).

Thermo-fluid dynamics simulations carried out for the hottest day of 2022 (28 June) showed significant improvements in design conditions at 9:00 a.m., 12:00 a.m., and 3:00 p.m. For case E, decreases of up to 15% in PMV (predicted mean value), up to 1.5 °C in potential air temperature, and up to 3.4 °C in surface temperature were recorded. For case D, decreases of up to 21% in the PMV value, up to 2 °C in the potential air temperature, and up to 3 °C in the surface temperature were recorded.

The landscaping project involves the planting of 268 new trees in case E and 165 in case D. The increase of trees related to their strategic placement allows reducing the energy costs of adjacent buildings by EUR 1610/year for case E and EUR 521/year for the other. Trees allow their properties to achieve positive effects on the energy behavior of buildings. With a simplified calculation considering the transmittance of the envelope walls and the outdoor temperature, a natural reduction in indoor temperature of almost 1 °C is estimated. Further building interventions would reduce the indoor temperatures of buildings while diminishing the use of HVACs in the summer period.

Repeating the REC sizing phase through (23) and (24) results in a higher number of PODs than calculated with (14) and (15). The recalculating indicators (16), (17), and (18) allow new configurations to be determined. Defining flexible perimeters with variable sizes according to both urban and building energy and environmental retrofits identifies additional sharable resources within the community, beyond energy and environmental resources. Sometimes these are not equally distributed, as in the case of the community of school D. Such disparities may have negative consequences for the social cohesion and well-being of the involved communities.

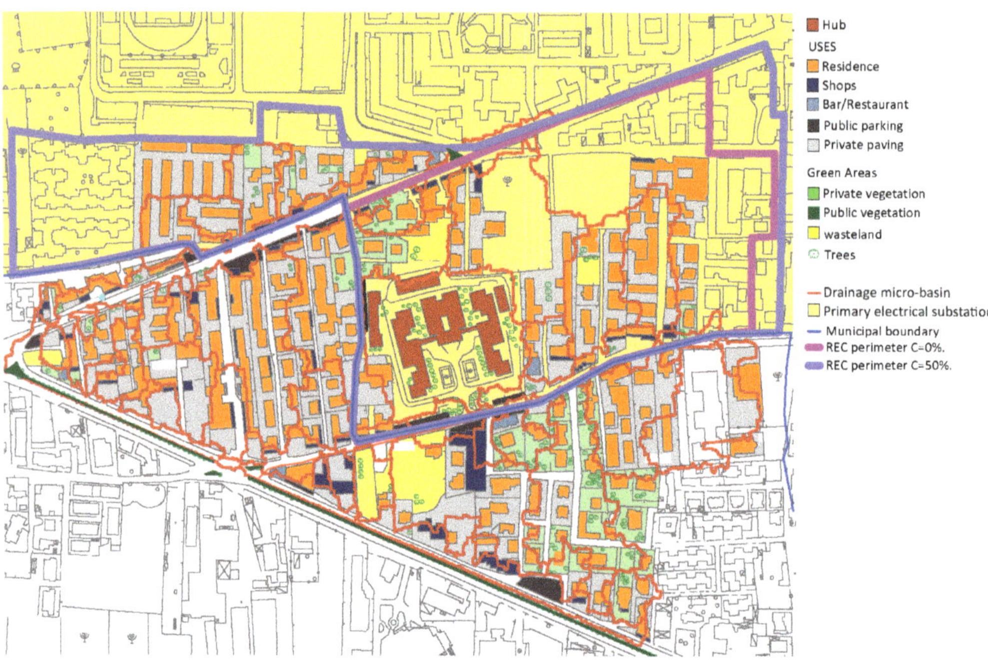

Figure 7. Comparison of water, energy, and urban boundaries in case D (source: authors).

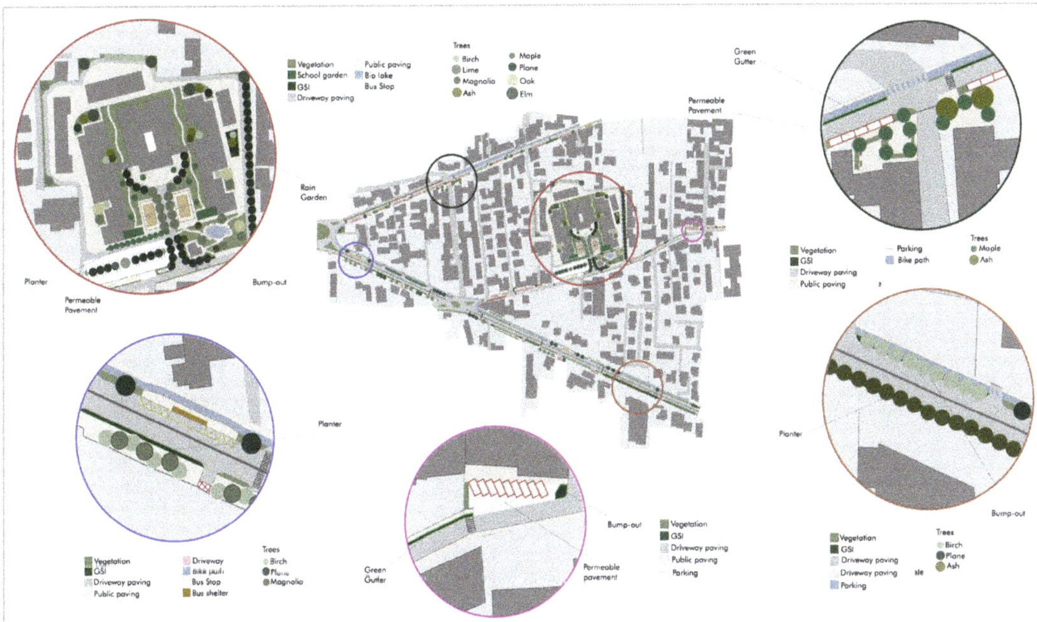

Figure 8. Pilot project D (source: authors).

Figure 9. Pilot project E (source: authors).

Therefore, the aim was to define an inclusive process for sharing community services even with those entities that, although belonging to the same urban catchment area, cannot be part of the REC. In this way, efficient resource management can be ensured, promoting equitable access to benefits and thus strengthening bordering communities.

The definition of a sustainable urban community involves sharing the immediate benefits of a REC together with improvements in:

- Microclimatic conditions;
- Resilience to extreme weather events;
- Real estate value;
- Sharing economy and pooling of additional community services (e.g., neighborhood laundries, internet points, and recycling of goods);
- Energy performance of the hub and buildings.

In conclusion, a sustainable urban community is a function of:

$$\text{S.U.C.} = f(FV,\ Trees,\ Microclimate,\ Social,\ Retrofit\ HUB,\ Retrofit\ Building). \quad (25)$$

This relationship can be represented by an increasing graph with upward concavity in which, as benefit values increase, the number of PODs that a community can accommodate increases (Figure 10).

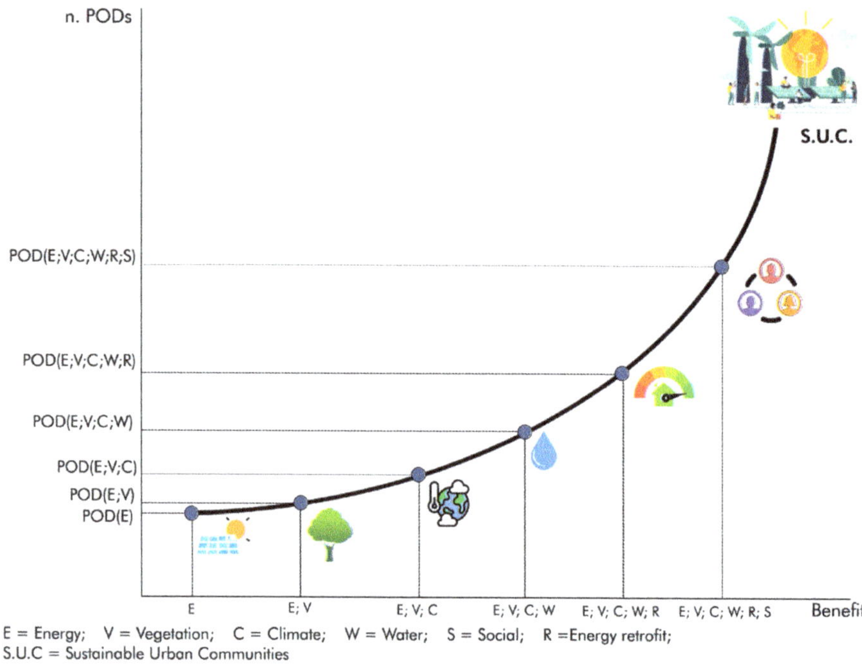

Figure 10. Graphical representation of the function (25) describing sustainable urban communities according to the balance between benefits and the number of PODs (source: authors).

The work focuses on defining an operational process model, starting with estimating the suitability of public buildings to become hubs of energy communities. The annual trend of energy consumption in a school shows that residential units can benefit from almost all the energy produced during periods of interruption of educational activities, resulting in even more substantial savings.

We overlap thematic layers representing the distribution of other resources in the REC perimeters, predicting their induced effects, criticalities, and appropriate technical solutions. The proposal seeks to prompt bureaucrats and stakeholders to take an integrated view of urban and social problems. The concept is the extension of the technical energy model for communities to include sustainability, solidarity, and sociality in urban neighborhoods for activism projects underlying aggregation initiatives and the transformation of the energy pattern itself.

Keyword analysis of the selected contributions in the scoping review led to the identification of the following five areas of inquiry: renewable energy, energy management, energy policy, sustainable urban development, and built environment management. The last three present the most numerous connections, suggesting a strong correlation between urban and energy policy issues. The research related to each one can have a positive impact on the others. The analysis suggests a possible line of investigation about the integration of more strictly technological issues with the social and management nature of the built environment. In particular, the research focuses on how energy communities formed around public buildings can represent effective organizational models.

The conditions of the existing buildings suggest the need for a renovation effort set at the level of territorial government, encouraging its transformation into a resource capable of producing energy alongside other community services. Public-interest endowments constitute a material and symbolic space where influential city regeneration processes can be triggered [53]. Rethinking the concept of standards through the lens of accessibility, sustainability, and people's health [54] fosters the transition to energetic community hubs. This means embracing a systemic vision where the facilities become dynamic and responsive supports whose spatial relationships shape people's practices on principles of mutualism and subsidiarity [55]. Considering energy in the same way as other unalienable community assets (water, health, and education), it is natural to think of its distribution and management as a public duty, unified by general regulations.

The reflection is about an integrated and complex consideration of urban resources, considering an effectively ecosystemic and intra-scale vision under which the relations of coexistence to re-design are based on tangible opportunities and shared characteristics. These become easily conveyed through the intrinsic effects of environmental education and social empowerment, which can be developed in living labs and urban action structures.

Maps of potential public energy hubs with their perimeters of influence provide tools of considerable utility to local governments for envisioning and assessing renewable energy production using public rooftops and open spaces. Around such pivots of physical, cultural, energy, and social aggregation, the perimeters connect the community in relation to the amount of shareable renewable energy, urban settlement types, technical-constructive characteristics, green areas, and runoff water sub-basins. In addition to environmental aspects, the protocols and indicators will include socio-technical considerations, from job creation to addressing energy poverty.

The logical systematization of evaluation criteria in dedicated scorecards allows easy integration of data obtained and processed for multi-purpose interventions (Figure 11). The proposed implementation of existing evaluation methods for the redevelopment of the built environment provides a useful starting point for planning interventions to improve the existing conditions, explicating the needs related to the individual areas under investigation.

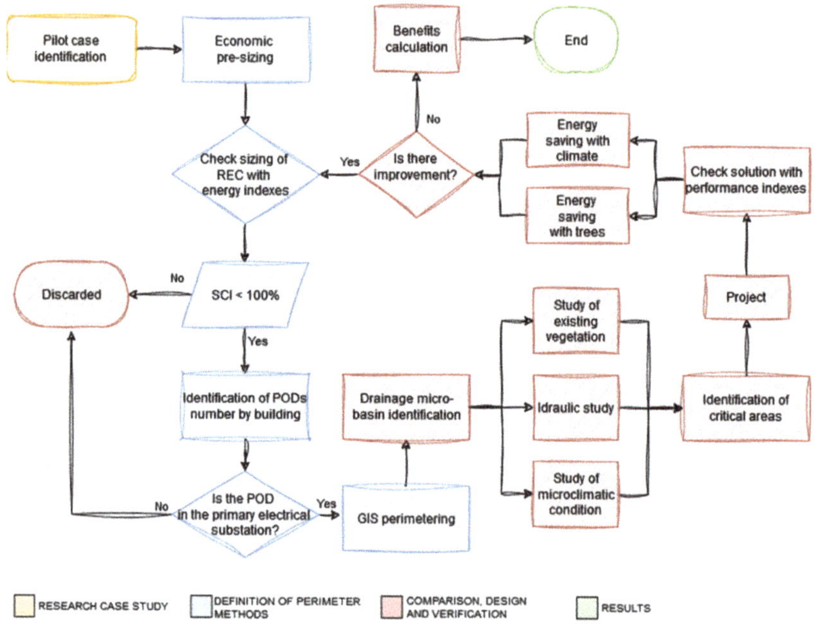

Figure 11. Flow chart relating to the perimeter phase and project definition and verification of the research process (source: authors).

6. Concluding Remarks

Transforming neighborhoods into renewable energy communities within a larger urban context is best addressed in physical space planning, involving communities and stakeholders in decision-making. This approach also benefits from the engagement of stakeholders with direct administrative responsibilities, such as regions, provinces, metropolitan cities, and communes. These decisions can be systematized in the form of a strategic-structural urban plan for the programming stage, an operational plan for implementation, a sub-municipal detailed plan, and a building-regulating plan. Municipal urban plans should integrate analyses of the areas' energy capacities, adding them to the traditional knowledge framework to regulate the use of available resources and programming transformations considering these aggregate capacities.

The change in existing electricity infrastructure and the implementation of a project network can accelerate an equitable energy transition that distributes these resources even to places that could never become self-sufficient by location and configuration. To this end, energy-environmental prescriptions could assign different performance levels depending on territorial areas, apply energy equalization, like urban planning, and regulate fair use rights.

Extending the scope of REC to the neighborhood scale, as the economic savings in the electricity bill decrease, the potential human capital of the community will increase. The presence of solar systems in public schools can also help develop local groups beyond the energy aspect, extending community benefits to different areas and sectors, such as health, safety, environment, culture, and economic activities. A further advancement for the valorisation of the idea of community consists in overcoming the dichotomy between 'prosumer' and 'consumer' by extending the possibility of sharing to goods, services, individual capabilities, or their own time. Identifying the REC with criteria that are not exclusively technical and of immediate economic return can initiate the return of long-standing good social practices. Furthermore, the pandemic accentuated the economic gap between the different strata of the population in many cities. In response, the Italian

Bishops' Conference (Conferenza Episcopale Italiana—CEI) has set up a technical table for the establishment of an energy community for each parish [56] to offer some concrete, though not exhaustive, responses to energy poverty. In smaller municipalities, these initiatives arise from local efforts involving the community. In the case of large cities, models that can be adapted to the urban territory are also being developed [57].

The establishment of appropriately sized RECs can significantly improve the living conditions of socially disadvantaged urban enclaves characterized by high population density, poor financial resilience, and difficulty accessing essential services. For these reasons, the first step in building an adequate evaluation method is to appropriately delimit the REC's perimeter.

For this research, the schools' needs were considered through the theoretical parameter of the annual energy consumption per student. For a more refined calibration of the processed data, the next step will consider the actual consumption of the schools. The study can provide valid indications to technicians and local administrators to implement sustainable technologies by preparing a map of potential energy communities. The proposed calculation parameters represent a starting point for defining administration strategies to integrate the environmental requalification process.

To evaluate the chosen methodology, our case study is the stock of buildings for higher school education managed by the Metropolitan City of Naples. However, in such a vast field of research, this constitutes a partial test of the obtainable findings, as we will extend the considerations to other types of public buildings and equipment.

The development of the research involves the activation of multidirectional channels of exchange between universities, local governments, and citizen groups. The university's relationship with the territory is of paramount importance, as the various departments, in their respective fields and through interdisciplinary collaborations, produce public engagement activities that can influence the decisions of central and local administrations [58]. The identification of pilot cases in the local area can induce a strengthening of the connection between research and management entities. This implies that administrations, technical bureaucrats, and citizens should themselves be equipped with appropriate technical, technological, and cultural tools. The updated critique reflects on the unequal conditions under which participation can take place by different stakeholders, i.e., the frequent technical and financial unpreparedness constitutes a major handicap.

The ongoing study experiments with environmental analyses, simulation models, calculations, and descriptions of the resulting perimeters and contents for critical evaluations of pilot cases. For their identification, one criterion is the selection of locations where there is correspondence between different thematic perimeters to maximize possible outcomes. In this regard, protocols and indicators for evaluating such initiatives will need to include socio-technical considerations. In addition, it will be necessary to verify the adequacy of grid infrastructure to integrate increasing shares of renewables efficiently and without compromising the quality of service. Finally, quantitative benchmark metrics need to be defined with respect to parameters such as the percentage of integrated renewable resources, emission reductions, savings guaranteed to members, and territorial energy resilience. Integrated approaches are needed for the full development of the potential of energy communities, considering technical, economic, and social factors and identifying replicable best practices.

Supplementary Materials: The following supporting information can be downloaded at: https://www.mdpi.com/article/10.3390/en17102247/s1. File S1: Energy calculations for case studies A, B, C, D, and E; File S2: 2.1 I-Tree Eco and 2.2 ENVI-met software modeling results.

Author Contributions: Conceptualization, R.V. and L.A.M.; methodology, R.V., R.B. and S.G.; validation, R.V. and L.A.M.; formal analysis, S.G. and R.B.; investigation, R.B., S.G., R.V. and S.L.; resources, R.B., S.G. and S.L.; data curation, S.G.; writing—original draft preparation, R.B. and S.G.; writing—review and editing, R.V.; visualization, R.B.; supervision, L.A.M.; project administration,

R.V.; funding acquisition, R.V. All authors have read and agreed to the published version of the manuscript.

Funding: This research and the APC were funded by European Union EU—Next Generation EU, Ministero dell'Università e della Ricerca (Italy), Italia domani PIANO NAZIONALE DI RIPRESA E RESILIENZA, Missione 4 Istruzione e Ricerca, grant number PRIN2022 A4XFPR. Title: 1. "FASTECH a model for rapid technological building retroFit to cut energy demAnd and GHG emiSsions of housing Toward rEnewable energy self-Consumption and tHe smart energy communities", Università degli Studi della Campania Luigi Vanvitelli.

Data Availability Statement: The original contributions presented in the study are included in the Supplementary Materials, further inquiries can be directed to the corresponding author.

Acknowledgments: We are grateful to the Architects of Schools' Technical Direction for the Metropolitan City of Naples, Annarita Marciano, Daniela D'Esposito, and Paolo Parravicini for their cooperation.

Conflicts of Interest: The authors declare no conflicts of interest. The funders had no role in the design of the study, in the collection, analysis, or interpretation of data, in the writing of the manuscript, or in the decision to publish the results.

Appendix A. Questionnaire Format (Edited in Microsoft Office Forms)

Question	Answer Mode
Name of the community	open-ended answer
Who oversaw the direction of the RESC design process?	open-ended answer
What is the administrative process that was used to establish the RESC?	open-ended answer
What were the main difficulties in implementing the RESC activities?	multiple-choice answers: • Administrative • Bureaucratic • Fund raising • Citizens' involvement • Technical planning
What funds financed the establishment of the RESC?	open-ended answer
What is the legal form of the RESC?	single-choice answers: • Social enterprise • Non-recognized association • Cooperative • Legal form provided for publicly held companies

Question	Answer Mode
Who are the participants in the RESC legal entity?	multiple-choice answers: • Citizens • Companies • Craftsmen • Associations • Public entities • Religious institutions
How many utilities are connected to the REC? (No. of households, No. of businesses, No. of entities . . .)	open-ended answer
How are expenses and revenues allocated within the community?	open-ended answer
Who is the management of the RESC?	open-ended answer
By what criteria were the energy community utilities selected?	multiple-choice answers: • Proximity • Belonging to the primary electrical substation • Belonging to the secondary electrical substation • Socio-economic conditions • Previous knowledge • Other
How much of self-generated energy is actually consumed?	0-to-100% scale
To optimize energy consumption, have improved interventions (energy class reduction, appliance renovation, etc.) been considered on the HUB and other buildings that are part of the REC?	single-choice answers: • Yes, and have been carried out/are being carried out • No, but they are scheduled • No
Has a plan been implemented to raise awareness of efficient energy consumption?	single-choice answers: • Yes, and have been carried out • No, but they are scheduled • No
Have any special technological interventions been implemented at the infrastructure level? (e.g., installation of new meters)	single-choice answers: • Yes • No
How relevant were the organizational skills and expertise provided by the local authority or possibly other private organizations?	0-to-10 scale
The new decree allows the establishment of RECS with an installed capacity of up to 1 MWp. Are there any plans to expand the RECS?	single-choice answers: • Yes, but participation will not be free • Yes, and participation will be free • No

Question	Answer Mode
What institutional (public) actors and/or private entities or organizations were involved? Among them, who played a leading role?	multiple-choice answers: • Municipality • Province • Region • Associations • Cooperatives • Electric service providers • Privates • Other
How many institutional (public) actors and/or private entities or organizations were involved?	multiple-choice answers: • Municipality • Province • Region • Associations • Cooperatives • Electric service providers • Privates • Other
What does "solidarity" mean to you within the Renewable and Solidarity Energy Community?	open-ended answer
In addition to economic benefits, in establishing the RESC, were sharing strategies considered that would give additional benefits to the neighborhood?	multiple-choice answers: • Rainwater harvesting through green systems • Vegetation increases • Shared services • Benefits for REC members
Once the incentives and assured maintenance for 25 years are over, what are the community management plans?	open-ended answer
Is it possible to obtain technical data related to REC from a web portal?	open-ended answer
Is it possible to obtain technical data on REC from a certified engineer?	open-ended answer
Is it possible to obtain the technical data related to REC from the ECUs?	open-ended answer
Can REC-related technical data be obtained from other resources?	open-ended answer
Which entities or organizations provided the capital needed to start the initiative?	multiple-choice answers: • Banks • Cooperative credit banks • REC members • Public authorities • Local energy service provider • Non-energy private sector

Question	Answer Mode
Beyond the funds in the previous question, have additional resources been allocated?	multiple-choice answers: • Self-financing (reinvesting profits) • No external funding • Crowdfunding • Donations • Loans from private banks • Loans from the public sector • Incentives and grants • Loans from other similar private initiatives (for energy-saving purposes) • Other forms of loans
Has the initiative obtained financial support from the public sector? If so, which entities are the main supporters?	multiple-choice answers: • No public support • Municipalities • Regions • State • Funds from EU institutions • Funds from non-EU institutions
Why do consumers and prosumers find it interesting to be part of the REC?	multiple-choice answers: • Getting more equitable energy prices • Obtaining below-market prices • Making a safe investment • Making a profitable investment • Be involved in the management of energy services • Be involved in the management of non-energy services • Invest in sustainable energy • Invest in projects with ethical purposes other than sustainable energy • A sense of belonging to the community • Tax deductions

References

1. European Commission. *Going Climate-Neutral by 2050. A Strategic Long-Term Vision for a Prosperous, Modern, Competitive and Climate-Neutral EU Economy*; European Commission: Brussels, Belgium, 2019.
2. Lennon, B.; Dunphy, N.; Gaffney, C.; Revez, A.; Mullally, G.; O'Connor, P. Citizen or Consumer? Reconsidering Energy Citizenship. *J. Environ. Policy Plan.* **2020**, *22*, 184–197. [CrossRef]
3. Barberis, L. *Comunità Energetiche Rinnovabili. Un'opportunità Di Sviluppo Per Il Territorio*; Terzjus: Rome, Italy, 2021.
4. Italia Domani. *Piano Nazionale Di Ripresa E Resilienza (Pnrr)*; Italia Domani: Rome, Italy, 2021.
5. Ceglia, F.; Marraso, E.; Martone, C.; Pallotta, G.; Roselli, C.; Sasso, M. Towards the Decarbonization of Industrial Districts through Renewable Energy Communities: Techno-Economic Feasibility of an Italian Case Study. *Energies* **2023**, *16*, 2722. [CrossRef]
6. Menniti, D.; Pinnarelli, A.; Sorrentino, N.; Vizza, P.; Barone, G.; Brusco, G.; Mendicino, S.; Mendicino, L.; Polizzi, G. Enabling Technologies for Energy Communities: Some Experimental Use Cases. *Energies* **2022**, *15*, 6374. [CrossRef]
7. Moncecchi, M.; Meneghello, S.; Merlo, M. Energy Sharing in Renewable Energy Communities: The Italian Case. In Proceedings of the 2020 55th International Universities Power Engineering Conference (UPEC), Turin, Italy, 1–4 September 2020; IEEE: Piscataway, NJ, USA, 2020.
8. Petrichenko, L.; Sauhats, A.; Diahovchenko, I.; Segeda, I. Economic Viability of Energy Communities versus Distributed Prosumers. *Sustainability* **2022**, *14*, 4634. [CrossRef]
9. Jasinski, J.; Kozakiewicz, M.; Sołtysik, M. Determinants of Energy Cooperatives' Development in Rural Areas—Evidence from Poland. *Energies* **2021**, *14*, 319. [CrossRef]
10. Leone, M.F.; Amirante, R.; Sferratore, A. Renewable Energy Communities as Public Architectures and Socio-Ecological Infrastructures. *TECHNE J. Technol. Arch. Environ.* **2023**, *26*, 173–183. [CrossRef]

11. Otamendi-Irizar, I.; Grijalba, O.; Arias, A.; Pennese, C.; Hernández, R. How Can Local Energy Communities Promote Sustainable Development in European Cities. *Energy Res. Soc. Sci.* **2022**, *84*, 102363. [CrossRef]
12. Butera, F.M. Emissioni Nette Zero Al 2050: Transizione Tecnologica O Rivoluzione Culturale. *Techne* **2023**, *26*, 22–24. [CrossRef]
13. De Vidovich, L.; Tricarico, L.; Matteo, Z. Modelli Organizzativi Per Le Comunità Energetiche. Riflessioni Dalla Ricerca 'Community Energy Map'. *Impresa Soc.* **2023**, *1*, 122–137.
14. Walker, G.; Devine-Wright, P. Community Renewable Energy: What Should it Mean. *Energy Policy* **2008**, *36*, 497–500. [CrossRef]
15. Bernardoni, A.; Borzaga, C.; Sforzi, J. Comunità Energetiche Rinnovabili. Una Sfida Per Le Imprese Sociali E Di Comunità. *Impresa Soc.* **2022**, *2*, 77–82.
16. Lode, M.L.; te Boveldt, G.; Coosemans, T.; Ramirez Camargo, L. A Transition Perspective on Energy Communities: A Systematic Literature Review and Research Agenda. *Renew. Sustain. Energy Rev.* **2022**, *163*, 112479. [CrossRef]
17. Birch, E.; Wachter, S.M. *Growing Greener Cities: Urban Sustainability in the Twenty-First Century*; University of Pennsylvania Press: Philadelphia, PA, USA, 2008.
18. Vecchi, F.; Stasi, R.; Berardi, U. Modelling tools for the assessment of Renewable Energy Communities. *Energy Rep.* **2024**, *11*, 3941–3962. [CrossRef]
19. Lazzeroni, P.; Lorenti, G.; Moraglio, F.; Repetto, M. Modeling of Renewable Energy Communities: The RECoupled approach. In Proceedings of the IEEE 46h Annual Computers, Software, and Applications Conference, Los Alamitos, CA, USA, 27 June–1 July 2022.
20. CENSIS, Centro Studi Investimenti Sociali. *45° Rapporto Censis Sulla Situazione Sociale Del Paese*; Territorio e Reti: Rome, Italy, 2011; pp. 303–378.
21. Yeang, K. *Ecomasterplanning*; John Wiley and Sons Ltd.: London, UK, 2009.
22. Farr, D. *Sustainable Urbanism: Urban Design with Nature*; Wiley: Hoboken, NJ, USA, 2008.
23. Agencia d'Ecologia Urbana de Barcelona. *Plan De Indicadores De Sostenibilidad Urbana De Vitoria-Gasteiz*; Agencia d'Ecologia Urbana de Barcelona: Barcelona, Spain, 2010.
24. Kriken, J.L.; Enquist, P.; Rapaport, R. *City Building. Nine Planning Principles for the Twenty-First Century*; Princeton Architectural Press: New York, NY, USA, 2010.
25. Colombo, G. *Dizionario Di Urbanistica, Voci E Locuzioni D'Uso Corrente in Urbanistica Applicata*; Pirola Editore: Milan, Italy, 1991.
26. Alexander, C. *"Note Sulla Sintesi Della Forma." A City Is Not a Tree*; Il Saggiatore: Milan, Italy, 1967.
27. UN General Assembly. *Convention-Cadre sur les Changements Climatiques*; UN General Assembly: New York, NY, USA, 2016.
28. Perry, C. The Neighborhood Unit, Regional Survey of New York and Its Environs. In *Neighborhood and Community Planning*; Regional Plan of New York and Its Environs: New York, NY, USA, 1929; Volume 7.
29. Parlamento Europeo e Consiglio dell'Unione Europea. *Direttiva (Ue) 2018/12 Sulla Promozione Dell'Uso Dell'Energia Da Fonti Rinnovabili*; Gazzetta Ufficiale dell'Unione Europea (1028): Luxembourg, 2018.
30. Decreto Legislativo, 8 novembre 2021/199. *Attuazione Della Direttiva (Ue) 2018/2001 Del Parlamento Europeo E Del Consiglio, Dell'11 dicembre 2018, Sulla Promozione Dell'Uso Dell'Energia Da Fonti Rinnovabili. (21g00214) (Gu Serie Generale N.285 Del 30-11-2021-Suppl. Ordinario N. 42)*; Parlamento Italiano: Rome, Italy, 2021.
31. Pichetto Fratin, G. *Decreto-Legge 9 Dicembre 2023, N. 181. Disposizioni Urgenti Per La Sicurezza Energetica del Paese*; Parlamento Italiano: Rome, Italy, 2023.
32. UN General Assembly. *Transforming Our World: The 2030 Agenda for Sustainable Development*; A/RES/70/1; UN General Assembly: New York, NY, USA, 2015.
33. GSE & Legambiente, Comunità Energetiche Rinnovabili. *Il Punto Della Situazione in Italia*; Eroe, K., D'Agostino, O., Imparato, M., Eds.; GSE & Legambiente, Comunità Energetiche Rinnovabili: Rome, Italy, 2024.
34. Valentini, G. *Il Ricorso Al Finanziamento Tramite Terzi*; ENEA Editore: Milan, Italy, 1997.
35. NREL. Transforming Energy. Community Solar. 2021. Available online: https://www.nrel.gov/state-local-tribal/community-solar.html (accessed on 28 March 2024).
36. Chavin, S. *Microgrid Regulation Challenges and Opportunities*; Kleinman Center for Energy Policy: Philadelphia, PA, USA, 2022.
37. Heeter, J.; Xu, K.; Chan, G. *Sharing the Sun: Community Solar Deployment, Subscription Savings, and Energy Burden Reduction*; NREL Transforming Energy: Golden, CO, USA, 2021.
38. Bosco, A.; Scognamiglio, A. *Fotovoltaico E Riqualificazione Edilizia*; ENEA: Rome, Italty, 2005.
39. Vivoli, F.P.; Castello, S.; De Lia, F.; Graditi, G.; Scognamiglio, A.; Zingarelli, L.; Schioppo, R. *Progettare e Installare un Impianto Fotovoltaico*; ENEA Editore: Milan, Italy, 2008.
40. GSE. *Gruppi Di Autoconsumatori Di Energia Rinnovabile Che Agiscono Collettivamente E Comunità Di Energia Rinnovabile. Modalità Di Profilazione Dei Dati Di Misura E Relative Modalità Di Utilizzo Ai Sensi Dell'articolo 9 Dell'allegato a Alla Delibera 318/2020/r/eel*; GSE: Tbilisi, GA, USA, 2022.
41. Bosco, R.; Giacobbe, S.; Valente, R. *The Chance for Resc in Italy: Study for a Sizing Model. 11*; Kurt, E., Ed.; European Conference on Renewable Energy Systems ECRES: Riga, Latvia, 2023; pp. 276–282.
42. Cielo, A.; Margiaria, P.; Lazzeroni, P.; Mariuzzo, I.; Repetto, M. Renewable Energy Communities Business Models Under the 2020 Italian Regulation. *J. Clean. Prod.* **2021**, *316*, 128217. [CrossRef]
43. Fatin, A.; Franzese, A.; Magnabosco, G.; Nicoletto, L. La Città Come Risorsa Rinnovabile. Il Ruolo Dello Spazio Nella Riscrittura Del Territorio. In Proceedings of the Atti XXIII Conferenza Nazionale SIU. I Torino, Turin, Italy, 17–18 June 2021; pp. 38–47.

44. Ascione, F.; De Masi, R.F.; de Rossi, F.; Fistola, R.; Sasso, M.; Vanoli, G.P. Analysis and diagnosis of energy performance of buildings and districts: Methodology, validation and development of Urban Energy Maps. *Cities* **2013**, *35*, 270–283. [CrossRef]
45. Ceglia, F.; Esposito, P.; Faraudello, A.; Marrasso, E.; Rossi, P.; Sasso, M. An energy, environmental, management and economic analysis of energy efficient system towards renewable energy community: The case study of multi-purpose energy community. *J. Clean. Prod.* **2022**, *369*, 133269. [CrossRef]
46. Fischera, A.; Marraso, E.; Sasso, M.; Volpe, R. Energy, Environmental and Economic Performance of an Urban Community Hybrid Distributed Energy System. *Energies* **2020**, *13*, 2545. [CrossRef]
47. ARERA. *Delibera 04 Agosto 2020 318/2020/r/eel. Regolazione delle Partite Economiche Relative All'energia Elettrica Condivisa da un Gruppo di Autoconsumatori di Energia Rinnovabile che Agiscono Collettivamente in Edifici e Condomini Oppure Condivisa in una Comunità di Energia Rinnovabile*; ARERA: Milan, Italy, 2020.
48. Albert, C.; Brillinger, M.; Guerrero, P.; Gottwald, S.; Henze, J.; Schmidt, S.; Ott, E.; Schröter, B. Planning Nature-Based Solutions: Principles, Steps, and Insights. *Ambio* **2021**, *50*, 1446–1461. [CrossRef] [PubMed]
49. McPherson, E.G.; Simpson, J.R.; Peper, P.J.; Xiao, Q. Benefit-Cost Analysis of Modesto's Municipal Urban Forest. *J. Arboric.* **1999**, *25*, 235–248. [CrossRef]
50. Barlow, J.F. Progress in Observing and Modelling the Urban Boundary Layer. *Urb. Clim.* **2014**, *10*, 216–240. [CrossRef]
51. Valente, R.; Mozingo, L.A.; Bosco, R.; Cappelli, E.; Donadio, C. *Green Street Framework for Mediterranean Urban Fringe Areas. From Urban Districts to Eco-Districts Knowledge Methodologies, Strategic Programmes, Pilot Projects for Climate Adaptation*; Losasso, M., Bologna, R., Mussinelli, E., Tucci, F., Eds.; Maggioli: Santarcangelo di Romagna, Italy, 2021; pp. 206–249.
52. Campiotti, C.A.; Giagnacovo, G.; Latini, A.; Margiotta, F.; Nencini, L.; Pazzola, L.; Scoccianti, M. L'uso Della Vegetazione Per Aumentare L'efficienza Energetica Degli Edifici E L'impiego Di Sistemi Di Climatizzazione Rinnovabile. 2018. Available online: https://www2.enea.it/it/Ricerca_sviluppo/documenti/ricerca-di-sistema-elettrico/adp-mise-enea-2015-2017/edifici-intelligenti/report-2017/rds-par2017-084.pdf (accessed on 22 April 2024).
53. Legambiente. *Scuole E Università a Zero Emissioni: La Sfida Delle Comunità Energetiche Rinnovabili E Solidali*; Legambiente: Rome, Italy, 2022.
54. Basso, S.; Marchigiani, E. *Questioni Di Accessibilità: Gli Standard Per Un Progetto Di Formazioni Urbane Più Sane E Inclusive. Diritti in Città. Gli Standard Urbanistici in Italia Dal 1968 Ad Oggi*; Baioni, M., Basso, S., Caudo, G., Marchigiani, A.F., Munarin, S., Renzoni, C., Savoldi, P., Tosi, M.C., Vazzoler, N., Eds.; Donzelli ed.: Rome, Italy, 2021.
55. Chaudhry, S.; Surmann, A.; Kühnbach, M.; Pierie, F. Renewable Energy Communities as Modes of Collective Prosumership: A Multi-Disciplinary Assessment, Part I—Methodology. *Energies* **2022**, *15*, 8902. [CrossRef]
56. Molinari, A. Un'indagine su Dottrina Sociale della Chiesa, ecologia integrale, obiettivi di sviluppo sostenibile: Le ACLI, prospettiva di impegno civile. In *Laudato si' +5, Fratelli Tutti. Human Development. A Great Cultural, Spiritual Political, Economic, Educational Challenge*; Bornatici, S., Galeri, P., Gaspar, Y., Malavasi, P., Vacchelli, O., Eds.; Pensa MultiMedia Editore s.r.l.: Lecce, Italy, 2020; pp. 295–310. Available online: https://publicatt.unicatt.it/handle/10807/177846 (accessed on 22 April 2024).
57. Miniaci, R.; Valbonesi, P. La povertà energetica in Italia. In *Rivista Trimestrale di Cultura e di Politica*; Il Mulino: New York, NY, USA, 2022; pp. 182–190. [CrossRef]
58. Anzivino, M.; Ceravolo, F.A.; Rostan, M. *Il Public Engagement Degli Accademici Italiani: Un'Opportunità di Rapporto tra Università e Territorio*; Il Mulino: Bologna, Italy, 2018.

energies

MDPI

Review

Smart City and Energy: A Bibliometric Review of the Smart City and Smart Energy Concept from the Perspective of the Bioclimatic Approach

Saveria Olga Murielle Boulanger

Department of Architecture, University of Bologna, 40136 Bologna, Italy; saveria.boulanger@unibo.it

Abstract: Smart Cities have emerged as a promising approach for transforming urban living into more sustainable and resilient systems through technology-driven innovations and data-driven governance. Despite its growing implementation and diffusion around the globe, many questions surrounding this topic have emerged. Many critics have emerged since its first conceptualization in the first decade of the current century. Smart Cities have been criticized for their utopian objectives and the security, safety, people's freedom, and privacy within these systems. There are also capitalistic and neoliberalism-related critiques. Other critiques also highlight the current climate cost of Smart City initiatives. In the context of those critiques, bioclimatic and passive strategies might provide an interesting evolution of the concept but seem to be left in the background. This paper aims to contribute to the understanding of the linkages between environmental design approaches and the Smart City discourse. The contribution will explore to which extent bioclimatic and environmental design principles are present in the Smart City discourse and what the patterns are inside the current literature. The methodology of the research included a quali-quantitative analysis of the body of literature in Scopus and a bibliometric analysis using the VOS Viewer tool.

Keywords: Smart City; smart districts; bioclimatic; environmental design

1. Introduction

The conceptualization of Smart Cities has evolved into a visionary paradigm, offering a glimpse into the trajectory of urban life over the past two decades [1–3]. Since its inception at the onset of the preceding century, cities have undergone diverse transformations, most of them aligning with the Smart City vision, especially within big cities and metropolis [1,4]. Numerous urban centers have embraced this paradigm by implementing pioneering services across many domains, such as mobility, governance, healthcare, and education. The energy sector, in particular, has been strongly hit by the Smart City paradigm, developing approaches in many different paths, such as the energy regeneration of entire districts and cities through smart grids [5–9], the increase of clean energy production [10,11], and the involvement of communities in energy production in the form of energy communities or energy citizenship [12–15]. In all these approaches, ICT and data have a central role, constituting the underlying digital infrastructure that seems to enable many innovative approaches.

Within the realm of Smart Cities, a pronounced attention to digital integration is evident, particularly in the service sector, where many innovative solutions, tools, and connectivity modalities are emerging. At the same time, novel concepts such as the "digital twin" [16], the "15-minute city" [17–19], and the "metaverse city"[20] are surfacing. These concepts reveal a current tendency for continuous evolutions, especially in city-related definitions, supported by digitalization and, in general, by a technocratic approach.

Contemporary challenges, exemplified by the COVID-19 pandemic and energy crises, accentuate the pressing need for substantive, systemic changes rather than isolated innovations. The consensus prevails that Smart Cities can serve as a viable approach to address these challenges, leveraging technology to support policies and urban management across diverse domains, including climate change mitigation efforts [21]. Nevertheless, a universal strategy for transforming a city into a smart entity remains puzzling. On one hand, each city exhibits unique requirements, while on the other, shared impediments and challenges are discernible. The quest for a collective trajectory is propelled by the shared aspirations of cities to enhance efficiency, address societal issues, embrace ecological sustainability, and fortify resilience, among other objectives such as economic ones.

Inside this framework, critics of the Smart City paradigm have surfaced and continue to do so, highlighting its potential drawbacks in areas such as security, personal privacy, and economic disparities. Scholars raise concerns about the paradigm's alignment with a capitalist society, accentuating the widening gap between socioeconomic classes due to these approaches [22–28]. This critical discourse sheds light on the potential side effects of current city strategies, necessitating reflective adjustments. Within the energy field, in fact, these considerations are even more advanced if considering, for example, the phenomenon of energy poverty [29–32]. However, it is important to note that the correlations between the Smart City concept and energy poverty have been scarcely investigated; some first reflections are shown in [33,34]. How can cities plan their sustainable strategies, taking advantage of the possibilities that are given by the new technologies, but at the same time taking into consideration the possible side effects and backlashes? This is a key question that scholars around the world will need to address in the future.

Moreover, critics underscore the capital-intensive nature of the paradigm, emphasizing the pervasive reliance on costly technologies for concrete transformation into a Smart City. Drawing parallels to historical moments in architectural and building science, where the widespread availability of cooling and heating supplies led to standardized construction overlooking local climates, this critique advocates for a shift in approach. Similarly, the field of bioclimatic design arose from addressing the oversight, delineating energy into active and passive components. Active energy encompasses energy produced using equipment like heating systems, while passive energy exploits local climates via strategies such as natural ventilation and indirect heating systems, forming the core of environmental or bioclimatic design approaches [35–41]. In addition, some scholars are also highlighting the sustainable costs (e.g., land use) of energy generation from renewables [11].

This critique prompts a question: can a similar correlation between the Smart City paradigm and environmentally conscious bioclimatic design emerge? This paper aims to explore this question in the context of the energy sector, investigating the potential links between the Smart City discourse and environmentally friendly bioclimatic approaches.

Two hypotheses underscore this contribution. The first one is that the current trajectory of Smart City development relies excessively on costly technocratic approaches, not considering or including the passive methodologies provided by the bioclimatic approach. The second one is that there are potentialities of applying bioclimatic design concepts from architecture to the Smart City discourse.

In the next sections, these hypotheses are explored by performing a bibliometric analysis of the Scopus database. To provide a comprehensive analysis, the paper will use both a quantitative and qualitative approach, as deepened in Section 3.

The paper is structured into five sections. Section 1 provides an introduction to the topics and highlights the research hypothesis. Section 2 deepens a background qualitative review of the key aspects of Smart Energy, bioclimatic approaches in architecture, and Smart City's main critics, suggesting potential gaps to be covered. Section 3 highlights the main methodology used in this contribution both for the qualitative and quantitative parts. Section 4 will show the results and Section 5 will provide their discussion. Finally, Section 6 will provide a conclusion and future potential paths of research.

2. Background Qualitative Literature Review, Objectives, and Research Questions

The discourse surrounding Smart Cities has significantly escalated, especially post-2015, owing to influential global agendas and reports [42,43]. This attention stimulated different interdisciplinary contributions, revealing cities' nature as multi-layered systems [44–46]. Nonetheless, a prevailing challenge persists in the absence of a cohesive definition, resulting in fragmented viewpoints very linked within specific domains [28,42,44,47–52].

Sectors like the smart economy, governance, mobility, and environment have notably framed scholarly conversations around the Smart City, expanding to emergent areas like security, education, health, energy, and performance [53–80]. Despite this diversity, the literature predominantly centers on highlighting and dissecting technological innovations, which remains one of the main aspects of the Smart City discourse. At the same time, there is an increasing trend in exploring future trajectories and potential developments [42,44,50,75,81–93], especially with methodologies such as scenarios, forecasting, roadmaps, and pathways [94,95]. In addition, the advent of COVID-19 accelerated the adoption of digital applications, spotlighting the potential of Smart City solutions in fortifying urban resilience [96,97].

Scholars identify three primary drivers as the ones framing the Smart City concept—community, technology, and policy—emphasizing outcomes such as productivity, sustainability, accessibility, well-being, livability, and governance [93]. Notably, there has been a significant shift from exclusively conceptual discussions toward an approach focusing more on the interrelations with climate change and sustainability, showcasing an evolving connection between Smart City concepts and sustainable practices [42,92]. Proposals advocating for the new concept of Smart Sustainable Cities show the need for technological integration, adherence to sustainability principles, and respect for environmental constraints [92].

Despite this attention to the topic, several critiques have emerged for the Smart City concept. Among these, the most frequent critiques consider the actual development of the Smart City discourse as a reflection of neoliberal, capitalistic, and, thus, political and economic approaches more than place-based necessities of evolution.

In 2013, Vanolo [22] conducted a critical analysis of the Smart City concept, with a particular focus on its implications for power dynamics and knowledge within the contemporary urban environment. The author argued that the Smart City is not just a technological innovation but also a disciplinary strategy that shapes and governs urban spaces. He highlighted how the Smart City discourse is embedded in neoliberal ideologies, reinforcing social inequalities. Similarly, in 2014, Hollands [27,98] intervened critically in the dominance of the corporate Smart City model, highlighting the need for a critical perspective on this approach. The article argued that the corporate Smart City model is driven by profit motives and corporate interests, often neglecting the needs and voices of urban residents. Already at the time, he called for a more inclusive and participatory approach to Smart City development, emphasizing the importance of addressing social justice and equity concerns.

Kitchin's articles [23,24] identified present shortcomings in critical scholarship on Smart Cities, emphasizing the need for detailed genealogies, empirical case studies, and collaborative engagement with stakeholders. The author argued that existing critical perspectives on Smart Cities often lack in-depth empirical research and comprehensive analysis of specific Smart City initiatives. He called for a more nuanced and context-specific understanding of Smart Cities, highlighting the importance of engaging with diverse stakeholders to address the complex social, political, and ethical dimensions of urban technology. His articles provided a comprehensive overview of the limitations of current critical scholarship on Smart Cities and offered recommendations for future research and practice.

More recently, in 2021, Struver et al. [25] emphasized the absence of a critical and emancipatory debate on the right to a Smart City, calling for socially smart, just, and sustainable ideas and strategies. The article critically examined the corporate storytelling and narratives surrounding the smart urbanism concept, highlighting the need to question and challenge the dominant discourses that often prioritize technological solutions over social and environmental concerns. In line with Kitchin's approach, the authors advocated for a more inclusive and participatory approach to urban development, emphasizing the importance of empowering communities and promoting equitable access to urban resources. Finally, Lynch's article [26], in 2019, explored the values, beliefs, and practices of the movement for technological freedom in Barcelona, engaging in a critical conversation with work on urban political movements and alternative economies. The author examined how grassroots movements in Barcelona are contesting the dominant narratives of digital futures and Smart City development. He highlighted the importance of alternative economies and community-driven initiatives in shaping urban technology and governance. The article provides valuable insights into the potential for grassroots movements to challenge the top-down approaches to Smart City development and advocate for more democratic and inclusive urban futures.

Despite the presence of multiple critics also in more recent years, Smart City applications and evolutions are predominant. A lot of attention has been paid to the energy theme as a key component of the Smart City topic and as a sector able to provide substantial impacts, especially within the climate transition and the decarbonization of cities. Smart Energy is included in the principal subdivision of the topic [53]. The concept of Smart Energy has evolved over time, but there is still no widely accepted formal definition, as there is for the broader Smart City concept [99]. The term "Smart Energy" encompasses various aspects, including energy flexibility, intelligent building clusters, and the integration of small-scale energy sources within the structures of virtual power plants [100,101]. Smart Energy involves the application of advanced technologies and innovative approaches to improve energy performance, occupant comfort, and sustainability in buildings [102,103]. It also encompasses the management, coordination, and control of multiple energy subsystems, including renewable energy sources, to enhance energy efficiency and address the challenges of electrification [104,105]. Within the broader context of Smart City, Smart Energy can be clustered into various subdivisions, each addressing specific aspects of energy management and sustainability. These subdivisions include—but are not limited to—intelligent building clusters, energy flexibility, virtual power plants, and smart grid technologies. The concept of intelligent building involves electrically interconnected smart buildings within the same microgrid, aiming to optimize energy usage and enhance sustainability [100,106]. Energy flexibility, on the other hand, focuses on the adaptive and responsive energy management strategies within buildings, contributing to the overall concept of Smart Energy [100,106]. Virtual power plants represent an innovative approach to integrating small-scale energy sources and managing energy distribution in a decentralized manner, contributing to the resilience and reliability of energy systems [101,105]. Additionally, smart grid technologies play a crucial role in enabling the integration of renewable energy sources, demand response mechanisms, and advanced energy management solutions within the urban energy infrastructure [5,107].

As it is possible to see from this synthetic overview, the Smart Energy concept is devoted to "positive" forms of energy aiming to support climate transition, mainly renewable sources of energy and, within these, electricity. The "passive" aspects are completely left in the Smart City discourse. To understand the difference between "passive" and "active" energy, it is essential to consider the concepts of energy transfer and production. Passive systems do not produce energy internally, and strictly passive systems do not consume or dissipate input energy [108]. On the other hand, active systems involve generating or consuming energy [109]. Generally, the distinction between passive and active energy lies in the generation, consumption, and transfer of energy within a system. Passive systems do not produce energy internally and have limitations on energy extraction, while active systems involve energy generation, consumption, and fluctuations. In the context of sustainable building design and the urban dimension, the role of passive and active energy transitions has been recognized, emphasizing the interdependence between urban processes and energy transitions [110]. Furthermore, the study of passive design and alternative energy in building energy optimization has demonstrated that passive measures in building envelope design can significantly reduce overall energy consumption without compromising occupants' comfort [111]. Moreover, the optimal combination of passive design strategies has been identified as crucial for improving building energy performance while ensuring indoor thermal comfort [112]. The integration of passive and active technology packages has been proposed as a means to improve building energy efficiency, particularly in the context of zero-energy buildings [113]. Within this context, the bioclimatic approach is relevant. The bioclimatic approach in architecture involves integrating the natural climate conditions of a specific location into the architectural design process. This approach focuses on creating buildings that harmonize with their environment, utilizing passive design strategies to enhance energy efficiency and occupant comfort. By considering factors such as solar orientation, natural ventilation, thermal mass, and shading, architects can design buildings that effectively respond to the local climate, leading to reduced energy consumption and improved indoor environmental quality. Bioclimatic architecture aims to optimize the use of natural resources like sunlight, wind, and vegetation to minimize reliance on mechanical systems for heating, cooling, and lighting [37,114,115]. By exploring the seamless integration of bioclimatic principles from individual buildings to urban planning, the bioclimatic approach not only enhances the sustainability of structures but also extends its impact on entire urban environments. Some authors have recently started to investigate the potentialities of extending the bioclimatic approach of buildings to cities. By incorporating these principles into urban design, cities can strategically address challenges such as the urban heat island effect, elevate air quality, and cultivate inviting outdoor spaces that prioritize the well-being and comfort of residents [116,117].

The long history of inclusion of passive and bioclimatic design in the context of building and urban design makes the absence of this concept in the mainstream Smart City discourse surprising. It seems that the Smart City topic only revolves around active forms of energy (mainly in the realm of energy generation from renewables and smart distribution), as supported by the most common definitions of Smart Energy.

This is the main objective of the present contribution: to verify if passive design and bioclimatic approaches are present in the framework of Smart City in the current literature. An additional objective is to verify the presence of evolutionary trends in this direction. The paper aims to fill this gap by verifying the hypothesis through a bibliometric review of the Smart City topic concerning two specific elements:

- the understanding of the concept of Smart Energy and passive and bioclimatic design applied to the district dimension and, in particular, their relations and clusterization
- the understanding of potential trends and future research paths for a more passive-oriented and bioclimatic Smart City evolution.

This paper seeks to explore recent literature and provide thoughtful considerations for advancing research on Smart Cities. As such, the hypothesis that the paper aims to verify is the following:

- Passive and bioclimatic design strategies seem not to be included in the current debate about Smart Cities.

Based on this hypothesis, the two underlying research questions for this review are as follows:

- RQ1: Are there existing or potential correlations between the Smart City concept and passive and bioclimatic strategies, and what are those correlations?
- RQ2: Is it possible to foresee a potential future development of the Smart City discourse into more passive-oriented and bioclimatic approaches, or is this development completely outside this research field?

3. Materials and Methods

This review paper uses the methodology of bibliometric research with the intention of identifying patterns and trends through bibliometric analysis. To perform this analysis a systematic literature review was performed in the Scopus database. Data were analyzed with the VOS Viewer tool, an open-source tool able to create clusterization from bibliographic data [118].

The Scopus database was chosen for its reputation for indexing high-standard and peer-reviewed papers and for the substantial presence of the Smart City topic in the collections. Additionally, Scopus allows downloading formats which are compatible with VOSviewer, and it also provides some analysis of citations and results that are useful to direct the search and immediately have insights on emerging aspects. As referred to in Section 6, the inclusion of other databases, such as Web of Science, will be considered in further publications. In order to include all the relevant contributions and to answer the research questions, several entries have been considered in Scopus, as detailed in Table 1. The literature search was conducted between August 2023 and February 2024, with a final check in April 2024.

Table 1. Detail of the literature search on Scopus.

Query Wording	Details	No. of Appearances
"Smart city" OR "Smart cities" AND "energy"	In abstract, keywords, and title	8615 total 5043 after refinement
"Smart Energy"	In abstract, keywords, and title	4924 total 3590 after refinement
"Bioclimatic" AND "design" OR "architecture"	In abstract, keywords, and title	1170 941 after refinement
"Smart city" OR "Smart cities" AND "energy" AND "bioclimatic"	In abstract, keywords, and title	20

The literature search had different phases. The first one was a search in the general Smart City and energy framework. It included the wordings of "smart city" and "energy" in the topic. This was important to understand the general response of the database and check the big picture of the topic. This first search resulted in 8615 papers that were refined to a final number of 5043. The filtering included the exclusion of research fields not in line with the research question (e.g., mathematics, medicine), the type of contribution (excluded editorials, collections of papers), and the limitation to papers written in English. The second search aimed to better identify the keywords inside the specific Smart Energy sector. The identification of clusters has been derived from this search. The query included the wording "Smart Energy", and it resulted in 4924 papers, which became 3590 papers after refinement.

To understand the role of bioclimatic approaches, a two-step approach was then taken. The first one aimed at identifying the main keywords and clustering in the topic of bioclimatic. This query included the wording of "bioclimatic" and "design" or "architecture". This search resulted in 1170 papers, which became 941 papers after refinement. In this case, the refinement also regarded the exclusion of research fields not in line with the research question (e.g., mathematics, medicine, nursing) and the type of contribution (excluded editorials, collections of papers) and was limited to papers written in English.

A last step was carried out to check the presence of direct relationships between the Smart City topic and the bioclimatic one. The search included the query "smart city" with "energy" and "bioclimatic". However, only 20 papers emerged, and the results were not necessarily relevant. For this reason, this part has only been used for the literature background and the qualitative part of the research.

The first, second, and third searches were used to perform a clustering analysis with the VOS Viewer tool. In contrast, the last search was used qualitatively. Table 1 shows the details of the literature search on Scopus, while Figure 1 shows the flow chart of the methodology.

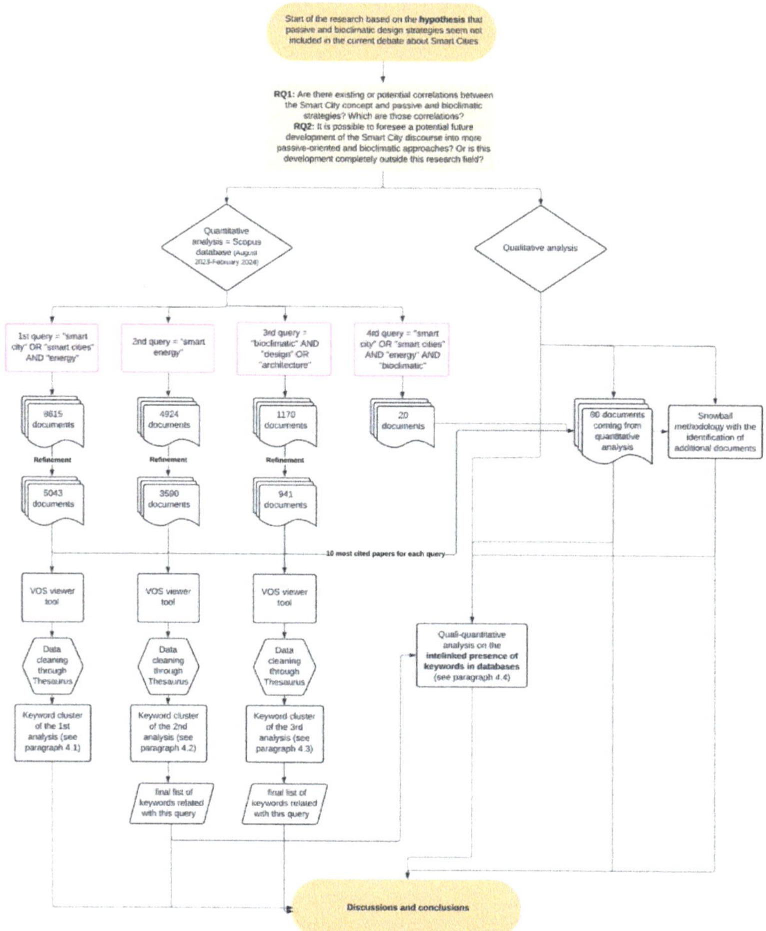

Figure 1. Flowchart of the methodology used, created by the author.

A total of 80 papers (including the 10 most cited papers for each query and the 20 papers for the final search) underwent a detailed qualitative analysis, while 9574 papers were included in the general quantitative analysis. The initial screening involved reviewing the titles and abstracts of the 80 papers, and a more thorough examination was conducted for the most important ones (included in the background literature review and in Section 5). These in-depth readings were crucial to ensure that key aspects of the topic were not missed in the co-occurrence maps and to avoid overlooking essential details in the analyses. Additionally, other contributions were identified through a qualitative approach during the database scan and were incorporated for further qualitative reviews.

For the bibliometric analysis, both Scopus's embedded tools and VOS Viewer were utilized. VOS Viewer, an open-source software available for all operating systems [118], has been widely adopted in recent papers across various research fields [42,119–121]. This software facilitates the correlation of terms found in literature, including those in titles, abstracts, authors, and journals. Moreover, VOS Viewer allows for a cleaning process involving the incorporation of thesaurus files created for each analysis. This step is crucial to prevent duplications, double counting, and normalization of abbreviations. After obtaining initial results, adjustments were made to clusterization parameters to generate a meaningful number of clusters, resulting in a more relevant map. Details regarding this procedure are delineated at the beginning of every subsection in Section 4. By default, all VOS Viewer analyses were conducted using textual data to showcase the co-occurrence of terms found in titles and abstracts. The input files utilized were bibliographic database files obtained from Scopus. A binary counting methodology was chosen to prevent duplicate counting within individual papers, with a minimum occurrence threshold of five documents set to exclude terms occurring in fewer than five documents.

However, as emphasized by Zheng et al. [87], drawing on the insights of numerous authors [122,123], hybrid techniques were employed in the analysis, integrating both qualitative and quantitative approaches. For the qualitative approach, a selection of 80 papers was meticulously reviewed and carefully read, starting with the top 10 most cited papers for each analysis. This selection has then been enlarged using a snowball methodology [124,125]. The findings from this qualitative reading are incorporated into Sections 2, 5 and 6.

4. Results

This section provides a comprehensive yet concise overview of the primary findings from the analysis. It is segmented into sub-paragraphs, with each subsection detailing the outcomes of a specific query.

4.1. The Smart City Concept in the Energy Context

This analysis aimed to provide an updated insight into the Smart City topic within the energy sector. The complete collection of papers was entered into the VOS Viewer software 1.6.18 using a suitable format, emphasizing word co-occurrence in both titles and abstracts. Subsequently, binary counting was employed, with a minimum threshold of five term co-occurrences set to be depicted on the map. The normalization process involved utilizing the association strength method.

A cleaning phase was executed, scrutinizing all considered words and incorporating a thesaurus file. This file guided the system to exclude non-relevant terms, preventing duplications (e.g., singular and plural words) and averting double counting (e.g., abbreviations), as elaborated in Section 2. Further optimizations were carried out in the number of clusters, aiming for the most optimal configuration. Each cluster comprised a minimum of 100 words. The outcomes are presented in Figure 2.

The map illustrated in Figure 2 highlights the emergence of four primary clusters concerning the Smart City topic within the energy framework. The largest cluster, denoted by the color red, comprises 315 items. This cluster encompasses keywords associated with various energy aspects intertwined with the Internet of Things, sensors, power management, and computing. Notably, the two principal keywords, "energy utilization" and "energy efficiency", demonstrate a robust connection to the query itself ("smart city" AND "energy"), establishing a strong link with all other branches. Within the red cluster, noteworthy connections stem from these two keywords. Firstly, the "Internet of Things" node leads to interconnected branches, including security-related keywords, sensor-related terms, energy computing, and, ultimately, power management and resource allocation.

Of significance within the red cluster are keywords associated with sensors, which manifest in multiple forms with substantial frequency. Additionally, the keyword related to "blockchain" surfaces prominently, affirming the growing significance of this topic within the field, as previously highlighted in 2022 [97].

The second cluster, containing 287 items, is identified by the color green. It encompasses keywords related to sustainability, covering various perspectives such as environmental sustainability and protection, sustainable cities, urban planning, climate change, decision-making, population-related aspects, and, lastly, buildings and architectural design. In this preliminary analysis, it appears that this cluster is more linked with architectural and sustainable approaches. Interestingly, even though the precise term is absent, bioclimatic design seems to have some elements in common with this cluster. Within this green cluster, the principal keywords include "sustainable development", "sustainability", "decision-making", "urban planning", "buildings", "artificial intelligence", and "information management". In contrast to the preceding cluster, this one seems to include more parallel keywords, hinting at a more varied and interconnected thematic landscape.

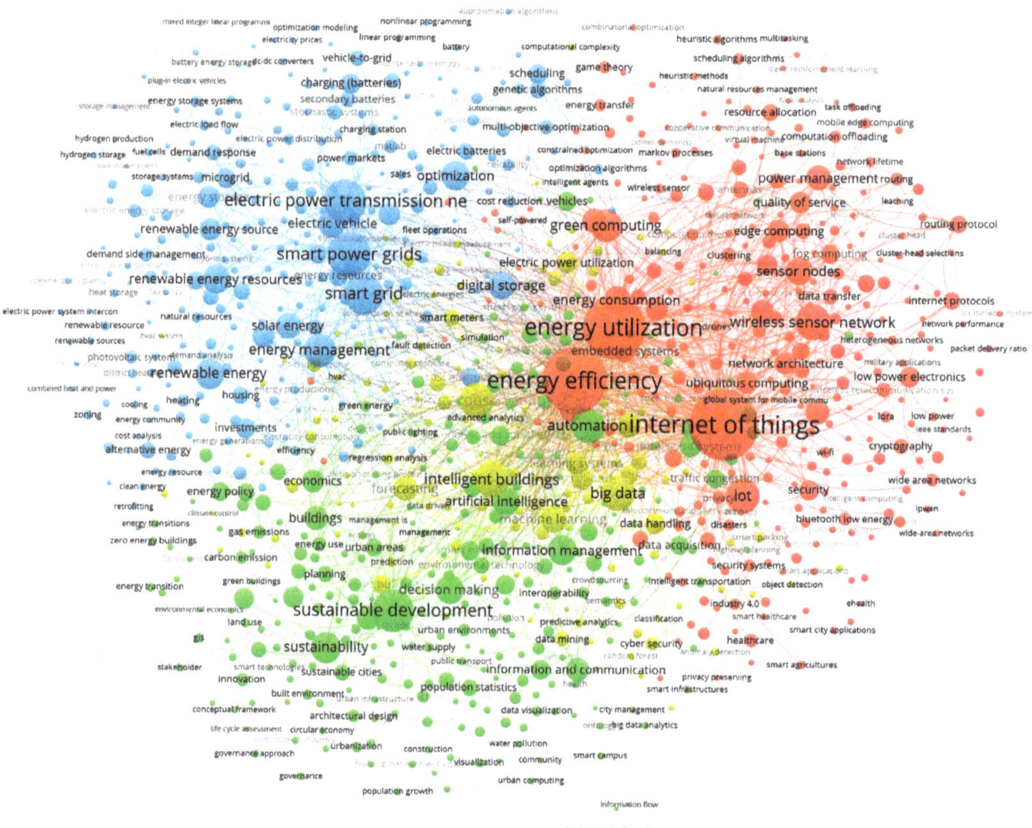

Figure 2. The output of the co-occurrence analysis of terms for the Smart City topic in the energy framework performed with VOS Viewer. The figure shows the physical distances among terms present in titles and abstracts of the literature analyzed, providing clusters of networks of co-occurring terms.

The third cluster, encompassing 264 items and characterized by the color blue, primarily revolves around topics related to the smart grid, storage, and renewable energies. This cluster includes several essential keywords, exploring various facets of the energy sector. However, a prevalent focus within this cluster is on the role of renewable energy sources and their management, emphasizing innovative distribution methods like the smart grid.

Concluding the analysis, the fourth cluster, represented by the color yellow and comprising 134 items, encompasses a diverse array of keywords. These range from "energy conservation" to "big data", "intelligent buildings", "machine learning", "smart homes", and "deep learning". Despite the higher diversity observed in this cluster compared to others, it appears that these clusters converge around the data aspects of the Smart City concept. This involves both methodologies, such as "forecasting", "learning systems", and "monitoring", and specific objects like "smart home" and "intelligent building".

4.2. Smart Energy Keywords Clusters

This analysis aims to provide an overview of the keyword clusters around the specific term of Smart Energy. As defined in the introduction, Smart Energy is intended to be one of the recognized components of a Smart City. Figure 3 shows the results of the cluster analysis.

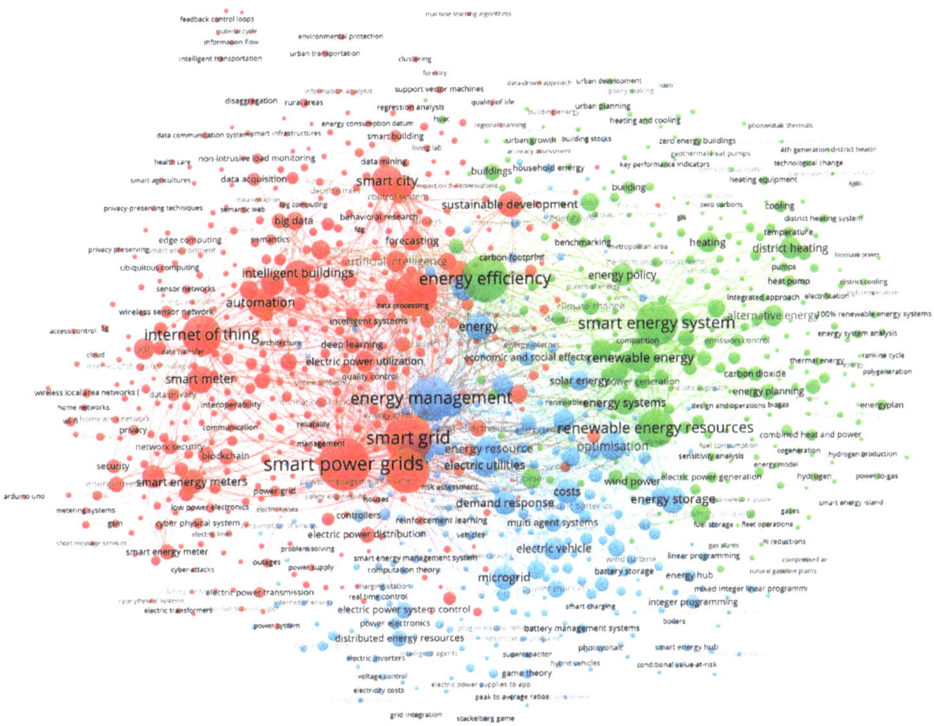

Figure 3. The output of the co-occurrence analysis of terms for the Smart Energy topic performed with VOS Viewer. The figure shows the physical distances among terms present in titles and abstracts of the literature analyzed, providing clusters of networks of co-occurring terms.

The image reveals three main clusters, each seemingly originating from a core keyword. Specifically, the red cluster is shaped by "energy utilization", the green cluster is shaped by "energy efficiency", and the blue cluster is shaped by "energy management".

The largest cluster, depicted in red and containing 461 items, revolves around keywords associated with the smart aspect of energy. It emphasizes terms like "smart grid", "internet of things", and "smart meters". Additionally, it incorporates specific keywords related to big data, smart monitoring, automation, and smart buildings. Two notable subtopics emerge within this cluster, visible as substantial nodes in the image. The first, located in the lower part, revolves around smart grids and electric power, showcasing strong connections with nearly all other keywords. This underscores the pivotal role of this aspect in the Smart City, particularly within the Smart Energy domain. The second subtopic, linked to the Internet of Things node, extends to smart metering, measuring aspects, smart buildings, and the broader Smart City theme.

The green cluster, with 280 items, is the second cluster, incorporating keywords linked with "energy efficiency". However, the nodes within this cluster extend beyond the "energy efficiency" theme, covering keywords related to energy generation, especially through renewable resources. The cluster appears divided into two core sub-groups.

The first centers around the general term "smart energy systems", connecting keywords such as "heating", "cooling", "district heating", and "buildings". The second sub-group, intrinsically linked with the overarching "smart energy system", emphasizes "renewable energy resources", interconnecting with various types of renewable resources. In contrast to the red cluster, which seems like the juxtaposition of two parallel sub-groups, the green one appears more interconnected, resembling a matryoshka where each main keyword contains the subsequent one.

The final cluster, depicted in blue and comprising 259 items, is initiated by the keyword "energy management". This keyword establishes links with numerous smaller nodes, differing from the more concentrated links in the red and green clusters. Despite the presence of the management keyword, the other words within this cluster seem more aligned with a range of subtopics, spanning vehicles, microgrids, distributed energy resources (proximate to the red cluster), storage-related terms, and economic-related words.

In contrast to the prior analysis, which focused on locating the energy topic within the Smart City context, this analysis reveals a broader array of keywords outside the robust connections. These peripheral keywords, visible at the edges of the image, exhibit minimal or no connections with the main network. This suggests that the topic remains dynamic, encompassing various subtopics and specifications that delve into significant specialization and detail.

4.3. Environmental Design and Bioclimatic Keyword Clusters

This analysis aims to provide an overview of the keyword clusters around the specific terms of bioclimatic design and architecture. As recalled in Section 3, the objective of this analysis is to better understand which main keywords are included in this research field in order to detect their position within the Smart City and Smart Energy topic. The results of the VOS Viewer analysis are synthetized in Figure 4.

Figure 4 shows the presence of three core clusters. The first one is the red one, with 131 items. It collects keywords pertaining to the theoretical concept around bioclimatic design and bioclimatic architecture. In particular, it is possible to see words such as "sustainable development", "architectural design", "energy efficiency", and "housing" but also "environmental impact", "sustainable buildings", "intelligent buildings", "environmental performance", "climate control", and "zero energy buildings". The cluster includes many synonyms linked with more sustainable and passive types of architectural practice, together with keywords related to the generic concepts of sustainability.

Three main keywords seem to create relative subgroups. The first one is "sustainable development", which is mainly linked with the broad topic of sustainability. The second one is "architectural design", strongly linked with the "housing" keywords. Both words refer to the more practical aspects of bioclimatic implementation in the architectural field, which are also linked with the building dimension (embodied by the green cluster). Links to this subgroup are, for example, keywords such as "sustainable construction", "intelligent buildings", "participatory design", "economic and social effects", "energy", and "building envelope". It is interesting to note that this cluster also contains the words "energy" and "energy efficiency", highlighting the presence of a clear connection between the two topics. The third subgroup is related to the keyword "energy efficiency", which is linked with "zero energy buildings" and the literature sector devoted to environmental and energy performance analysis ("life cycle", "environmental comfort", "environmental performance", and "environmental impact").

The second cluster is the green one, and it contains 82 items. It mainly contains keywords related to the building dimension and passive energy. It is interesting to see that the main keywords in this cluster are "buildings" and "energy utilization". The first one is linked with both the design part of the bioclimatic approach ("structural design", and design strategy") and the energy one ("heating", "cooling", and "bioclimatic chart"). The second one, "energy utilization", is connected with all the energy-related words such as "energy management", "energy performance", "passive design", "passive solar", "energy use", "energy savings", "energy consumption", and "energy conservation". The second cluster seems very linked with the energy sector, especially within the building dimension.

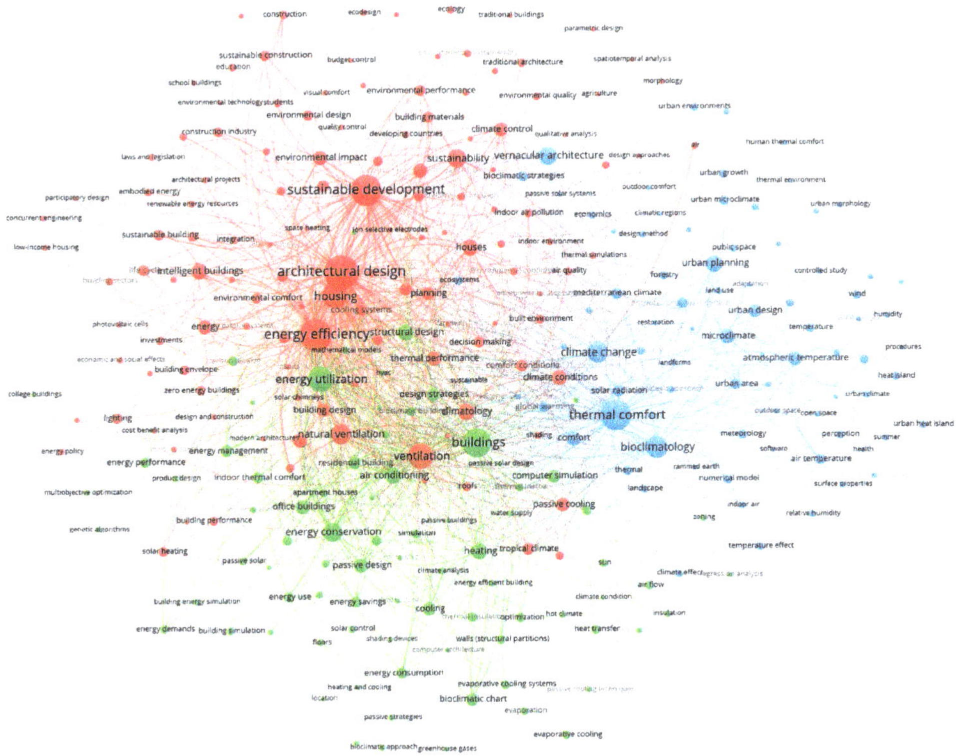

Figure 4. The output of the co-occurrence analysis of terms for the bioclimatic design and bioclimatic architecture field performed with VOS Viewer. The figure shows the physical distances among terms present in titles and abstracts of the literature analyzed, providing clusters of networks of co-occurring terms.

The third cluster is the blue one, and it contains 82 items as well. It has different subgroups mainly related to bioclimatology and climate. Strong keywords are "thermal comfort" with "bioclimatology" as a first sub-group, "climate change" with "urban planning" as a second sub-group, and "vernacular architecture" as a last small one.

Specific attention can be paid to the links of the "energy" word. This map clearly shows a link between the topic of the bioclimatic approach and the energy sector. As shown in Figure 5, the "energy" keyword seems to be mainly linked with "sustainable building" and "intelligent building" for the building-related part of the red cluster, then to the three core keywords of the same cluster ("sustainable development", "architectural design", and "energy efficiency"), and then to the words "planning" and ventilation for what concern the red cluster. It is then linked with the green cluster and, in particular, with "energy utilization", "buildings", "air conditioning", and "energy conservation". A final link is also present with the blue cluster, with the words "climate change", "thermal comfort", and "comfort".

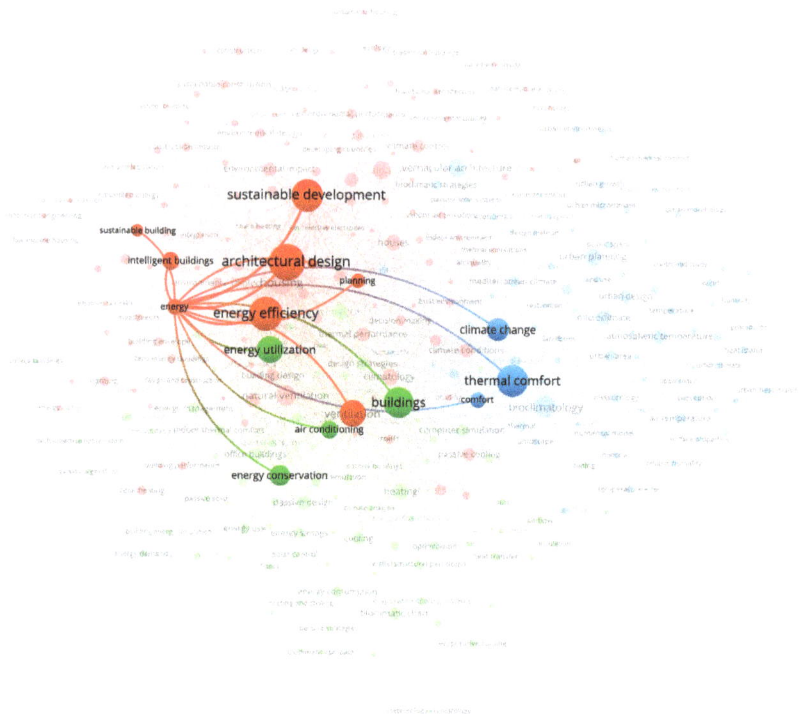

Figure 5. The output of the co-occurrence analysis of terms for the bioclimatic design and bioclimatic architecture field. The image shows, in particular, the connections of the word "energy" with other keywords. The figure shows the physical distances among terms present in titles and abstracts of the literature analyzed, providing clusters of networks of co-occurring terms.

For the objective of the paper, it is also crucial to note that the words linked with the Smart City or Smart Energy fields are not present in the results of this analysis, showing the absence of a strict connection between the topics of bioclimatic design and the Smart City discourse.

4.4. Interlinks between Environmental Design Keywords and the Smart Energy and Smart City Fields

This section aims to detect the interlinked presence of the most recurrent keywords in the three searches. The analysis was done in two steps: the first was searching the keywords of the bioclimatic network (see Section 4.3) into the Smart City and Smart Energy ones (Table 2), and the second was searching the keywords of the Smart City and Smart Energy sectors into the bioclimatic one (Table 3). Both analyses are intended to qualitatively verify the results detected within the background literature review and the VOS Viewer-based analyses. The selection of the keywords is made qualitatively and quantitatively considering the following: (1) the qualitative relevance in the field (selected based on the background literature review) and (2) their dimension in the correspondent energy map (the ten most recurrent keywords in the database).

The verification analysis strongly supports the outcomes from the VOS Viewer analysis. Specifically, it appears that all themes fall within the same broad framework of studies, which is sustainability. However, there's currently no clear link between the bioclimatic approach and the realms of Smart Cities or Smart Energy. Notably, the ten most frequently used keywords in the bioclimatic sector are largely mirrored in the Smart City and Smart Energy domains, emphasizing the commonality of the overall framework.

Table 2. Interlinks among keywords (k. in the table) in the three searches performed with VOS Viewer (n.m. stands for "network map").

Keywords from the Bioclimatic Network	Selection Methodology	Smart City n. m.	Smart Energy n. m.
Bioclimatic design	Main topic	No	No
Bioclimatic architecture	Main topic	No	No
Architectural design	Ten most recurrent k.	Yes	Yes
Energy efficiency	Ten most recurrent k.	Yes	Yes
Sustainable development	Ten most recurrent k.	Yes	Yes
Thermal comfort	Ten most recurrent k.	Yes	Yes
Buildings	Ten most recurrent k.	Yes	Yes
Ventilation	Ten most recurrent k.	Yes	Yes
Energy utilization	Ten most recurrent k.	Yes	Yes
Housing	Ten most recurrent k.	Yes	Yes
Bioclimatology	Ten most recurrent k.	No	no
Climate change	Ten most recurrent k.	Yes	Yes
Ecodesign	Qualitative	No	No
Zero-energy buildings	Qualitative	Yes	Yes
Comfort	Qualitative	No	No
Bioclimatic building	Qualitative	No	No
Environmental quality	Qualitative	No	No
Sustainable building	Qualitative	No	No
Passive energy	Qualitative	No	No
Embodied energy	Qualitative	No	No
Climate control	Qualitative	Yes	Yes
Microclimate	Qualitative	No	No

Table 3. Interlinks among keywords (k. in the table) in the three searches performed with VOS Viewer. n.m. stands for "network map", SC stands for "smart city", and SE stands for "smart energy".

Keywords from the Smart City and Smart Energy Networks	Selection Methodology	Bioclimatic n. m.
Internet of things	Ten most recurrent k. (SC + SE)	No
Energy efficiency	Ten most recurrent k. (SC + SE)	Yes
Energy utilization	Ten most recurrent k. (SC + SE)	Yes
Smart power grids	Ten most recurrent k. (SC + SE)	No
Electric power transmission networks	Ten most recurrent k. (SC + SE)	No
Sustainable development	Ten most recurrent k. (SC)	Yes
Smart grid	Ten most recurrent k. (SC + SE)	No
Intelligent buildings	Ten most recurrent k. (SC)	Yes
Wireless sensor network	Ten most recurrent k. (SC)	No
Big data	Ten most recurrent k. (SC)	No
Energy management	Ten most recurrent k. (SE)	Yes
Smart energy system	Ten most recurrent k. (SE)	No
Renewable energy resources	Ten most recurrent k. (SE)	Yes
Energy management system	Ten most recurrent k. (SE)	No

This alignment is also evident in Table 3, where the most general keywords from the Smart City and Smart Energy searches overlap with those in the bioclimatic domain. Nevertheless, the unique, qualitatively selected keywords in the bioclimatic field are mostly absent in the Smart City and Smart Energy sectors. Similarly, the most precise keywords associated with the smart discourse are not evident in the bioclimatic field.

5. Discussion

This paper aims to contribute to the study of the Smart City concept, making some small progress in its connection to the Smart Energy sector and, specifically, the bioclimatic approach. The focus is on exploring a specific aspect of literature, particularly contributions related to integrating passive and bioclimatic approaches into the Smart City field. A range of search terms was used on Scopus to map out different identified domains. Both qualitative and bibliometric analyses were conducted using the Scopus database and VOS Viewer tools. The results reveal some interesting points, which will be discussed in relation to the research questions.

Research Question 1: Are there any existing or potential correlations between the Smart City concept and passive and bioclimatic strategies, and what are they?

According to the analysis, the answer is twofold. On the negative side, the findings indicate that direct connections between the Smart City and the bioclimatic approach are not emerging. The Smart City discourse primarily focuses on active forms of energy rather than incorporating considerations about the passive use of natural resources, as seen in the bioclimatic debate (e.g., passive solar gains, natural ventilation, and passive cooling). This is also quite evident in some new publications that show some potential future trends in the Smart City discourse in the direction of artificial intelligence and blockchain [126,127]. However, both domains fall under the broader framework of sustainable design and the broader field of climate change-related studies, suggesting a potential correlation in the future, given the increasing costs, including land use, associated with generating new energy, even from renewables. This aspect has been widely observed and analyzed by Ahveniemmi et al. [128] in 2017 in a highly cited paper. The authors, in particular, observe the presence of an overlapping of the two concepts, proposing a new definition of "smart sustainable cities", to underline the importance of including sustainability in the Smart City discourse, also from a monitoring perspective. Furthermore, the approach taken by Kylili et al. [129] shows some interesting approaches to the link between the Smart City and sustainable building design. In their contribution, they see this link mainly through the BIM (Building Information Modeling) technologies and the life-cycle approach. As it is possible to see, in this case, the relationship is not direct between Smart City and bioclimatic but between Smart City and more nuanced approaches linked with the more general sustainable design.

Finally, both fields involve two main scales of analysis: the city (including the intermediate dimension of districts/neighborhoods) and the building. However, it is possible to note how the few papers trying to create a link between Smart Cities and the bioclimatic approach are mainly looking at the outdoor dimension, especially at the district scale [130], and linking the topic to resilience strategies [131,132].

Even if, quantitatively, the interlink is not evident, from a qualitative perspective, some specific publications are investigating this potentiality. In particular, Serghides et al. [116] focus on the bioclimatic approach in developing sustainable cities, particularly within the EU ERANET project "Smart bioclimatic low-carbon urban areas as innovative energy isles in the sustainable city" (SUI). The main objective of the project is to develop sustainable cities by balancing energy systems at a local level, with a focus on urban energy and CO_2 reduction. The bioclimatic approach is one of the three cornerstone procedures of the project, alongside Smart Grids and Management Platform. Even if the Smart City is not directly cited, it is clear that the project's direction is aligned with the Smart City approach.

Research Question 2: Can we foresee a potential future development of the Smart City discourse toward more passive-oriented and bioclimatic approaches, or is this development beyond this research field?

Based on the analysis, there are no clear existing overlaps, and no elements (quantitative or qualitative) support an evolution of the Smart City discourse toward more passive-oriented approaches. However, considering the shared general framework of both fields, it is plausible to suggest that future interrelations may emerge. Some studies reflect on the cost of energy generation, even from renewables, and the bioclimatic approach has long considered this, especially at the building scale. This potential future development is not entirely beyond the research field but is not yet present.

6. Conclusions

In summary, the review has shed light on aspects of the Smart City concept and proposed potential lines of further research. The methodology utilized has involved two main aspects: firstly, conducting a qualitative review of recent contributions on the Smart City topic, with a specific focus on their connections with both active and passive forms of energy—highlighting the significance of bioclimatic approaches; and secondly, performing a bibliometric analysis utilizing the VOS Viewer tool. This analysis has identified some literature gaps that could be addressed in future research and has raised questions that could guide future reviews. The key findings are as follows:

1. There is a lack of substantial emphasis on bioclimatic approaches or more nuanced and passive forms of energy within the discourse surrounding Smart Cities and Smart Energy. However, both fields share the overarching goal of sustainable cities and sustainability, suggesting the potential for future connections to emerge.
2. Even if there are no contributions creating a direct link between the two concepts, some reflections on the costs (e.g., in land use) of active forms of energy are arising, poising doubt on the long-term sustainability of these technologies and paving the way for the inclusion of different approaches.

The consequences of this study mainly revolve around affirming and spotlighting prevailing trends in Smart City development, especially considering the connection between two important fields of investigation such as the Smart City and the Bioclimatic one, while also putting forward new hypotheses for further exploration. However, a notable limitation of this study is the scarcity of actual contributions that recognize a connection between these two fields of inquiry. Future research will focus on expanding the literature base, including querying the Web of Science database using similar search terms, to determine whether there is a larger body of literature that either supports or modifies the findings presented in this study.

Funding: This research received no external funding.

Conflicts of Interest: The author declares no conflict of interest.

References

1. Boulanger, S. *Smarter and Greener: A Technological Path for Urban Complexity*; Ricerche di Tecnologia dell'Architettura; Franco Angeli: Milano, Italy, 2020; ISBN 978-88-917-9061-3.
2. Boulanger, S.O.M. Smart City: Utopia o realtà? Comprendere l'evoluzione per comprendere la trasformazione. *FAMagazine Ric. Progett. Sull'architettura Città* **2015**, *1*, 24–33. [CrossRef]
3. Valdez, A.-M.; Cook, M.; Potter, S. Roadmaps to utopia: Tales of the smart city. *Urban Stud.* **2018**, *55*, 3385–3403. [CrossRef]
4. Crysler, C.G.; Cairns, S.; Heynen, H. Metropolis, megalopolis and metacity. In *The Sage Handbook of Architectural Theory*; Sage: London, UK, 2012.
5. Tricoire, A. Uncertainty, Vision, and the Vitality of the Emerging Smart Grid. *Energy Res. Soc. Sci.* **2015**, *9*, 21–34. [CrossRef]
6. Bruckner, D.; Haase, J.; Palensky, P.; Zucker, G. Latest trends in integrating building automation and smart grids. In Proceedings of the IECON 2012-38th Annual Conference on IEEE Industrial Electronics Society, Montreal, QC, Canada, 25–28 October 2012; pp. 6285–6290.
7. Lyulyov, O.; Vakulenko, I.; Pimonenko, T.; Kwiliński, A.; Dźwigoł, H.; Dźwigoł-Barosz, M. Comprehensive Assessment of Smart Grids: Is There a Universal Approach? *Energies* **2021**, *14*, 3497. [CrossRef]
8. Mollah, M.B.; Zhao, J.; Niyato, D.; Lam, K.-Y.; Zhang, X.; Ghias, A.M.Y.M.; Koh, L.H.; Yang, L. Blockchain for Future Smart Grid: A Comprehensive Survey. *IEEE Internet Things J.* **2021**, *8*, 18–43. [CrossRef]
9. Butt, O.M.; Zulqarnain, M.; Butt, T.M. Recent Advancement in Smart Grid Technology: Future Prospects in the Electrical Power Network. *Ain Shams Eng. J.* **2021**, *12*, 687–695. [CrossRef]
10. Hoang, A.T.; Pham, V.V.; Nguyen, X.P. Integrating Renewable Sources into Energy System for Smart City as a Sagacious Strategy towards Clean and Sustainable Process. *J. Clean. Prod.* **2021**, *305*, 127161. [CrossRef]
11. Vuković, N.; Koriugina, U.; Illarionova, D.; Pankratova, D.; Kiseleva, P.; Gontareva, A. Towards Smart Green Cities: Analysis of Integrated Renewable Energy Use in Smart Cities. *Strat. Plan. Energy Environ.* **2023**, *40*, 75–94. [CrossRef]
12. Devine-Wright, P. Towards zero-carbon: Citizenship, responsibility and the public acceptability of sustainable energy technologies. *Sol. Energy Soc.-Energy Inst. Jt. Conf. Proc.* **2004**, *21*, 51–62.

13. Lennon, B.; Dunphy, N.; Gaffney, C.; Revez, A.; Mullally, G.; O'Connor, P. Citizen or consumer? Reconsidering energy citizenship. *J. Environ. Policy Plan.* **2020**, *22*, 184–197. [CrossRef]
14. Olivadese, R.; Alpagut, B.; Revilla, B.P.; Brouwer, J.; Georgiadou, V.; Woestenburg, A.; van Wees, M. Towards Energy Citizenship for a Just and Inclusive Transition: Lessons Learned on Collaborative Approach of Positive Energy Districts from the EU Horizon2020 Smart Cities and Communities Projects. *Proceedings* **2021**, *65*, 20. [CrossRef]
15. Massari, M.; Boulanger, S.O.M.; Longo, D. Collective energy actions to pursue a just transition. A Southern European observation. *Eur. J. Spat. Dev.* **2024**, *21*, 1–32. [CrossRef]
16. Ahn, E.-Y. Analysis of Digital Twin Technology Trends Related to Geoscience and Mineral Resources after the Korean New Deal Policy in 2020. *Econ. Environ. Geol.* **2021**, *54*, 659–670. [CrossRef]
17. Allam, Z.; Nieuwenhuijsen, M.; Chabaud, D.; Moreno, C. The 15-minute city offers a new framework for sustainability, liveability, and health. *Lancet Planet. Health* **2022**, *6*, e181–e183. [CrossRef] [PubMed]
18. Moreno, C.; Allam, Z.; Chabaud, D.; Gall, C.; Pratlong, F. Introducing the "15-Minute City": Sustainability, Resilience and Place Identity in Future Post-Pandemic Cities. *Smart Cities* **2021**, *4*, 93–111. [CrossRef]
19. Pozoukidou, G.; Chatziyiannaki, Z. 15-Minute City: Decomposing the New Urban Planning Eutopia. *Sustainability* **2021**, *13*, 928. [CrossRef]
20. Allam, Z.; Sharifi, A.; Bibri, S.E.; Jones, D.S.; Krogstie, J. The Metaverse as a Virtual Form of Smart Cities: Opportunities and Challenges for Environmental, Economic, and Social Sustainability in Urban Futures. *Smart Cities* **2022**, *5*, 771–801. [CrossRef]
21. Shahrabani, M.M.N.; Apanavičienė, R. Towards Integration of Smart and Resilient City: Literature Review. *IOP Conf. Series: Earth Environ. Sci.* **2022**, *1122*, 012019. [CrossRef]
22. Vanolo, A. Smartmentality: The Smart City as Disciplinary Strategy. *Urban Stud.* **2013**, *51*, 883–898. [CrossRef]
23. Kitchin, R. The real-time city? Big data and smart urbanism. *GeoJournal* **2014**, *79*, 1–14. [CrossRef]
24. Kitchin, R. Making Sense of Smart Cities: Addressing Present Shortcomings. *Camb. J. Reg. Econ. Soc.* **2014**, *8*, 131–136. [CrossRef]
25. Strüver, A.; Saltiel, R.; Schlitz, N.; Hohmann, B.; Höflehner, T.; Grabher, B. A Smart Right to the City—Grounding Corporate Storytelling and Questioning Smart Urbanism. *Sustainability* **2021**, *13*, 9590. [CrossRef]
26. Lynch, C.R. Contesting Digital Futures: Urban Politics, Alternative Economies, and the Movement for Technological Sovereignty in Barcelona. *Antipode* **2019**, *52*, 660–680. [CrossRef]
27. Hollands, R.G. Critical Interventions into the Corporate Smart City. *Camb. J. Reg. Econ. Soc.* **2014**, *8*, 61–77. [CrossRef]
28. Bria, F. *Morozov, Evgeny Ripensare la Smart City, Codice Edizioni*; Codice: Torino, Italy, 2018; ISBN 978-88-7578-765-3.
29. Boeri, A.; Gianfrate, V.; Boulanger, S.O.M.; Massari, M. Future Design Approaches for Energy Poverty: Users Profiling and Services for No-Vulnerable Condition. *Energies* **2020**, *13*, 2115. [CrossRef]
30. Longo, D.; Olivieri, G.; Roversi, R.; Turci, G.; Turillazzi, B. Energy Poverty and Protection of Vulnerable Consumers. Overview of the EU Funding Programs FP7 and H2020 and Future Trends in Horizon Europe. *Energies* **2020**, *13*, 1030. [CrossRef]
31. Papada, L.; Katsoulakos, N.; Kaliampakos, D. Fighting Energy Poverty by Going Underground. *Procedia Eng.* **2016**, *165*, 49–57. [CrossRef]
32. Bouzarovski, S.; Thomson, H.; Cornelis, M. Confronting Energy Poverty in Europe: A Research and Policy Agenda. *Energies* **2021**, *14*, 858. [CrossRef]
33. Parks, D.; Rohracher, H. From Sustainable to Smart: Re-Branding or Re-Assembling Urban Energy Infrastructure? *Geoforum* **2019**, *100*, 51–59. [CrossRef]
34. Jagganath, G. The Transforming City: Exploring the Potential for Smart Cities and Urban Agriculture in Africa. *Orient. Anthr. A Bi-Annu. Int. J. Sci. Man* **2021**, *22*, 24–40. [CrossRef]
35. Esfehankalateh, A.T.; Farrokhzad, M.; Tamaskani Esfehankalateh, F.; Soflaei, F. Bioclimatic passive design strategies of traditional houses in cold climate regions. *Environ. Dev. Sustain.* **2022**, *24*, 10027–10068. [CrossRef]
36. Ameur, M.; Kharbouch, Y.; Mimet, A. Optimization of passive design features for a naturally ventilated residential building according to the bioclimatic architecture concept and considering the northern Morocco climate. *Build. Simul.* **2020**, *13*, 677–689. [CrossRef]
37. Aghimien, E.I.; Li, D.H.W.; Tsang, E.K.-W. Bioclimatic architecture and its energy-saving potentials: A review and future directions. *Eng. Constr. Arch. Manag.* **2022**, *29*, 961–988. [CrossRef]
38. Bestara, A.K. Bioclimatic Architecture as a Design Approach for High School in Gedebage, Bandung City. In Proceedings of the International Webinar on Digital Architecture 2021 (IWEDA 2021), Yogyakarta, Indonesia, 13–14 December 2021.
39. Michael, A.; Phocas, M.C. Construction design and sustainability in architecture: Integrating environmental education in the architectural studies. *Renew. Energy Power Qual. J.* **2012**, *1*, 190–195. [CrossRef]
40. Michael, A.; Phocas, M.C. Bioclimatic approaches of modern residential architecture in Cyprus, 1952-1974. *Renew. Energy Power Qual. J.* **2010**, *1*, 27–32. [CrossRef]
41. Nguyen, A.T.; Reiter, S. Bioclimatism in architecture: An evolutionary perspective. *Int. J. Des. Nat. Ecodynamics* **2017**, *12*, 16–29. [CrossRef]
42. Sharifi, A.; Allam, Z.; Feizizadeh, B.; Ghamari, H. Three Decades of Research on Smart Cities: Mapping Knowledge Structure and Trends. *Sustainability* **2021**, *13*, 7140. [CrossRef]

43. Catriona Manville, G.; Kotterink, B. *Mapping Smart Cities in the EU*; EPRS—European Parliamentary Research Service: Brussels, Belgium, 2014; p. 200.
44. Kirimtat, A.; Krejcar, O.; Kertesz, A.; Tasgetiren, M.F. Future Trends and Current State of Smart City Concepts: A Survey. *IEEE Access* **2020**, *8*, 86448–86467. [CrossRef]
45. Antonini, E.; Boulanger, S.O.M.; Gaspari, J. Multi-Layered Urban Strategies to Foster the Smart Cities Development. *WIT Trans. Ecol. Environ.* **2015**, *194*, 23–34. [CrossRef]
46. Silva, B.N.; Khan, M.; Han, K. Towards sustainable smart cities: A review of trends, architectures, components, and open challenges in smart cities. *Sustain. Cities Soc.* **2018**, *38*, 697–713. [CrossRef]
47. Camero, A.; Alba, E. Smart City and information technology: A review. *Cities* **2019**, *93*, 84–94. [CrossRef]
48. Cocchia, A. Smart and Digital City: A Systematic Literature Review. In *Smart City, Progress in IS*; Dameri, R.P., Rosenthal-Sabroux, C., Eds.; Springer International Publishing: Cham, Switzerland, 2014. [CrossRef]
49. Dall'O, G. *Smart City. La Rivoluzione Intelligente delle Città*; Il Mulino: Bologna, Italy, 2014.
50. Kim, J. Smart city trends: A focus on 5 countries and 15 companies. *Cities* **2022**, *123*, 103551. [CrossRef]
51. Bonomi, A.; Masiero, R. *Dalla Smart City alla Smart Land, Marsilio Editore*; Marsilio Editore: Venezia, Italy, 2014; ISBN 978-88-317-1884-4.
52. Batty, M. Big data, smart cities and city planning. *Dialog-Hum. Geogr.* **2013**, *3*, 274–279. [CrossRef] [PubMed]
53. Giffinger, R.; Fertner, C.; Kramar, H.; Kalasek, R.; Pichler-Milanović, N.; Meijers, E. *Smart Cities. Ranking of European Medium-Sized Cities*; Centre of Regional Science Vienna UT: Vienna, Austria, 2007.
54. Anttiroiko, A.V.; Valkama, P.; Bailey, S. J Smart cities in the new service economy: Building platforms for smart services. *AI Soc* **2014**, *29*, 323–334. [CrossRef]
55. Intille, S.S. The goal: Smart people, not smart homes. In Proceedings of the International Conference on Smart Homes and Health Telematics (ICOST2006), Belfast, UK, 26–28 June 2006; IOS Press: Amsterdam, The Netherlands, 2006; pp. 3–6.
56. Barab, S.A.; Plucker, J.A. Smart People or Smart Contexts? Cognition, Ability, and Talent Development in an Age of Situated Approaches to Knowing and Learning. *Educ. Psychol.* **2002**, *37*, 165–182. [CrossRef]
57. Deakin, M. Smart cities: The state-of-the-art and governance challenge. *Triple Helix* **2014**, *1*, 7. [CrossRef]
58. Papa, R.; Gargiulo, C.; Battarra, R. *Città Metropolitane e Smart Governance: Iniziative di Successo e nodi Critici verso la Smart City*; FedOA-Federico II University Press: Napoli, Italy, 2016; Volume 1, ISBN 978-88-6887-005-8.
59. Meijer, A.; Bolivar, M.P.R. Governing the smart city: A review of the literature on smart urban governance. *Int. Rev. Adm. Sci.* **2016**, *82*, 417–435. [CrossRef]
60. Bakogiannis, E.; Siti, M.; Kyriakidis, C.; Vassi, A. Using Traditional and New Digital Technology Tools to Promote Sustainable Mobility: Current Trends in the Evolving Transformation of the Smart City. In *Smart Cities in the Mediterranean: Coping with Sustainability Objectives in Small and Medium-Sized Cities and Island Communities*; Stratigea, A., Kyriakides, E., Nicolaides, C., Eds.; Springer: Berlin/Heidelberg, Germany, 2017; pp. 113–133. ISBN 2196-8705.
61. Orlowski, A.; Romanowska, P. Smart Cities Concept: Smart Mobility Indicator. *Cybern. Syst.* **2019**, *50*, 118–131. [CrossRef]
62. Paiva, S.; Ahad, M.A.; Tripathi, G.; Feroz, N.; Casalino, G. Enabling Technologies for Urban Smart Mobility: Recent Trends, Opportunities and Challenges. *Sensors* **2021**, *21*, 2143. [CrossRef]
63. Schmitter-Edgecombe, M.; Cook, D.J. Assessing the Quality of Activities in a Smart Environment. *Methods Inf. Med.* **2009**, *48*, 480–485. [CrossRef]
64. Fazio, M.; Celesti, A.; Puliafito, A.; Villari, M. Big Data Storage in the Cloud for Smart Environment Monitoring. *Procedia Comput. Sci.* **2015**, *52*, 500–506. [CrossRef]
65. Vinod Kumar, T.M. (Ed.) *Smart Environment for Smart Cities*; Springer: Berlin/Heidelberg, Germany, 2020; pp. 1–53. [CrossRef]
66. Ullo, S.L.; Sinha, G.R. Advances in Smart Environment Monitoring Systems Using IoT and Sensors. *Sensors* **2020**, *20*, 3113. [CrossRef] [PubMed]
67. Valcárcel-Aguiar, B.; Murias, P.; Rodríguez-González, D. Sustainable Urban Liveability: A Practical Proposal Based on a Composite Indicator. *Sustainability* **2019**, *11*, 86. [CrossRef]
68. Nugent, C.; Augusto, J.C. *Smart Homes and Beyond: ICOST 2006*; IOS Press: Amsterdam, The Netherlands, 2006; ISBN 978-1-60750-178-7.
69. Collotta, M.; Pau, G. An Innovative Approach for Forecasting of Energy Requirements to Improve a Smart Home Management System Based on BLE. *IEEE Trans. Green Commun. Netw.* **2017**, *1*, 112–120. [CrossRef]
70. Haque, A.K.M.B.; Bhushan, B.; Dhiman, G. Conceptualizing smart city applications: Requirements, architecture, security issues, and emerging trends. *Expert Syst.* **2022**, *39*, e12753. [CrossRef]
71. Liu, Z.; Seo, H.; Sun, H.; Huang, C. Special Issue on Emerging Information Security and Privacy Trends for Smart City Foreword. *J. Inf. Sci. Eng.* **2017**, *33*, 873–1100.
72. Diaz-Parra, O.; Fuentes-Penna, A.; Barrera-Camara, R.; Trejo-Macotela, F.; Ramos-Fernandez, J.; Ruiz-Vanoye, J.; Zezzatti, A.; Rodriguez-Flores, J. Smart Education and future trends. *Int. J. Comb. Optim. Probl. Inform.* **2022**, *13*, 65–74.
73. Bin Ahmad, K.A.; Khujamatov, H.; Akhmedov, N.; Bajuri, M.Y.; Ahmad, M.N.; Ahmadian, A. Emerging trends and evolutions for smart city healthcare systems. *Sustain. Cities Soc.* **2022**, *80*, 103695. [CrossRef]
74. Amin, S.U.; Hossain, M.S.; Muhammad, G.; Alhussein, M.; Rahman, A. Cognitive Smart Healthcare for Pathology Detection and Monitoring. *IEEE Access* **2019**, *7*, 10745–10753. [CrossRef]

75. Angelidou, M.; Politis, C.; Panori, A.; Bakratsas, T.; Fellnhofer, K. Emerging smart city, transport and energy trends in urban settings: Results of a pan-European foresight exercise with 120 experts. *Technol. Forecast. Soc. Change* **2022**, *183*, 10745–10753. [CrossRef]
76. Boeri, A.; Boulanger, S.O.M.; Turci, G.; Pagliula, S. Enabling strategies for mixed-used PEDs: Energy efficiency between smart cities and Industry 4.0. *Techne* **2021**, *22*, 170–180. [CrossRef]
77. Matheri, A.; Ngila, J.; Njenga, C.K.; Belaid, M.; van Rensburg, N.J. Smart City Energy Trend Transformation in the Fourth Industrial Revolution Digital Disruption. In Proceedings of the IEEE International Conference on Industrial Engineering and Engineering Management (IEEM), Macao, China, 15–18 December 2019; pp. 978–984.
78. Manitiu, D.N.; Pedrini, G. Smart and Sustainable Cities in the European Union. An Ex Ante Assessment of Environmental, Social, and Cultural Domains; Sustainability Environmental Economics and Dynamics Studies. 2015. Available online: https://boa.unimib.it/retrieve/handle/10281/106251/155701/wp%20seeds.pdf (accessed on 10 March 2024).
79. Kourtit, K.; Deakin, M.; Caragliu, A. An advanced triple helix network framework for smart cities performance. In *Smart Cities: Governing, Modelling and Analysing the Transition*; Routledge: London, UK, 2013; Volume 196, ISBN 978-0-203-07622-4.
80. Lombardi, P.; Giordano, S.; Farouh, H.; Yousef, W. Modelling the smart city performance. *Innov. Eur. J. Soc. Sci. Res.* **2012**, *25*, 137–149. [CrossRef]
81. Juma, M.; Shaalan, K. Cyber-Physical Systems in Smart City: Challenges and Future Trends for Strategic Research. In Proceedings of the International Conference on Advanced Intelligent Systems and Informatics 2019; Hassanien, A., Shaalan, K., Tolba, M., Eds.; Springer: Berlin/Heidelberg, Germany, 2020; Volume 1058, pp. 855–865. [CrossRef]
82. Katalin, F. Smart city trends and concepts according to the most popular collaborative documentation. *Inf. Tarsad.* **2017**, *17*, 25–38.
83. Kemec, A. Analysis of smart city global research trends with network map technique. *Manag. Res. Pract.* **2022**, *14*, 46–59.
84. Neirotti, P.; De Marco, A.; Cagliano, A.C.; Mangano, G.; Scorrano, F. Current trends in Smart City initiatives: Some stylised facts. *Cities* **2014**, *38*, 25–36. [CrossRef]
85. Mora, L.; Bolici, R.; Deakin, M. The First Two Decades of Smart-City Research: A Bibliometric Analysis. *J. Urban Technol.* **2017**, *24*, 3–27. [CrossRef]
86. Guo, Y.-M.; Huang, Z.-L.; Guo, J.; Li, H.; Guo, X.-R.; Nkeli, M.J. Bibliometric Analysis on Smart Cities Research. *Sustainability* **2019**, *11*, 3606. [CrossRef]
87. Zheng, C.; Yuan, J.; Zhu, L.; Zhang, Y.; Shao, Q. From digital to sustainable: A scientometric review of smart city literature between 1990 and 2019. *J. Clean. Prod.* **2020**, *258*, 120689. [CrossRef]
88. Pérez, L.M.; Oltra-Badenes, R.; Gutiérrez, J.V.O.; Gil-Gómez, H. A Bibliometric Diagnosis and Analysis about Smart Cities. *Sustainability* **2020**, *12*, 6357. [CrossRef]
89. Moradi, S. The scientometrics of literature on smart cities. *Libr. Hi Tech* **2019**, *38*, 385–398. [CrossRef]
90. Zhao, L.; Tang, Z.-Y.; Zou, X. Mapping the Knowledge Domain of Smart-City Research: A Bibliometric and Scientometric Analysis. *Sustainability* **2019**, *11*, 6648. [CrossRef]
91. Samuel, F.W.F.; Queiroz, M.M. A Bibliometric Analysis and Research Agenda on Smart Cities. In Proceedings of the International Working Conference on Transfer and Diffusion of IT (TDIT), Accra, Ghana, 21–22 June 2019; Dwivedi, Y., Ayaburi, E., Boateng, R., Effah, J., Eds.; Springer International Publishing: Accra, Ghana, 2019; Volume AICT-558, pp. 325–335. [CrossRef]
92. Janik, A.; Ryszko, A.; Szafraniec, M. Scientific Landscape of Smart and Sustainable Cities Literature: A Bibliometric Analysis. *Sustainability* **2020**, *12*, 779. [CrossRef]
93. Yigitcanlar, T.; Kamruzzaman; Buys, L.; Ioppolo, G.; Sabatini-Marques, J.; da Costa, E.M.; Yun, J.J. Understanding 'smart cities': Intertwining development drivers with desired outcomes in a multidimensional framework. *Cities* **2018**, *81*, 145–160. [CrossRef]
94. Boulanger, S.O.M. Running after pathways: A critical reflection on climate change roadmaps. In *Architectures for Next Generation EU Cities*; Franco Angeli: Milan, Italy, 2022; pp. 47–58.
95. Antonini, E. Uncertainty, fragility, resilience. *Agathon-Int. J. Archit. Art Des.* **2019**, *6*, 6–13. [CrossRef]
96. Sharifi, A.; Khavarian-Garmsir, A.R.; Kummitha, R.K.R. Contributions of Smart City Solutions and Technologies to Resilience against the COVID-19 Pandemic: A Literature Review. *Sustainability* **2021**, *13*, 8018. [CrossRef]
97. Boulanger, S.O.M. The Roadmap to Smart Cities: A Bibliometric Literature Review on Smart Cities' Trends before and after the COVID-19 Pandemic. *Energies* **2022**, *15*, 9326. [CrossRef]
98. Hollands, R. Will the real smart city please stand up. In *The Routledge Companion to Smart Cities*; Routledge: London, UK, 2020; pp. 303–320.
99. Callcut, M.; Agliozzo, J.-P.C.; Varga, L.; McMillan, L. Digital Twins in Civil Infrastructure Systems. *Sustainability* **2021**, *13*, 11549. [CrossRef]
100. Sibilla, M.; Manfren, M. Envisioning Building-as-Energy-Service in the European Context. From a literature review to a conceptual framework. *Arch. Eng. Des. Manag.* **2021**, *18*, 495–520. [CrossRef]
101. Kucęba, R. Dimensions and Factors that Determine Integration of Small-Scale Sources in the Structures of Virtual Power Plants. *Prod. Eng. Arch.* **2022**, *28*, 185–192. [CrossRef]
102. Newton, S.; Shirazi, A.; Christensen, P. Defining and Demonstrating a Smart Technology Configuration to Improve Energy Performance and Occupant Comfort in Existing Buildings: A Conceptual Framework. *Int. J. Build. Pathol. Adapt.* **2021**, *41*, 182–200. [CrossRef]

103. Benavente-Peces, C. On the Energy Efficiency in the Next Generation of Smart Buildings—Supporting Technologies and Techniques. *Energies* **2019**, *12*, 4399. [CrossRef]
104. Hussain, H.M.; Narayanan, A.; Nardelli, P.H.J.; Yang, Y. What is Energy Internet? Concepts, Technologies, and Future Directions. *IEEE Access* **2020**, *8*, 183127–183145. [CrossRef]
105. Cioara, T.; Antal, M.; Mihailescu, V.T.; Antal, C.D.; Anghel, I.M.; Mitrea, D. Blockchain-Based Decentralized Virtual Power Plants of Small Prosumers. *IEEE Access* **2021**, *9*, 29490–29504. [CrossRef]
106. Vigna, I.; Pernetti, R.; Pasut, W.; Lollini, R. New Domain for Promoting Energy Efficiency: Energy Flexible Building Cluster. *Sustain. Cities Soc.* **2018**, *38*, 526–533. [CrossRef]
107. Ma, L.; Liu, N.; Wang, L.; Zhang, J.; Lei, J.; Zeng, Z.; Wang, C.; Cheng, M. Multi-Party Energy Management for Smart Building Cluster with PV Systems Using Automatic Demand Response. *Energy Build.* **2016**, *121*, 11–21. [CrossRef]
108. Antoulas, A.C.; Ionutiu, R.R.; Martins, N.; ter Maten, E.J.; Mohaghegh, K.; Pulch, R.; Rommes, J.; Saadvandi, M.; Striebel, M. Model Order Reduction: Methods, Concepts and Properties. In *Coupled Multiscale Simulation and Optimization in Nanoelectronics*; Springer: Berlin/Heidelberg, Germany, 2015. [CrossRef]
109. Abreu-Vieira, G.; Xiao, C.; Gavrilova, O.; Reitman, M.L. Integration of Body Temperature into the Analysis of Energy Expenditure in the Mouse. *Mol. Metab.* **2015**, *4*, 461–470. [CrossRef] [PubMed]
110. Huang, P.; Broto, V.C. Interdependence Between Urban Processes and Energy Transitions: The Dimensions of Urban Energy Transitions (DUET) Framework. *Environ. Innov. Soc. Transitions* **2018**, *28*, 35–45. [CrossRef]
111. Gondal, I.A.; Athar, M.S.; Khurram, M. Role of Passive Design and Alternative Energy in Building Energy Optimization. *Indoor Built Environ.* **2019**, *30*, 278–289. [CrossRef]
112. Lu, H.-P.; Chen, C.-S.; Yu, H. Technology roadmap for building a smart city: An exploring study on methodology. *Futur. Gener. Comput. Syst.* **2019**, *97*, 727–742. [CrossRef]
113. Sung, U.-J.; Kim, S.-H. Development of a Passive and Active Technology Package Standard and Database for Application to Zero Energy Buildings in South Korea. *Energies* **2019**, *12*, 1700. [CrossRef]
114. Okoye, P.U.; Ogbuagu, O.G.; Ohaedeghasi, C.I.; Ngwu, C. Bioclimatic Practices in Modern Residential Building Design and Construction in South-Eastern Nigeria. *J. Constr. Eng. Manag. Innov.* **2020**, *3*, 158–178. [CrossRef]
115. Ahriz, A.; Mesloub, A.; Elkhayat, K.; Alghaseb, M.A.; Abdelhafez, M.H.; Ghosh, A. Development of a mosque design for a hot, dry climate based on a holistic bioclimatic vision. *Sustainability* **2021**, *13*, 6254. [CrossRef]
116. Serghides, D.K.; Dimitriou, S.; Kyprianou, I.; Papanicolas, C. The Bioclimatic Approach in Developing Smart Urban Isles for Sustainable Cities. *Renew. Energy Environ. Sustain.* **2019**, *4*, 2. [CrossRef]
117. Chiesa, G.; Li, Y. Including Urban Heat Island in Bioclimatic Early-Design Phases: A Simplified Methodology and Sample Applications. *Sustainability* **2021**, *13*, 5918. [CrossRef]
118. Van Eck, N.J.; Waltman, L. Visualizing bibliometric networks. In *Measuring Scholarly Impact*; Springer: Cham, Switzerland, 2014; pp. 285–320.
119. Ranjbari, M.; Esfandabadi, Z.S.; Ferraris, A.; Quatraro, F.; Rehan, M.; Nizami, A.-S.; Gupta, V.K.; Lam, S.S.; Aghbashlo, M.; Tabatabaei, M. Biofuel supply chain management in the circular economy transition: An inclusive knowledge map of the field. *Chemosphere* **2022**, *296*, 133968. [CrossRef]
120. Shah, S.H.H.; Lei, S.; Ali, M.; Doronin, D.; Hussain, S.T. Prosumption: Bibliometric analysis using HistCite and VOSviewer. *Kybernetes* **2019**, *49*, 1020–1045. [CrossRef]
121. Yu, Y.; Li, Y.; Zhang, Z.; Gu, Z.; Zhong, H.; Zha, Q.; Yang, L.; Zhu, C.; Chen, E. A bibliometric analysis using VOSviewer of publications on COVID-19. *Ann. Transl. Med.* **2020**, *8*, 816. [CrossRef] [PubMed]
122. King, J. A review of bibliometric and other science indicators and their role in research evaluation. *J. Inf. Sci.* **1987**, *13*, 261–276. [CrossRef]
123. Cobo, M.J.; López-Herrera, A.G.; Herrera-Viedma, E.; Herrera, F. Science mapping software tools: Review, analysis, and cooperative study among tools. *J. Am. Soc. Inf. Sci. Technol.* **2011**, *62*, 1382–1402. [CrossRef]
124. Greenhalgh, T.; Peacock, R. Effectiveness and Efficiency of Search Methods in Systematic Reviews of Complex Evidence: Audit of Primary Sources. *BMJ* **2005**, *331*, 1064–1065. [CrossRef]
125. van Haastrecht, M.; Sarhan, I.; Ozkan, B.Y.; Brinkhuis, M.; Spruit, M. SYMBALS: A Systematic Review Methodology Blending Active Learning and Snowballing. *Front. Res. Metrics Anal.* **2021**, *6*, 685591. [CrossRef] [PubMed]
126. Alaeddini, M.; Hajizadeh, M.; Reaidy, P. A Bibliometric Analysis of Research on the Convergence of Artificial Intelligence and Blockchain in Smart Cities. *Smart Cities* **2023**, *6*, 764–795. [CrossRef]
127. Szum, K. IoT-based smart cities: A bibliometric analysis and literature review. *Eng. Manag. Prod. Serv.* **2021**, *13*, 115–136. [CrossRef]
128. Ahvenniemi, H.; Huovila, A.; Pinto-Seppä, I.; Airaksinen, M. What are the differences between sustainable and smart cities? *Cities* **2017**, *60*, 234–245. [CrossRef]
129. Kylili, A.; Georgali, P.-Z.; Christou, P.; Fokaides, P. An integrated building information modeling (BIM)-based lifecycle-oriented framework for sustainable building design. *Constr. Innov.* **2024**, *24*, 492–514. [CrossRef]
130. Dunichkin, I.V.; Ilina, I.N. Climate adaptation of Smart City by assessing bioclimatic comfort for UBEM. *Smart Innov. Syst. Technol.* **2020**, *163*, 519–529. [CrossRef]

131. Chondrogianni, D.V.; Stephanedes, Y.J. Assessment of the Bioclimatic Index for resilient urban spaces in Mediterranean cities. *Environ. Plan. B Urban Anal. City Sci.* **2024**, *51*, 234–258. [CrossRef]
132. Chondrogianni, D.V.; Stephanedes, Y.J. Evaluation of urban planning methods toward bioclimatic and resilient urban spaces. *Environ. Plan. B Urban Anal. City Sci.* **2022**, *49*, 1354–1370. [CrossRef]

energies

MDPI

Article

Integrated Energy and Environmental Modeling to Design Cost-Effective Building Solutions at a Regional Level

Mariana Januário, Ricardo Gomes, Patrícia Baptista and Paulo Ferrão *

IN+, Centre for Innovation, Technology and Policy Research, LARSyS—Laboratory for Robotics and Engineering Systems, Instituto Superior Técnico, Universidade de Lisboa, Av. Rovisco Pais, 1049-001 Lisboa, Portugal; mariana.januario@tecnico.ulisboa.pt (M.J.); ricardo.a.gomes@tecnico.ulisboa.pt (R.G.); patricia.baptista@tecnico.ulisboa.pt (P.B.)

* Correspondence: ferrao@tecnico.ulisboa.pt

Abstract: This study introduces a computationally efficient urban building energy model (UBEM) to assess decarbonization strategies for the residential sector at the regional level. The model considers a range of inputs, including building characteristics, climate data, technology penetration, and occupant behavior. The model provides an economic analysis associating emission reduction potential with economic returns through an abatement cost curve, which is critical to designing cost-effective solutions. The model was validated at its full scale in Portugal, using actual consumption data from all municipalities. Key findings showed that lighting upgrades (100% LEDs) are the most cost-effective measure, offering the lowest abatement cost (−521 EUR/tonCO_2eq) and a low discounted payback period of 2 years, while heat pumps for water heating provide the highest emission reduction potential, with an annual reduction of 863 tonnes of CO_2eq annually, equivalent to a 20% reduction in national emissions. Additionally, behavioral measures achieved an annual reduction of 147 tonnes of CO_2eq. The analysis further reveals that, while some measures might have a negative abatement cost at the national level, their economic viability varies locally, with certain municipalities incurring positive abatement costs, highlighting how local context affects the economic viability of decarbonization strategies.

Keywords: decarbonization; residential energy model; CO_2eq emission reduction; abatement cost

1. Introduction

In recent years, the effects of global warming have become increasingly worrying for humanity as climate change continues to intensify. It has led to the creation of plans and agreements by governments worldwide to address the issue, such as the Paris Agreement, which aims to limit global warming by limiting global temperature increase to well below 2 degrees Celsius [1]. Achieving these climate targets requires coordinated efforts across multiple sectors, each facing its own set of challenges. While sectors like transportation struggle with the complexity of energy consumption and carbon reductions, the building sector has been recognized as having the potential to make a greater contribution to energy and emission reductions [2].

Within the European Union (EU), buildings account for approximately 40% of the total energy consumption and 36% of greenhouse gas (GHG) emissions [3,4]. As a result, the residential sector holds significant potential for energy savings and emission reductions. Addressing this potential is essential to reach the EU's ambitious climate targets. Policies promoting energy efficiency in buildings, combined with the transition to renewable energy, are seen as crucial steps towards reducing emissions and improving energy performance.

In this context, urban building energy models (UBEMs) are essential tools for analyzing residential energy consumption and emissions, supporting the development of effective policies at scales ranging from individual buildings to neighborhoods or cities [5]. UBEMs enable researchers and policymakers to simulate alternative scenarios for evaluating decarbonization measures, such as energy efficiency upgrades, retrofitting, and renewable integration [6,7].

Several UBEM tools have become widely adopted for urban applications [5]. For example, CityBes, developed by Hong et al. [8], is a platform intended to simulate the energy performance of building stocks across cities, supporting the evaluation of retrofit strategies and energy-saving interventions. In a recent study, Li et al. [9] expanded CityBes capabilities to include district heating and cooling systems, enabling city planners to model district-level energy configurations and assess their impact on energy demand and emissions. Additionally, Fonseca and Schlueter [10] developed CEA, a UBEM designed to simulate energy consumption and emissions at the urban scale, with a case study in Zurich. Their model integrates spatial and temporal data, providing detailed insights into urban districts' potential for reducing emissions through a range of energy efficiency measures and interventions. Other widely used UBEM platforms include UMI, SimStadm, and CitySIM [5].

While these UBEM platforms offer valuable insights at the neighborhood, district, and city levels, extending them to regional or national applications remains a significant challenge due to several key factors. Johari et al. [11] highlight significant data requirements and computational complexity for UBEM even at the city scale. These challenges suggest that scaling such models to a national level would be substantially more difficult. Additionally, Ali et al. [12] point out that models based on detailed simulations require substantial computational resources, thereby limiting their scalability to larger regions or national levels. Sola et al. [13] similarly note that, while detailed UBEM simulation engines predict the operational energy use of individual buildings very well, it is common for difficulties to appear when scaling up to modeling at larger scales. These difficulties, along with the increased computational requirements, make it challenging for traditional UBEMs, designed primarily for single cities or districts, to maintain efficiency at larger scales, often leading to significantly longer run times.

For instance, Chen et al. [14] simulated 940 buildings in San Francisco using CityBES, which required hours to days, depending on the level of detail and available computational infrastructure. For larger simulations, like the 177,023 buildings across the entire city of San Francisco, the computational demands became even more pronounced.

While current UBEMs perform well at the city or district level, they often lack the computational efficiency necessary for national-scale deployment without the support of high-performance infrastructure. A study by Beltrán-Velamazán et al. [15] further underscores this limitation, highlighting that national-scale UBEMs, though valuable for estimating energy consumption across a country's building stock, demand supercomputing capabilities to handle simulations efficiently.

This paper addresses this gap by presenting a computationally efficient model capable of simulating multiple regions within a country, accounting for each region's specific context, and even extending to national-scale simulations. The model was designed to be computationally efficient, enabling simulations at a national scale in less than one minute while still maintaining the necessary detail for the analysis of cost-effective building solutions. This allows the model to bridge the gap between highly localized urban models and broader national models, providing insights into energy consumption and emissions across larger geographical areas without the typical computational constraints of more detailed models.

In addition to its computational efficiency, the model introduces an economic analysis through the use of abatement cost curves. This feature links environmental performance with economic feasibility, providing policymakers with critical insights into the cost-effectiveness of various building solutions designed to reduce CO_2eq emissions.

To demonstrate its applicability, the model was applied to Portugal with desegregated data at the municipal level, covering all 308 municipalities. Portugal was selected as a case study to highlight the model's adaptability in accounting for diverse regional factors, such as local climate variations, building characteristics, and penetration technologies. The simulation of decarbonization measures reveals valuable information about the energy savings, CO_2eq emission reductions, and cost-effectiveness of various strategies, helping to design the most effective approaches across different contexts. These findings are crucial for guiding resource allocation and informing targeted policy decisions.

2. Materials and Methods

The methodology for developing a regional residential energy model is outlined in this chapter. It includes the analytics used to determine energy consumption in each end use of a dwelling, which are derived from established research.

The model takes into account a variety of inputs, such as the building attributes, climate, and occupant behavioral trends. This hybrid approach is expected to provide an accurate estimate of residential energy consumption by combining both physical and behavioral factors. The model was implemented in Python (version 3.10).

Figure 1 showcases the diagram of the regional residential energy model, providing a visual representation of its various components and their interconnections.

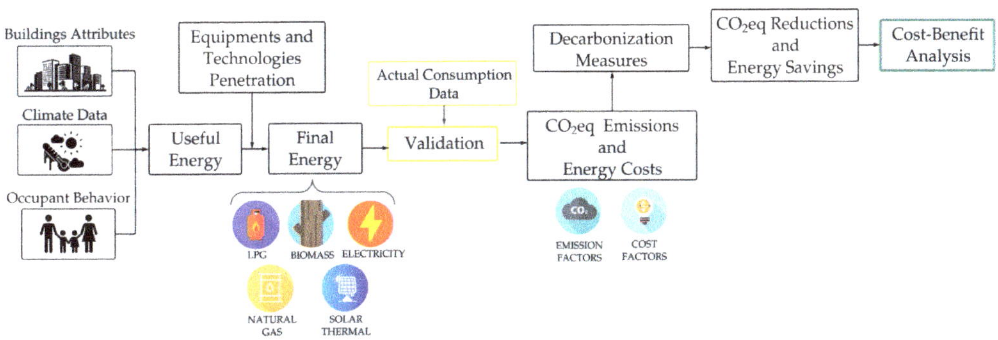

Figure 1. Regional residential building energy model diagram.

The model estimates the useful energy consumption of residential buildings using inputs such as building constructive solutions, climate data, and occupant behavior. Useful energy refers to the amount of energy required to meet a specific demand (e.g., heating a space or providing hot water), while final energy is the amount of energy consumed via the system to fulfill that demand, considering the efficiency of the equipment.

When evaluating energy consumption for cooking and household appliances, it is important to understand that these calculations are based on final energy consumption, not useful energy. To calculate the useful energy for these two end uses would require detailed information on factors like the types of food being prepared and the materials and efficiency of each appliance. These details are necessary because different foods and cooking methods consume varying amounts of energy, and appliances made from different materials operate at different efficiencies. Collecting these comprehensive data on a national scale is unfeasible, so final energy consumption was adopted for these two end uses as a more practical approach.

Space heating and water heating are often the biggest contributors to overall dwelling energy consumption [16]; according to the IEA, space and water heating accounts for 46% of the energy consumed in buildings [17]. In Portugal, water heating consumption accounts for 20% of the total residential energy consumption, while space heating accounts for 23% [18].

The thermal dynamics of a building involve heat-transfer processes that determine how energy is gained and lost. Heat gains include solar radiation, which heats building surfaces like windows, walls, and roofs, and internal gains from occupants, appliances, lighting, and cooking. The solar heat gains can reduce space heating needs but may cause thermal discomfort during warmer periods.

Heat losses occur through the building envelope (e.g., walls, windows, and roofs), driven by the temperature gradient between the interior and exterior. In cold conditions, heat is lost through conduction, convection, and air infiltration, while in warm conditions, heat can flow inward, increasing cooling demand. Understanding these thermal dynamics is essential to accurately estimating the space heating and cooling demands of the residential building stock.

In the following sections, the model formulation is presented in detail, outlining the mathematical equations and methodologies used to estimate the final energy demand for various end uses in residential buildings. The model covers space heating, space cooling, water heating, lighting, household appliances, and cooking. Each subsection delves into the specific factors and parameters influencing energy consumption for these end uses, such as thermal losses, solar gains, equipment efficiency, and occupant behavior.

2.1. Space Heating

The equations required to compute space heating final energy were adapted from the Regulation on the Energy Performance of Residential Buildings (REH) [19] and other studies [20,21].

Equation (1) was used to calculate space heating's final energy demand.

$$Q_{SH} = \frac{\left(Q_{t,sh} + Q_{v,sh} - Q_{g,sh}\right)}{\eta_{T,sh}} \quad (1)$$

where $Q_{t,sh}$ (kWh) represents the losses through transmission, $Q_{v,sh}$ (kWh) is the quantity of heat lost as a result of the ventilation system of a building, $Q_{g,sh}$ (kWh) represents the heat gains, and $\eta_{T,sh}$ is the space heating technology efficiency.

The heat losses through transmission are calculated through Equation (2).

$$Q_{t,sh} = 0.024 \cdot HDD \cdot H_t \quad (2)$$

where HDD (°C· Day) represents the heating degree days, and H_t (W/°C) is the global heat coefficient.

The heating degree days are a measure of the severity of the outdoor temperature in a given location during a specified time period. The use of heating degree days from the REH ensures that the energy model represents the typical heating demands of buildings in Portugal.

The global heat coefficient is calculated through Equation (3).

$$H_t = U_{wall} \cdot A_{wall} + U_{window} \cdot A_{window} + U_{floor} \cdot A_{floor} + U_{roof} \cdot A_{roof} \quad (3)$$

where U_{wall}, U_{window}, U_{floor}, and U_{roof} (W/m^2·°C) are the overall heat transfer coefficient of the walls, windows, floors, and roofs of a dwelling, and A_{wall}, A_{window}, A_{floor}, and A_{roof} (m^2) are the respective areas of each of the elements.

The heat losses through ventilation are calculated with Equation (4).

$$Q_{v,sh} = 0.024 \cdot HDD \cdot \frac{c_p \cdot \rho}{3600} \cdot ACH \cdot Vol \quad (4)$$

where ACH is the air changes per hour (h^{-1}), C_p (1.005 kJ/kg· K) is the specific heat of air, ρ ($1.2\frac{kg}{m^3}$) is the density of air, and Vol (m^3) is the volume of the building under analysis.

The heat gains are calculated using Equation (5).

$$Q_{g,sh} = \eta_{gu} \cdot (Q_{in} + Q_{sol,sh}) \quad (5)$$

where Q_{in} (kWh) are the internal heat gains, $Q_{sol,sh}$ (kWh) is the solar irradiation heat gains, and η_{gu} is the gain utilization factor.

The gain utilization factor is a ratio of the thermal gains and losses of a building in the heating season, and it is calculated through the following Equation (6).

$$\eta_{gu} = \begin{cases} if \gamma \neq 1 \land \gamma > 0 & \frac{1-\gamma^{a}}{1-\gamma^{a+1}} \\ if \gamma = 1 & \frac{a}{a+1} \\ if \gamma < 0 & \frac{1}{\gamma} \end{cases} \tag{6}$$

$$\gamma = \frac{Q_g}{Q_t + Q_v} \tag{7}$$

where "a" can take the values of 1.8, 2.6, or 4.2, according to buildings' weak, average, and strong thermal inertia [19].

Thermal inertia, which represents a building's capacity to store and gradually release heat, is influenced by the U-value of the building envelope. Lower U-values, typically found in well-insulated buildings, enhance thermal inertia by reducing the heat transfer through the envelope [22]. A study by Giama et al. [23] showed that older buildings tend to have lower U-values due to outdated construction materials and methods, such as less insulation and single-pane windows, resulting in poorer thermal performance and lower thermal inertia.

Internal heat gains can significantly affect a building's energy balance and indoor thermal environment, and they are calculated through Equation (8).

$$Q_{in} = 0.72 \cdot q_{int} \cdot M_{sh} \cdot A_{\bar{floor}} \tag{8}$$

where q_{int} (W/m^2) are the internal gains, M_{SH} represents the duration of the heating season, measured in months, and $A_{\bar{floor}}$ (m^2)) is the average floor area.

The internal heat gains are assumed as 4 W/m^2 [24]; this value is used in energy performance calculations, and it represents the average heat generated via human occupation, cooking, and electrical and lighting equipment.

$Q_{sol,sh}$ (kWh) is the solar heat gains, and it is calculated with Equations (9)–(11).

$$Q_{sol,sh} = G_{south} \cdot \sum_{w} \left[X_i \cdot F_h \cdot F_o \cdot F_f \cdot A_w \cdot F_g \cdot g_i \right] \cdot M_{sh} \tag{9}$$

$$F_s = F_h \cdot F_o \cdot F_f \tag{10}$$

$$A_s = A_w \cdot F_g \cdot F_i \tag{11}$$

where G_{south} (kWh/m^2 $\cdot$ month) is referred to as the average monthly solar radiation through the heating season on a vertical south-facing surface, X_i is the window orientation coefficient, F_h is the horizon window obstruction factor, F_o and F_f are window obstruction factors from horizontal and vertical elements, respectively. A_w (m^2) is the window area, F_g is the glassed fraction of the window, and finally, g_i is the solar factor, a coefficient that accounts for properties of the glass and shading elements, while F_s is the total obstruction factor, and A_s (m^2) is the effective area.

The majority of the coefficients utilized were obtained from [19]. The coefficients that required calculations will be discussed in the following subsections.

2.2. Space Cooling

In the case study through which the model was tested, space cooling consumption accounted for a very small (<5%) [18] portion of the total energy consumption. However, since the intention was to create a replicable methodology, it was still important to include space cooling consumption in the analysis. In other contexts where the model may be applied, this end use may not be negligible, and modeling it ensures the model's adaptability to different contexts.

The space cooling final energy demand of a building is provided via Equation (12).

$$Q_{SC} = \frac{(1-\eta_{gu}) \cdot Q_{g,sc}}{\eta_{T,sc}} \tag{12}$$

where $\eta_{T,sc}$ depicts the efficiency of the cooling technology, η_{gu} is the gain utilization factor for the cooling season, and $Q_{g,sc}$ (kWh) are the solar gains.

Thus, the following Equation (13) is employed to calculate the solar gains:

$$Q_{g,sc} = \sum_{w}[G_{solar} \cdot F_s \cdot A_{s,w}] + \sum_{o}[G_{solar} \cdot F_s \cdot A_{s,o}] \tag{13}$$

where G_{solar} (kWh/m^2) is the solar gains coefficient, $A_{s,o}$ (m^2)) is the effective area of the opaque elements, and $A_{s,w}$ (m^2)) is the effective area of the glazed elements.

The first parcel of the formula refers to the windows, while the second parcel refers to the opaque elements. The solar gains of the window elements for space cooling are computed in exactly the same manner as for space heating.

The solar gains coefficient is an assessment of how effectively a window or other building material transmits solar radiation into a structure. It represents the percentage of incident solar radiation that passes through a medium and can be used to heat or cool the inside environment. This coefficient is dependent on a variety of parameters, such as the glazing's material characteristics, the angle at which solar radiation impacts the building, and the existence of shading devices or other building components that might reflect or obstruct solar radiation.

The effective areas are calculated with Equations (14) and (15).

$$A_{s,w} = A_w \cdot F_g \cdot g_i \tag{14}$$

where A_w (m^2)) is the window area, F_g is the glassed fraction of the window, and g_i is the solar factor.

$$A_{s,o} = U.\alpha \cdot A_o \cdot R_{es} \tag{15}$$

where A_o (m^2) is the surface area, α is the solar radiation absorption coefficient, R_{es} (m^2·°C/W) is the external surface thermal resistance of opaque element, and U (W/m^2·°C) is the overall heat transfer coefficient.

In order to compute the heat gain utilization factor, it is first necessary to calculate the heat transmissions using transmission, ventilation, and infiltration from Equations (16) and (17), respectively.

$$Q_{t,sc} = 0.72.M_{sc} \cdot H_t\left(T_{ref} - T_{ext}\right) \tag{16}$$

$$Q_{v,sc} = 0.72 \cdot M_{sc} \cdot ACH \cdot \frac{C_p \cdot \rho}{3600}\left(T_{ref} - T_{ext}\right) \cdot Vol \tag{17}$$

where M_{sc} is the duration of the cooling season, usually expressed in months, T_{ref} (°C) is a reference temperature set for the cooling season, and T_{ext} (°C) is the average outside temperature in the cooling season. The T_{ref} (°C) for space cooling is set to 25 °C, according to the REH [19].

2.3. Water Heating

The energy needed to heat the water used in a residential building is referred to as the water heating energy demand.

The water heating energy demand for the residential building stock is calculated through Equation (18).

$$Q_{WH} = \frac{C_{p,w} \cdot \rho_w \cdot \Delta T \cdot d \cdot n_r \cdot V_w \cdot f_w}{3600 \cdot \eta_{T,wh}} \tag{18}$$

where $C_{p,w}$ (4.186 kJ/kg·K) is the specific heat of water, ρ_w (1 kg/l) is the density of water, ΔT (°C) is the water temperature difference desirable, n_r is the number of residents in the dwelling, V_w (l/person.day) is the volume of water required daily for each resident, f_w is a factor for hydraulic efficiency, and finally, $\eta_{T,wh}$ represents the water heating technology efficiency.

2.4. Lighting

The lighting requirements for a dwelling can fluctuate, depending on the size of the home, how the rooms are organized, and the individual requirements of the occupants. The same formulation as in [21,25] was used. Equation (19) calculates the annual energy demand for lighting in a dwelling.

$$Q_L = 0.001 \cdot AFA \cdot \sum_{Tl} \left[\frac{S_{Tl}}{\eta_{Tl}} \right] \cdot L \cdot T \tag{19}$$

where Q_L (kWh) is the annual energy demand for lighting for a single dwelling, AFA (m^2) is the average floor area of the dwelling, S_{Tl} is the share of the lighting technology, and η_{Tl} (Lm/W) is the corresponding lighting efficiency.

Finally, L (Lux) is the lighting requirement, and T (hours/years) is the equivalent amount of lighting hours required per year. According to Daioglou et al. [26], the combination of the last two factors can be expressed as the average lighting requirements, measured in lumen hours per square meter.

2.5. Cooking

The quantity of energy required for cooking depends on an array of factors, such as the type of cooking appliance used, the frequency and duration of cooking, and the efficiency of the cooking appliances used.

Most dwellings include more than one piece of equipment for food preparation; thus, the total amount consumed for cooking is the sum of all the appliances' consumption [25], given by Equation (20).

$$Q_C = \sum P_{c,equip} \cdot S_{c,equip} \tag{20}$$

where Q_C (kWh) is the annual final energy consumption for cooking, $P_{c,equip}$ (units/dwelling) is the equipment penetration, and $S_{c,equip}$ (kWh/year) is the specific energy consumption of given cooking equipment.

2.6. Abatement Cost

The CO_2eq cost of abatement is the amount spent to reduce GHG emissions, usually expressed in terms of how much it costs per tonne of CO_2eq that is prevented from being released into the atmosphere. It represents the cost of implementing measures or technologies to reduce emissions. Measures with a negative abatement cost on an abatement curve are often referred to as "no-regret" measures since they simultaneously provide economic and environmental benefits. This can be the result of energy efficiency improvements that lead to cost savings that exceed the costs associated with implementing them [27].

Equation (21) was employed to calculate the net abatement cost of each measure under examination.

$$AC = \frac{I - \sum_1^n \frac{S_t}{(1+r)^n}}{\Delta E} \tag{21}$$

where AC (EUR/tonCO_2eq) represents the abatement cost, I (EUR) is the total investment cost to implement the given measure, S_t (EUR) denotes the annual energy cost savings achieved by the measure, and n is the lifetime of the measure, expressed in years; this represents the expected operational duration of the measure before it needs to be replaced or upgraded. In other words, it is the number of years during which the measure will generate energy savings and emission reductions, r is the discount rate applied to adjust future savings to their present value, and ΔE (tonCO_2eq) refers to the total emission reductions achieved through the measure over its lifetime.

The discount rate, r, is a key factor in evaluating the economic feasibility of measures, as it determines how future costs and benefits are valued in present terms [28]. Over time, the discount rate applied in energy efficiency and decarbonization projects has generally decreased, both at the European level and globally [29]. This reflects a growing recognition of the importance of long-term environmental sustainability and societal benefits. Historically, higher discount rates reflected short-term financial returns, often devaluing projects with significant future environmental and social gains. However, with an increasing focus on climate change and the transition to low-carbon economies, the rate has been progressively lowered, incentivizing investments in measures with longer payback periods but substantial future benefits.

According to the European Commission's Guide, Economic Appraisal Vademecum 2021–2027—General Principles and Sector Applications, the use of lower discount rates is recommended for evaluating decarbonization and energy efficiency projects [30]. This approach emphasizes the need to account for the long-term societal and environmental benefits that such projects deliver. Specifically, by applying discount rates between 3% and 5%, the EU ensures that the future benefits, including reduced greenhouse gas emissions and improved air quality, are adequately reflected in cost–benefit analyses [30]. These benefits often take several decades to materialize, particularly in projects with long lifetimes, such as building renovations and renewable energy systems.

2.7. Model Validation

A validation step in an energy model is essential to ensure the model's accuracy and reliability. The error (%) was calculated for each region with the following Equation (22).

$$\text{Error}_i(\%) = \frac{(\hat{C}_i - C_i)}{C_i} \times 100 \tag{22}$$

where i represents each of the regions modeled, $\hat{C}_i$ (kWh) represents the consumption estimated by the model for the i-th region, and C_i (kWh) is the actual consumption for the i-th region.

3. Case Study

In the case study section, the focus is on characterizing the specific context to which the model was applied, namely at the municipality level in Portugal. This section provides an overview of the building stock, climate data, and occupant behaviors, offering a detailed understanding of the conditions under which the model operates. However, while this case study focused on Portugal, the methodology is designed to be replicable for other contexts, provided that the necessary input data are available. This flexibility demonstrates the model's adaptability and its potential application to diverse geographical areas for energy and emissions assessments.

3.1. Demographic and Geographic Characteristics of Portugal

According to INE [31], as of 2021, the population of Portugal is distributed as 9,855,909 in the Portuguese mainland, 236,413 in the Azores autonomous region, and 250,744 in the Madeira autonomous region.

Therefore, the population of Mainland Portugal is by far the largest, accounting for 95.30% of the total population of the country. The remaining 4.7% is divided between the two autonomous regions of Portugal, with the Madeira autonomous region having a slightly larger population than the Azores autonomous region.

With a total area of approximately 92,212 km^2, the mainland region occupies approximately 89,015 km^2, while the two autonomous regions—Azores and Madeira—span 2322 and 801 km^2, respectively.

The mean annual temperature for mainland Portugal is approximately 17 °C, some variations may occur, depending on the location and season. Particularly in the Southern regions, such as Algarve, where mean temperatures can range between 18 °C and 20 °C throughout the year. The cooler climates like the ones observed in northern areas like Porto and the Douro Valley experience an average temperature of 13 °C to 17 °C [32].

3.2. Portugal's Residential Building Stock

A comprehensive characterization of the residential building stock in Portugal is critical for a successful energy model because the energy performance of buildings is largely determined by their physical and thermal properties. Factors such as building age, construction materials, insulation quality, window types, and heating or cooling systems directly impact how buildings gain and lose heat, and consequently, their energy consumption for space heating and cooling.

An accurate characterization of the building stock is the first step to developing an accurate energy model. By reflecting the diversity of building types and conditions across different regions, this characterization allows the model to perform precise simulations and provide reliable predictions of energy consumption, potential energy savings, and emission reductions.

In Portugal, there are currently 4,142,581 dwellings that are primary residences, according to the 2021 Census [31]. The term primary residence refers to a dwelling or housing unit where a resident usually lives or stays for the majority of the year.

Figure 2 represents the number of dwellings of primary residence per year of construction; the data were extracted from [31].

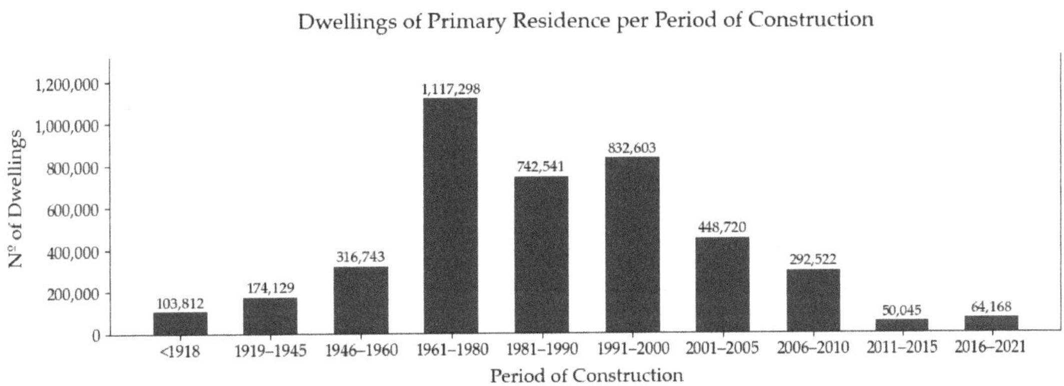

Figure 2. Dwellings as primary residences per year of construction.

From Figure 2, it is clear that the majority of primary dwellings in Portugal were constructed between 1961 and 1980. This period is described as the "construction boom", which occurred as a result of the country's economic and demographic growth. There was a high demand for housing, resulting in the construction of a large number of buildings that are now part of the existing building stock. Additionally, from Figure 2, it is possible to conclude that approximately 90% of primary dwellings were constructed before 2006.

The maintenance and repair needs of a building can also have substantial ramifications for energy consumption, resulting in an increase in energy usage. For example, if the building envelope (e.g., roof, walls, windows, and doors) is not properly maintained, it can lead to heat losses. This will cause heating and cooling systems to work harder to maintain comfortable indoor temperatures, resulting in an increase in energy usage.

Figure 3 represents the number of buildings that require maintenance or repairs per year of construction; the data were extracted from [31].

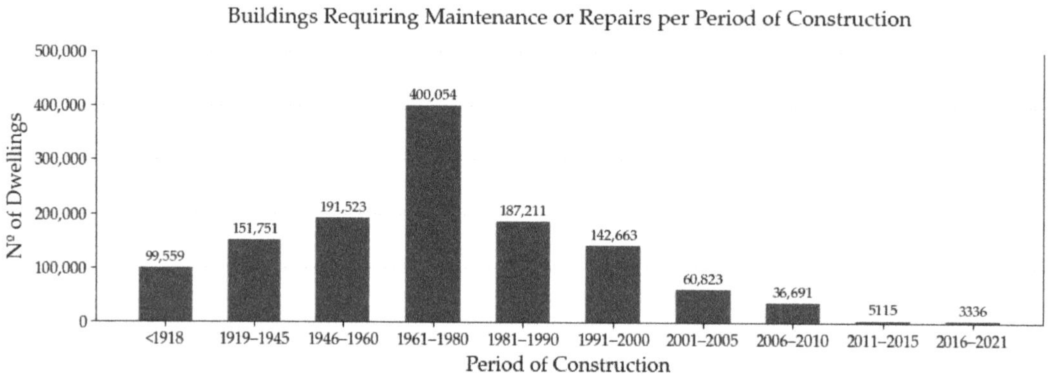

Figure 3. Buildings requiring maintenance or repair per period of construction.

From Figure 3, it is possible to conclude that the buildings built between 1961 and 1980 are the ones that require more maintenance. According to Almeida et al. [33], the 20th century saw marked growth in the implementation of reinforced concrete construction in Portugal, which was principally seen between 1960 and 1980. This increase reflected an effort to augment the longevity and resistance to earthquakes of such structures. However, it was later found that the reinforcement bars used in these constructions had inadequate protections against moisture and saltwater exposure, resulting in corrosion and swelling that caused damage to their structure.

3.3. Building Regulations and Legislation in Portugal

The REH is a set of regulations and technical specifications that establishes the energy performance requirements for residential buildings in Portugal [19]. The REH was first published in 2006 and has subsequently undergone multiple updates to reflect advancements in renewable energy and building energy efficiency technologies.

The REH establishes minimum energy performance standards for a number of building envelope components, including walls, roofs, floors, and windows, as well as for heating, cooling, ventilation, and lighting systems. It is suitable for both new and old residential buildings. It also creates guidelines for determining a building's energy performance and issuing energy performance certifications, both of which are necessary for all newly constructed buildings, as well as for any existing buildings that are sold or rented [19].

The Portuguese Regulation for Energy Performance of Buildings (RCCTE) [24] RCCTE specifies the standards for the thermal performance of buildings. This guidelines were initially established in 1990, and since then, multiple improvements have been made [24].

Weather plays a vital role in the energy performance of buildings, as it directly influences heating, cooling, and overall energy demands. In the Portuguese legislation, the country is classified into six climate zones: three for winter (I1, I2, and I3) and three for summer (V1, V2, and V3) [24]. Index 1 zones have a milder climate, whereas index 3 zones have a harsher climate [34].

The classification of winter climate zones is not based solely on HDD values; it is also influenced by the elevation of each municipality. While HDD captures the heating requirements over time, elevation plays a critical role in determining local temperature variations. Higher elevations tend to experience lower temperatures, which can significantly increase heating needs, even in regions where the HDD values alone might not reflect such extreme conditions.

To address this, the winter zone classification uses a weighted approach, combining both HDD values and the elevation of municipalities. This ensures a more accurate understanding of the heating demands in different areas. For instance, municipalities at higher elevations may fall into harsher winter zones due to colder temperatures that result from elevation, even if their HDD values suggest milder conditions. This classification acknowledges the dual influence of sustained cold temperatures and elevation, where milder zones (I1) are typically characterized by both lower HDD values and lower elevations, and harsher zones (I3) reflect both higher HDD values and greater elevations. The combination of these factors provides a more comprehensive reflection of heating demands across different regions, accounting for both long-term temperature patterns and the local impact of elevation.

The summer zones are determined on the basis of the average outside temperature for the cooling season. This classification helps in understanding the cooling requirements of buildings across different regions. In milder summer zones (V1), lower average temperatures reduce the need for cooling, while in harsher summer zones (V3), higher average temperatures indicate a greater need for cooling to maintain indoor thermal comfort.

3.4. Model Inputs

In this section, some of the required inputs for the energy model used in this study are presented and clarified. The simplifications and assumption made in the model are also presented throughout this section. This includes the simplifications and approximations made in the modeling process, as well as the uncertainties in the input data. Acknowledging these limitations is important for ensuring a transparent and thorough analysis of the results, helping to contextualize the model's findings within the scope of its design and the data available.

3.4.1. Residential Building Stock

In developing the energy model for this study, certain considerations were made regarding the physical and thermal characteristics of the residential stock. Specifically, it was assumed that all dwellings have a standard wall height of 2.7 m, which, according to Appolloni and D'Alessandro [35], is the minimum wall requirement in Portugal, and the ratio of a dwelling's length to its side is 1.

3.4.2. Average Floor Area

Based on census data, it is possible to determine the average floor area of the Portuguese residential stock in each municipality. The overall number of residences in each municipality, as well as the number of residential buildings classified according to their area, are both taken into consideration. The average floor area (m^2)) for every municipality is determined using Equation (23).

$$\bar{A}_{floor} = \frac{\sum_{i=1}^{n} \bar{A}_i \cdot N_i}{N_{dwellings}} \quad (23)$$

where $\bar{A}_i$ (m^2)) represents the average floor area for each interval, N_i represents the number of dwellings in each corresponding interval, and $N_{dwellings}$ is the total number of dwellings.

3.4.3. Number of Floors

The number of floors in residential buildings is also a crucial parameter for assessing the energy consumption in the Portuguese residential building stock. To this end, Equation (24) was developed to estimate the number of floors in each municipality.

$$N_{floors} = \sum_{1}^{x} x \cdot N_{RB,xf} \quad (24)$$

where x represents the number of floors in a given building, and $N_{RB,xf}$ is the number of buildings with x floors.

Data from the Portuguese 2021 census on the number of residential buildings categorized by the number of floors were used. These data indicate the number of buildings in each municipality, ranging from those with just one floor to those with seven or more floors.

Finally, Equation (25) calculates the average number of dwellings per floor in each municipality.

$$\bar{N}_{d/floor} = \frac{N_{dwellings}}{N_{floors}} \quad (25)$$

where $\bar{N}_{d/floor}$ is the average number of dwellings per floor.

3.4.4. Overall Heat Transfer Coefficient

The overall heat transfer coefficient (U-value) is a key factor in evaluating the energy performance of residential buildings. It measures how quickly heat is lost through the building's envelope, taking into account the thermal resistance of building components like walls, floors, roofs, and windows.

One significant factor that can influence the U-value is the period of construction of buildings. This is because building materials and construction practices have improved over time, largely due to advancements in regulations and laws. As a result, newer buildings have better building envelope performance and are better equipped to prevent heat losses. Conversely, older buildings have less thermal insulation or single-pane windows that allow heat to escape more easily.

Figure 4 presents the U-values considered for each envelope element based on the period of construction; the data were extracted from [36].

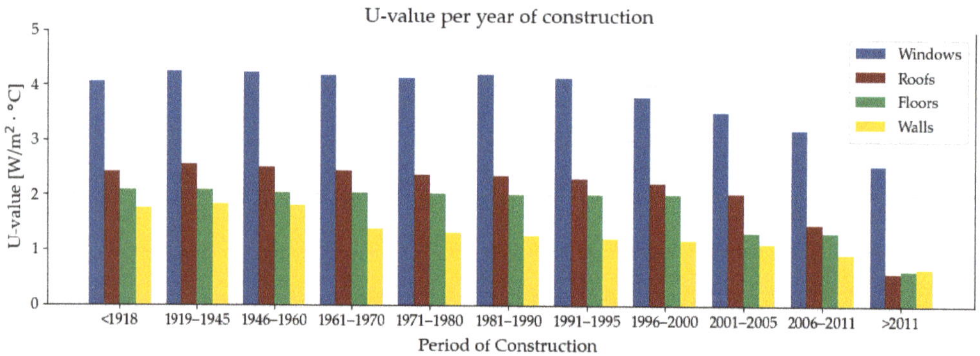

Figure 4. Overall heat transfer coefficient (W/m^2.°C) per year of construction. Adapted from [36].

In this case study, 11 archetypes were used to represent the building stock by construction period, simplifying the modeling process but assuming uniform characteristics within each period. This introduces uncertainty, as it may not fully capture the diversity of construction techniques and materials across buildings.

3.4.5. Glazing Area Percentage

The glazing area percentage (GAP) refers to the proportion of a building's exterior envelope that is made up of windows or glass surfaces. GAP is an important parameter to consider because it directly impacts a building's thermal performance, affecting both heat loss and solar gains.

In older buildings, there is a tendency to have less glazing area due to architectural styles and design preferences, which often prioritize solid walls over windows. These solid walls are also more prevalent in historical buildings. As a consequence, older buildings GAP is smaller than that of newer buildings [37].

The following figure, Figure 5, presents the GAP values based on the period of construction considered in this study; the data were extracted from [37].

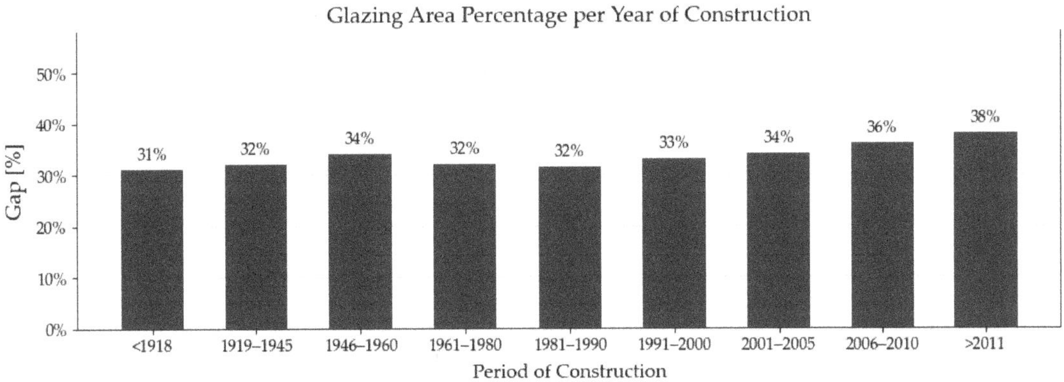

Figure 5. Glazing area percentage (%). Adapted from [37].

The variance in GAP across different periods of construction is not as pronounced as in some other building parameters. However, the trend of an increasing GAP over time is still noticeable, indicating a gradual shift toward incorporating more glazing in building designs, especially in post-2000 constructions.

3.4.6. Air Changes per Hour

The (ACH) is a measurement of how many times the air in a room is fully replaced with outdoor air in an hour. It is an essential metric for assessing indoor air quality and energy efficiency in buildings. The following figure, Figure 6, presents the evolution of the ACH based on the period of construction; the data were extracted from [21].

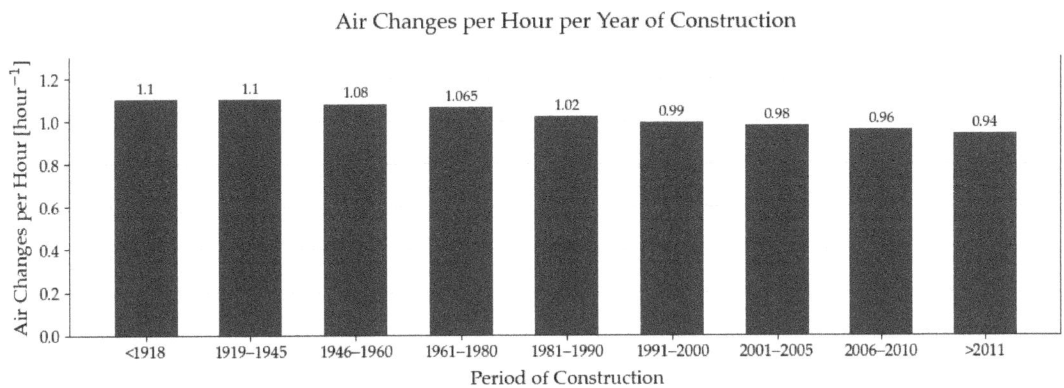

Figure 6. Air changes per hour [hour^{-1}] per period of construction. Adapted from [21].

3.4.7. Window Orientation Coefficient

The window orientation coefficient (X_i) is a parameter that, unlike some of the other factors presented in the model formulation section, was not directly obtained from the Portuguese legislation. Therefore, it is important to explain how this coefficient was calculated in order to ensure clarity and accuracy in the model's methodology.

To compute this parameter, it is first necessary to know the distribution of windows orientation for each climatic zone, which was obtained from [38] and is presented in Table 1.

Table 1. Window orientation per climate zone. Adapted from [38].

Orientation	I1V1	I1V2	I1V3	I2V1	I2V2	I2V3	I3V1	I3V2	I3V3
W	0.14	0.12	0.21	0.15	0.26	0.12	0.08	0.06	0.19
SW	0.08	0.18	0.08	0.12	0.03	0.20	0.09	0.07	0.05
S	0.06	0.11	0.08	0.15	0.10	0.06	0.17	0.25	0.24
SE	0.22	0.12	0.09	0.07	0.03	0.12	0.16	0.18	0.00
E	0.14	0.18	0.21	0.12	0.25	0.12	0.17	0.05	0.28
NE	0.08	0.17	0.08	0.11	0.05	0.20	0.00	0.17	0.00
N	0.07	0.12	0.17	0.15	0.12	0.07	0.16	0.11	0.24
NW	0.21	0.00	0.08	0.13	0.16	0.11	0.17	0.11	0.11

Secondly, the orientation factor for different exposures is required. This factor takes into account the orientation of the facade of a building with respect to the cardinal points, and it determines the amount of solar radiation received by the building. The values for this parameter were sourced from the RCCTE [24].

Finally, the window orientation coefficient by climate zone is calculated using Equation (26).

$$X_i = \sum_{j=1}^{n} D_{ij} \cdot F_j \tag{26}$$

where X_i is the window orientation coefficient for climate zone i, D_{ij} is the distribution of windows' orientation for climate zone i and orientation j, F_j is the orientation factor for orientation j, and n is the total number of orientations considered.

3.4.8. Solar Gains

Solar gains (G_{solar}) are the thermal energy a building receives from solar radiation, which affects its internal temperature. They can reduce heating needs in colder months but increase cooling demand in warmer periods, significantly influencing a building's overall energy performance. However, they are not directly obtainable from the legislation, requiring calculations based on solar irradiation data and building orientation to accurately estimate their contribution to the building's energy balance.

The equation used to calculate solar gains (G_{solar}) is as follows:

$$G_{solar} = \sum_{i=1}^{n} I_i \cdot D_i \tag{27}$$

where G_{solar} (kWh/m^2), the solar gains, represent the overall thermal energy received in the building from solar energy, and I_i (kWh/m^2) indicates the amount of solar energy incident on a surface for each direction (e.g., north and south) during the specified cooling period (June to September).

3.4.9. Emissions and Costs

In order to accurately calculate the CO_2eq emissions of the different sectors, it was necessary to select the CO_2eq emission factors associated with each type of energy vector used. Table 2 represents the emissions factors utilized in this work, as well as the price of each energy vector.

Table 2. CO_2eq emissions factors and energy costs.

Emission Factors ($kgCO_2eq/kWh$)	Electricity	Natural Gas	LPG	Biomass	Heating Oil	Solar
Energy Conversion [1]	0.220 [39]	0.041 [40]	0.028 [40]	-	0.047 [40]	-
Use Phase [2]	-	0.203 [18]	0.227 [18]	0 [18]	0.267 [18]	-
Total	0.220	0.244	0.255	0	0.314	0
Cost (EUR/kWh)	0.218 [41]	0.144 [42]	0.213 [42]	0.120 [42]	0.140 [43]	0.03

[1] Emissions resulting from the various stages of the energy supply chain, including extraction, separation, transport, distribution, and refining if it is applicable. [2] Emissions resulting from the combustion process itself, considering the chemical reactions involved and the release of pollutants into the atmosphere.

Biomass was considered to have no direct emissions, as it is often classified as carbon-neutral due to the assumption that the CO_2 released during combustion is offset by the carbon absorbed during the growth of biomass [44].

The CO_2eq emissions factors also take into account the emissions resulting from extracting, transporting, and using the energies considered in the model. These additional non-combustion factors are essential to ensure that a complete picture of energy-related CO_2eq emissions is taken into account when assessing its true environmental impact.

To obtain the price of the kWh produced via the solar thermal panels, Equation (28) was utilized.

$$\mathrm{LCOE_{solar}} = \frac{C_{\mathrm{solar}}}{E_{\mathrm{total}}} \tag{28}$$

where C_{solar} (EUR) is the capital cost of the solar thermal panel, and E_{total} (kWh) is the total energy output produced via the panel over its operational lifetime.

3.5. Equipment and Technology Penetration

In this section, the average penetration of various equipment and technologies across different end uses in residential dwellings in Portugal is presented. The specific penetration for each individual municipality is not shown, as such information would be too extensive, but the national average is provided to suggest an overall idea of the distribution.

3.5.1. Space Heating

The 2021 Census categorizes space heating technologies into four distinct categories: central heating, non-central heating (fixed and movable equipment), fireplaces, and no heating [31]. In the case study, among the total dwellings with access to space heating, movable equipment had the highest penetration at 40.7%, followed by fireplaces at 31.1%, central heating at 20.0%, and fixed equipment at 8.2%.

Additionally between the 2010 and 2021 census, an indicator that provided information on the distribution of space heating technologies by their corresponding energy source was eliminated. As a result, the 2021 census only included data on the distribution of space heating technologies. To obtain the corresponding distribution of space heating technologies by energy source for 2021, a projection was made using the data from the 2010 census [18,45].

3.5.2. Space Cooling

The space cooling technologies considered in the model include air conditioners, fans, and heat pumps. These technologies are commonly used in maintaining comfortable indoor temperatures during hot weather conditions. While heat pumps are a more energy-efficient option, air conditioners and fans are widely available, with fans being the most inexpensive option. It is important to note that, while fans can provide some level of comfort to individuals, they do not have the capability of decreasing the ambient temperature to the desired set point temperature.

The availability of space cooling technologies in residential buildings in Portugal was determined by analyzing the data gathered from the 2021 census, as well as the Energy Consumption Survey in the Domestic Sector 2020 [18,31].

Among the dwellings equipped with cooling technologies, which make up approximately 32.7% of the total, air conditioners are present in around 16.7% of the dwellings with cooling technologies in Portugal. Fans are more widely used, being found in about 58.8% of these dwellings, while heat pumps are available in approximately 45.4% of the dwellings with cooling systems.

3.5.3. Water Heating

The technologies considered for water heating in the model were gas water heaters, electric water heaters, boilers, and solar thermal systems. Each technology was assigned a distribution value between 0 and 1, with the sum of all values in a municipality equal to 1. This approach is based on the assumption that each dwelling has only one piece of water heating equipment, unlike the space cooling and space heating categories, where multiple systems can coexist in a single dwelling or a dwelling may not include any such system at all. This distinction is made because water heating is typically served through a dedicated system, whereas cooling and heating needs may vary, with some dwellings using a combination of devices or none at all, depending on the layout and usage patterns.

In the case study, gas water heaters' penetration rate was of 57.0%, electric water heaters' was 14.0%, gas boilers' was 16.0%, and solar thermal was present in 13% of the dwellings; the data were obtained from [18,46]. According to the data, gas water heaters remain the most widely used technology for water heating in Portugal, while solar thermal systems are the least commonly used.

3.5.4. Lighting

In assessing the lighting technologies present in Portuguese dwellings, several factors were considered. Notably, four different types of light bulbs were taken into account, including incandescent, halogen, fluorescent, and LED bulbs [18]. Notably, in dwellings that had these bulbs present, there were often multiple options available.

The Energy Consumption Survey in the Domestic Sector 2020 [18] indicates that LED lighting is the most widely used technology in the case study, being present in 80.10% of dwellings, with an average of 12.60 bulbs per household. Fluorescent lighting follows, with a presence in 56.60% of dwellings, although with a significantly lower average of 3.20 bulbs. Halogen bulbs are installed in 23.50% of dwellings, with an average of 5.80 bulbs per household, while incandescent bulbs are still found in 35.10% of dwellings, with an average of 5.70 bulbs.

3.5.5. Cooking

In this study, four major cooking technologies were considered: cookers with ovens, cookers without ovens, independent ovens, and cooking robots. Among the technologies assessed, cookers with ovens had the highest penetration at 52.00%, followed closely by cookers without ovens and independent ovens, both at 51.60%. Cooking robots, with a penetration rate of 13.90%, were the least common technology among the options considered in the case study. This distribution highlights the prominence of traditional cooking technologies, while newer technologies, such as cooking robots, are still less widespread. These results were extracted from the Energy Consumption Survey in the Domestic Sector 2020 [18].

3.5.6. Household Appliances

In this study, a diverse number of household appliances were considered, namely fridges, freezers, dishwashers, washing machines, and clothes dryers, among others. These are the appliances most common in a residential setting, and they have a greater impact on the total energy consumption of a dwelling.

It is important to also take into account the energy efficiency of electronic appliances, as it can vary significantly, impacting energy consumption. As the 2020 distribution of efficiency for electronic devices was not available, it was necessary to use the 2010 distribution presented in [45] in this study. It is important to note, however, that electronic devices are not frequently replaced, and thus, the 2010 distribution may still be relevant for many dwellings. Nevertheless, updated data must be made available to improve the accuracy of energy consumption models. Figure 7 presents the energy efficiency share for the case study; the data were extracted from [18,45].

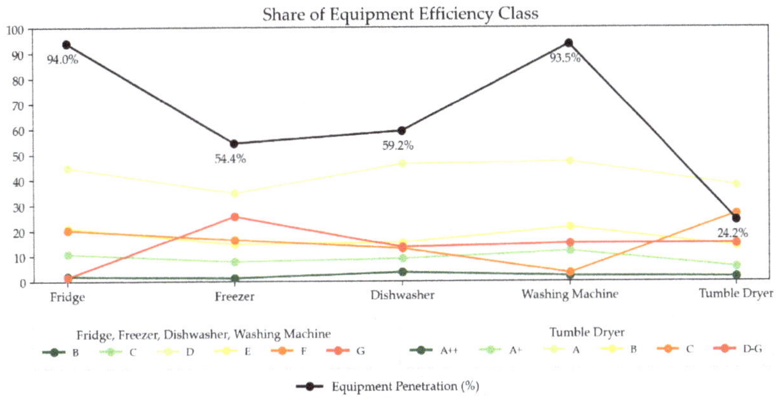

Figure 7. Energy efficiency share. Adapted from [18,45].

The analysis of household appliances in Portugal revealed that high-efficiency appliances make up a mere 2% of the total. The most common efficiency rating is D, accounting for 43% of appliances, followed by E class with 19%, Class F with 14%, and lower efficiency ratings representing 12% of appliances. This distribution does not account for dryers since they possess their own separate energy efficiency label.

It is also important to note that some of the data used in this study, particularly regarding appliance energy efficiency, originate from the 2010 distribution, prior to the introduction of the new energy efficiency classifications. To ensure consistency with current standards, the data for appliances were updated to match the latest energy classification brochure. As a result, the first four appliances in Figure 7 (fridges, freezers, dishwashers, and washing machines) do not show an "A" rating, reflecting the revised classifications.

The model additionally included electronics related to entertainment, such as television sets and boxes. It was important to include these devices in the model due to their high penetration rate (99.2%) and high usage across dwellings nationally.

Unlike certain end uses for which detailed data on equipment penetration are available at the municipal level, the information for cooking technologies, lighting, and household appliances is only accessible at the NUTS I level.

This presents a limitation because different regions may vary in their cooking habits, access to technologies, and income levels, leading to differences in the actual use of these technologies. However, due to the lack of disaggregated data, assuming a uniform penetration rate for all municipalities was a necessary simplification. This approach allowed for the inclusion of these technologies in the model but may have obscured local differences, potentially leading to less precise estimates at the municipal scale.

4. Results

This section presents the results and findings of the regional residential energy model, providing insights into the energy consumption patterns and dynamics within the residential sector. Firstly, the validation process conducted in this work will be presented, ensuring the accuracy and reliability of the regional residential energy model. Subsequently, an analysis of residential energy consumption, its associated emissions, and costs will be presented and discussed.

4.1. Validation

A comparison and a validation of the regional residential energy model's results were performed for electricity and natural gas.

Electricity and natural gas consumption data were used to ensure the model's accuracy and reliability using reference data from the DGEG website [42] for the year 2021. Validating the model using 2021 data is particularly important because the census data from 2021 were used extensively in the model, making it representative of the country's state during that year.

Figures 8 and 9 display the normal distributions of errors associated with the residential electricity and natural gas consumption estimated via the model, respectively.

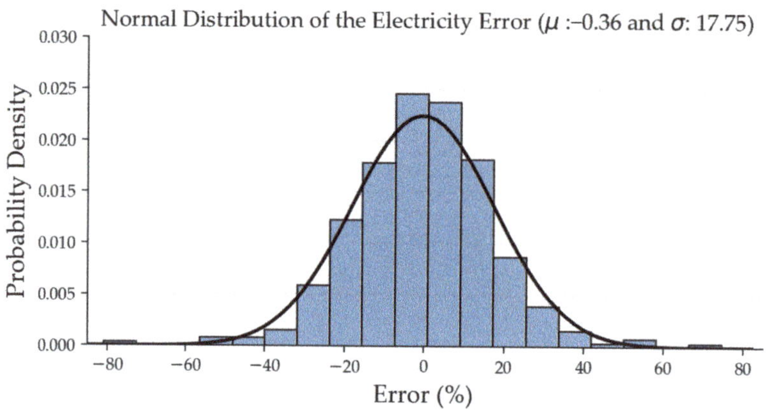

Figure 8. Normal distribution of the electricity error.

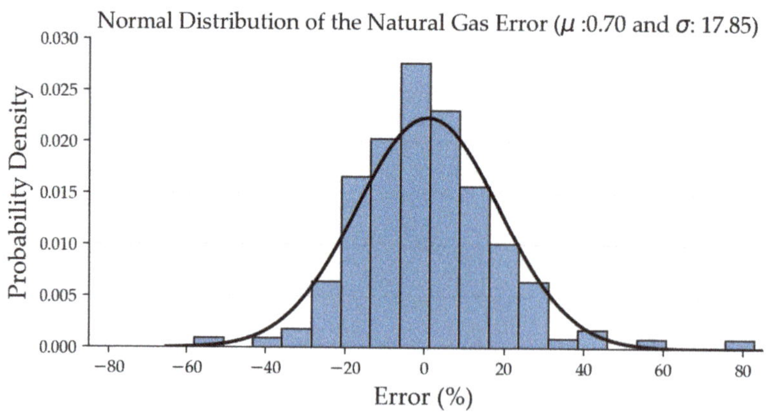

Figure 9. Normal distribution of the natural gas error.

Both models demonstrate strong performance, with a slight underestimation of electricity consumption (mean error of −0.36%) and a slight overestimation of natural gas consumption (mean error of 0.70%).

The standard deviation of errors for the electricity model is 17.75%, while the natural gas model involves a slightly higher standard deviation of 17.85%. This indicates that, although the models perform effectively on average, there is variability in their predictions. The electricity model's error typically ranges between −18.11% and +17.39%, and the natural gas model's error spans from approximately −17.15% to +18.55%. These ranges represent the spread of errors in the model's predictions, meaning that, while the average error is close to zero, individual predictions can vary considerably from the actual values. The variability is likely due to the complex and diverse factors that influence energy consumption across municipalities, such as differing household behaviors, appliance usage patterns, and the unpredictability of certain local factors that the model may not fully capture.

Although these models are not without limitations, they offer a robust representation of overall consumption trends. The relatively low mean errors suggest that the models are largely unbiased at a national scale, although some deviations may occur for individual municipalities. Given the complexities involved, the performance of both models is well within acceptable ranges for this type of analysis, as supported by other UBEM studies where acceptable error ranges for energy consumption predictions typically fall between 1% and 19%, with this range reflecting the absolute error calculated from comparisons of predicted energy use against real-world utility data [47].

Additionally, Chen et al. [14] noted that deviations at smaller scales, such as individual municipalities, are acceptable as long as UBEMs remain effective in capturing broader consumption trends. Their study emphasized that these deviations are considered a necessary trade-off, given the need to balance model complexity with computational efficiency.

In this case study, some data, such as the equipment penetration of cooking appliances, were only available at the multi-municipal level, and they were assumed to be homogeneous across all the municipalities at this level. Should more granular data become accessible in the future, they would enhance the precision of localized results and further strengthen the model ability to support decarbonization strategies tailored to municipal contexts.

4.2. As-Is Scenario

Characterizing the current state of the case study is important for understanding its energy consumption and emissions patterns. This baseline not only offers insight into how energy is being used and where emissions are originating from but also serves as a reference point for evaluating the impact of future interventions.

Figures 10 and 11 depict a characterization of energy consumption patterns across the various end uses in the residential dwellings within the case study, offering insights into both the proportional distribution of energy consumption and the energy use per dwelling.

The figures illustrate that, in this specific case study, space heating and cooking represent the largest portions of energy consumption. The high standard deviation observed in space heating (Figure 11) is largely a consequence of each municipality's specific weather conditions. More specifically, municipalities in the northern regions of the country exhibit the highest energy consumption for space heating, which is closely associated with higher HDD values.

Space heating consumption is also influenced by other factors, such as building age, envelope thermal insulation quality, and heating system efficiency, and this combination of factors, with HDD playing a significant role, directly contributes to the larger standard deviation observed in space heating consumption throughout the case study.

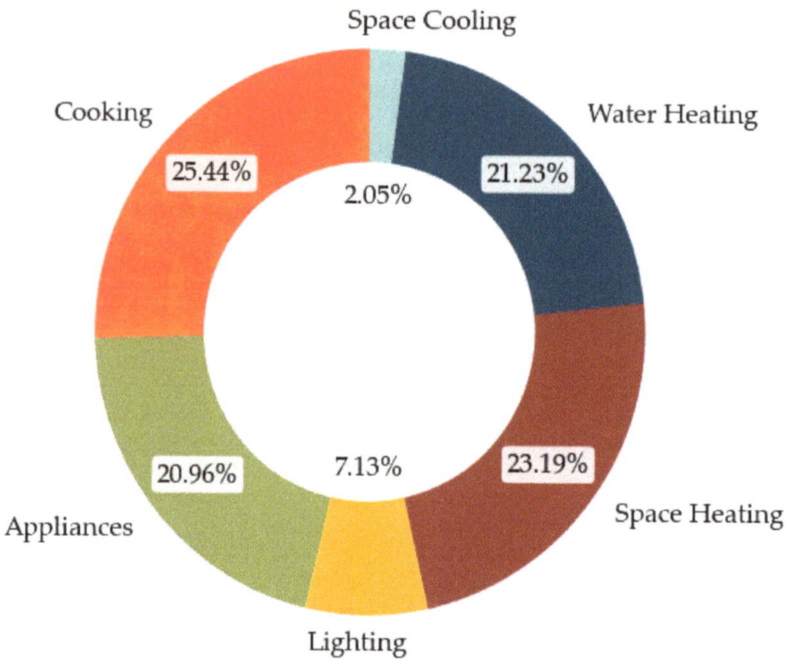

Figure 10. Energy consumption distribution by end use in dwellings.

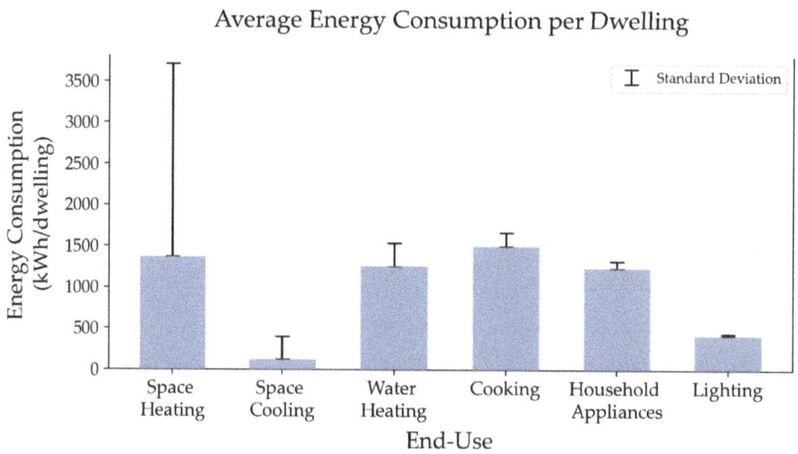

Figure 11. Energy consumption (kWh/dwelling) per dwelling and by end use.

For cooking, household appliances, and lighting, the low standard deviations can be explained by the lack of detailed municipal-level data for these end uses. In the absence of such data, assumptions had to be made at the national level. For instance, lighting requirements, including hours of usage and the types of appliances used, were standardized across all municipalities. These generalized inputs contribute to less variability in the model's outputs, as they assume similar consumption behaviors across regions. This results in relatively low standard deviation values, as there is little differentiation between municipalities in terms of these energy uses.

In the case of water heating, while detailed information on equipment penetration is available at the municipal level, the actual usage patterns—such as the amount of water consumed and the temperature settings—were assumed to be constant across all households. This decision, while necessary due to data limitations, contributes to the medium level of standard deviation observed.

For space cooling, despite the relatively low average energy consumption, there is a higher standard deviation compared to other end uses. This can be attributed to the climatic variability across the case study area. Municipalities in the interior of the country, which experience higher outside temperatures during the cooling season, tend to have greater energy consumption for space cooling, while coastal or cooler regions have lower space cooling needs. This geographic disparity introduces greater variability in the energy consumption for space cooling, even though the overall demand remains low compared to other end uses.

Finally, understanding the total emissions per end use in the case study is crucial for identifying which energy-consuming activities contribute the most to overall GHG emissions. This breakdown allows for a more targeted approach when developing decarbonization strategies, as it highlights the specific areas where emission reductions can be most impactful.

Figure 12 presents the energy source distribution by end use, and Figure 13 presents the total emissions per end use across the case study, providing a clearer picture of how energy consumption translates into CO_2eq emissions.

Although space heating is one of the most energy-intensive end uses, a significant portion of its consumption comes from biomass, which in this study is considered to produce no emissions. This assumption explains why space heating, despite its high energy consumption, contributes less to the total emissions when compared to other end uses.

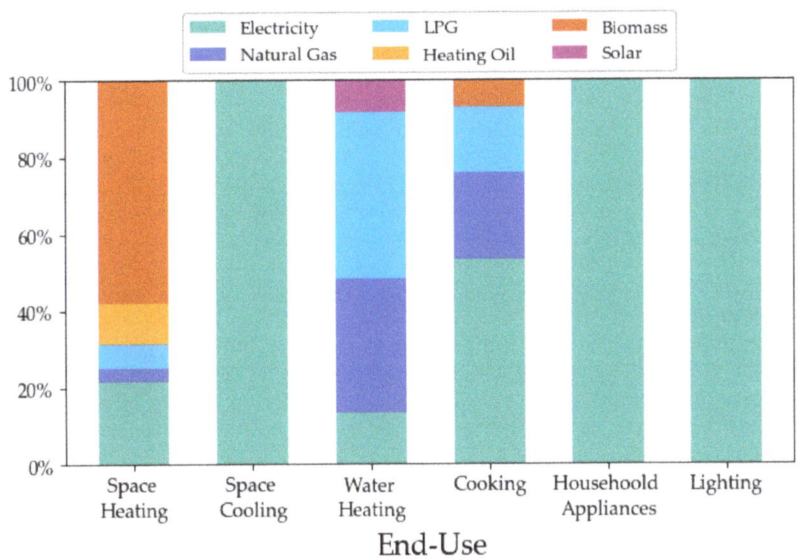

Figure 12. Energy source distribution by end use in dwellings.

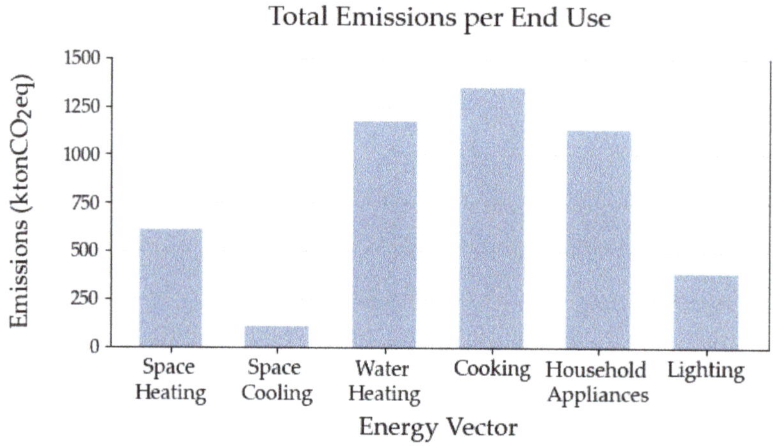

Figure 13. Total emissions (ktonCO_2eq) per end use.

Unlike other energy-intensive end uses like household appliances, water heating, and cooking, which exhibit consistently high consumption across all municipalities, space heating exhibits significant regional variation. As previously discussed, this is largely due to differences in HDD, where northern regions with higher HDD values experience much greater demand for space heating compared to their lower HDD counterpart. This regional disparity means that space heating plays a major role in energy consumption and emissions in colder areas while contributing relatively little in warmer regions.

Space cooling and lighting are the least-emitting sectors, which aligns with the fact that these end uses rely solely on electricity and represent a smaller share of overall energy use, as shown in Figures 10 and 12.

4.3. Decarbonization Scenarios

In this section, various decarbonization measures are explored with a focus on reducing emissions across different building types and energy systems. These measures are evaluated to highlight potential energy savings, environmental benefits, and economic considerations.

The next step is particularly important, as it transforms the model from only estimating emissions and energy consumption into a design tool. By providing detailed insights into the cost-effectiveness and environmental impact of different measures, the model enables stakeholders make informed decisions about solutions that balance sustainability and investment returns.

Although the number of potential measures to test is extensive, a set of scenarios has been carefully selected based on their appropriateness for the case study. These measures represent a diverse mix of passive design improvements, active technological upgrades, and behavioral changes tailored to the specific needs and context of the study area.

Table 3 provides a comprehensive overview of the measures and design scenarios evaluated in this study.

This study tested three distinct categories of scenarios: passive, active, and behavioral. Passive measures focus on improving the energy efficiency of buildings by enhancing the design and characteristics of the building envelope. The building envelope refers to the physical barrier between the conditioned indoor space and the external environment, including elements like walls, roofs, windows, and floors. These measures are generally non-mechanical and focus on reducing the building's energy demand by leveraging design principles that naturally maintain thermal comfort. Active measures, in contrast to passive ones, involve the use of mechanical or electrical systems to manage energy demand and improve the energy efficiency of a building. These systems typically include HVAC, and lighting, among others. Active measures often focus on upgrading equipment to more energy-efficient alternatives or integrating renewable energy sources into a building [48].

Table 3. Measures under study overview.

Type of Measure	Measure ID	Measure Description	Scope
Passive measures	E1	Window renovations	13 Municipalities
	E2	Thermal insulation renovations	13 Municipalities
Active measures	W1	Solar thermal	High-irradiation municipalities [1]
	W2	Electrification of water heating	Nationally
	W3	Electrification of water heating	High population density *
	H1	Electrification of space heating	Nationally
	H2	Electrification of space heating	Highest penetration of fossil fuels *
	L1	100% LED lighting	Nationally
	A1	More efficient appliances	Nationally
	A2	More efficient appliances	High population density *
	A3	More efficient appliances	High income *
Behavioral measures	B1	Shower water conservation	Nationally
	B2	Decreased TV consumption	Nationally

[1] Performed in the following NUTS III municipalities: Algarve, Alentejo Central, Baixo Alentejo, Alto Alentejo, and Alentejo Litoral. * Performed in the top 50 municipalities of this scope.

In addition to passive and active measures, behavioral measures focus on influencing the behavior of building occupants to promote energy-efficient practices. Unlike passive and active measures, which are related to physical upgrades or equipment, behavioral measures aim to reduce energy consumption by changing how people use energy.

Despite biomass being recognized as a renewable energy source, it was not included in the decarbonization measures. This decision was based on concerns that incentivizing its use could promote deforestation, as increased demand for biomass may lead to unsustainable harvesting of forests, ultimately undermining the environmental goals of the study.

For each measure, a base investment cost was established, which reflects the specific characteristics of the case study that are presented in Table 4. While tailored to the conditions of this case study, the costs can be adjusted and adapted for application in other geographical or socioeconomic contexts.

Table 4. Base investment costs.

Measure ID	Intervention	Investment
E1	EPS 150 (80 mm thickness)	45 EUR/m^2 [49]
E2	Window framing and double glass	380 EUR/m^2 [49,50]
W1	Solar thermal panel	1500 EUR/unit [49]
W2, W3	Heat pump water heater	2200 EUR/unit [49]
H1, H2	Heat pump space heating	2800 EUR/unit [49]
L1	LED bulbs	4 EUR/W [50]
A1, A2, A3	Fridge	950 EUR/unit [49]
	Washer	300 EUR/unit [49]
	Dryer	450 EUR/unit [49]
B1, B2	Education and awareness programs	3.30 EUR/resident [51]

5. Discussion

An abatement cost curve was utilized to support the design of cost-effective solutions by comparing decarbonization measures. The abatement cost curve illustrates the relationship between the cost per tonne of CO_2eq avoided (vertical axis) and the annual emission reductions achieved via each measure (horizontal axis). This enables a clear comparison of different strategies based on their financial cost and their ability to reduce emissions on a yearly basis.

By applying this method, we can identify measures that are both cost-effective and capable of contributing significantly to overall emission reductions.

When interpreting an abatement cost curve, measures can appear on either the negative or positive side of the cost axis, which has different implications for their cost-effectiveness. Measures on the negative axis (below zero on the cost axis) represent negative abatement costs. These measures not only reduce emissions but also save money over time. The savings they generate, typically through lower energy use or increased efficiency, more than offset their initial costs. Measures on the positive axis represent positive abatement costs, meaning they require a net investment for each tonne of CO_2eq reduced.

While the abatement cost curve is a valuable tool for assessing cost-effectiveness, it is important to recognize that it does not capture additional benefits that certain measures may provide. For instance, measures such as building insulation or renewable energy installations can improve thermal comfort, enhance indoor air quality, or offer energy security through local production. Additionally, measures like investments in renewable energy infrastructure might yield future financial gains, especially if energy prices fluctuate or carbon pricing becomes more stringent. Thus, while a measure with a positive abatement cost may seem less attractive from a purely financial perspective, its broader societal and environmental contributions should not be disregarded. These extra benefits can play a crucial role in long-term decarbonization strategies, making such measures essential in achieving comprehensive climate goals.

An abatement cost curve serves as a design tool to assist decision-makers in interpreting cost-effectiveness alongside other objectives, such as maximizing emission reductions or addressing societal benefits. Factors such as policy goals, long-term sustainability, and the specific context of the region or sector should all play a role in shaping decisions. Ultimately, the curve provides valuable insight into the relative costs of different strategies, but the final choice depends on the balance between economic feasibility, environmental impacts, and other strategic priorities that are unique to each decision-making scenario.

Figure 14 presents the abatement cost curve for the measures tested in the case study on a national level.

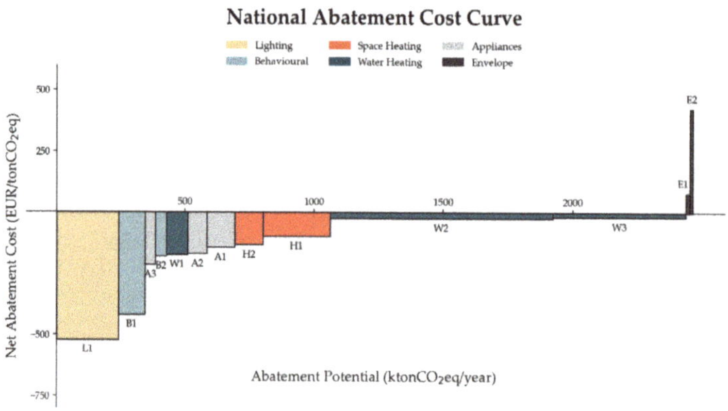

Figure 14. National abatement cost curve.

The passive measures, E1 and E2, which involve the installation of thermal insulation and the upgrading of windows in I3 winter-zone municipalities, both show a positive abatement cost on the abatement cost curve. Moreover, they demonstrate relatively marginal emission reductions when compared to some of the other measures. This can largely be attributed to the nature of passive measures, which typically involve high upfront investment costs for materials, labor, and installation, particularly for building-envelope upgrades.

Additionally, the effectiveness of these passive measures in reducing energy consumption for space heating and cooling can be limited in certain cases. For example, while thermal insulation and upgraded windows reduce heat transfer, their impact on actual energy consumption depends on the existing efficiency of the heating and cooling systems. If the heating and cooling systems in place are already efficient, the marginal energy savings from insulation upgrades can be lower, thus resulting in a higher cost per tonne of CO_2eq reduced. This is especially true in cases where the baseline energy consumption for space heating is not as high, making the reduction in consumption smaller relative to the investment.

It is important to note that space heating is largely dependent on biomass, which accounts for approximately 60% of the energy used. Biomass is often regarded as carbon-neutral, meaning measures aimed at reducing energy consumption for space heating, such as improved insulation or window upgrades, have a limited direct impact on emission reductions in this scenario. However, the remaining 40% of energy used for space heating comes from electricity, natural gas, LPG, and heating oil, which do contribute to emissions. As results passive measures, while they are beneficial in terms of energy efficiency, they have a reduced overall impact on emissions in cases where biomass dominates the heating mix.

However, this does not diminish the importance of such measures in improving thermal comfort and ensuring energy efficiency, even in situations where the direct emissions impact may be less significant. To assess whether these benefits are truly justified or worthwhile, an additional and more detailed analysis would be necessary.

All the active measures in this study demonstrated a negative abatement cost, indicating that they result in net economic gains over their lifetimes.

Among the active measures, W2, which involves the installation of heat pumps to replace fossil fuel-based water heating systems on a national level, is the most impactful in terms of direct emission reductions. Heat pumps are an energy-efficient technology that significantly reduces dependence on fossil fuels, offering substantial reductions in CO_2eq emissions; however, W2 presents a higher abatement cost compared to other measures in this study, primarily due to the high upfront costs associated with purchasing and installing heat pump systems. Despite these costs, the measure still offers a positive return on investment over time, making it financially viable in the long term.

Although W2 offers substantial emission reduction potential, implementing it on a national scale presents significant challenges. Replacing all fossil fuel-based water heating systems with heat pumps across the country would require a large financial investment and logistical coordination, making it unfeasible in the short term. However, this measure provides a valuable perspective on the long-term environmental gains that could be achieved.

Measure L2, which involves converting the entire residential sector to LED light fixtures, stands out as the measure with the lowest abatement cost and offers the best economic returns, along with considerable emission reductions. In the case study, a large portion of residential lighting still relies on non-LED lights, leaving substantial room for improvement. The low abatement cost associated with this measure can be attributed to the affordability of LED technology and its high efficiency, making it an attractive option for decarbonization efforts in this region.

The higher efficiency of LED lights, coupled with their relatively low cost, underscores their potential as a key technology in the future decarbonization plans for this case study. Although replacing all lighting fixtures on a national scale at once may not be feasible, a gradual approach could be taken. This could include phasing out non-LED fixtures from the market and incentivizing the adoption of LEDs. Given the already low price and long lifespan of LEDs, this transition would likely be easy to implement as older lighting systems naturally phase out, ensuring a smooth and cost-effective shift toward energy-efficient lighting.

Finally, it is important to closely examine the behavioral measures. In these measures, the investment required was based on the cost per resident, which represents the education needed to shift behaviors. These costs include expenses for education campaigns aimed at informing the population. Both B1 (water conservation) and B2 (reduced TV watch time) achieved a negative abatement cost, indicating that they result in net economic gains over time; however, the abatement cost for B1 is lower than that for B2. This difference arises because water conservation leads to more significant energy savings by reducing the amount of hot water needed, which lowers the energy demand for heating. In contrast, reducing TV watch time has a smaller impact since the energy used by TVs is much lower than the energy required for water heating.

These measures demonstrate that behavioral changes can have a substantial impact on the emissions of the residential sector. Encouraging shifts in daily activities, such as conserving water or reducing unnecessary energy use, can effectively lower household emissions. However, while promoting behavioral changes has clear benefits, achieving a nationwide shift in behavior is challenging, and such changes require persistent and well-designed education campaigns to ensure long-term adoption.

While the proposed decarbonization measures offer significant insights, certain practical aspects, such as physical space requirements for the installation of technologies, were not considered in this analysis. These limitations, although important for real-world implementation, fall outside the scope of this study.

In addition to analyzing the cost-effectiveness of the measures, the discounted payback period was also evaluated for each intervention. The discounted payback period refers to the amount of time it takes for the cumulative discounted savings from a measure to equal its initial investment. This period accounts for the time value of money, meaning that future savings are discounted to reflect their present value. By doing so, it provides a more accurate representation of how long it takes for the benefits of a measure to outweigh its costs.

Table 5 presents the discounted payback period for the decarbonization measures under study.

Table 5. Discounted payback period.

Measure ID	Discounted Payback Period
E1	41 years
E2	47 years
W1	10 years
W2	15 years
W3	15 years
H1	11 years
H2	10 years
L1	2 years
A1	16 years
A2	15 years
A3	13 years
B1	1 year
B2	1 year

The analysis in Table 5 reveals that the highest discounted payback periods were found in the passive measures, particularly in Measure E2. Notably, the discounted payback period for E2 is significantly longer than the expected lifetime of windows ($\approx$25 years), indicating that the financial returns from energy savings will not cover the initial investment within the useful lifespan of the measure. As discussed earlier, passive measures are generally less effective at reducing energy consumption, which means that their high upfront costs are harder to recover through savings.

Most of the active measures, with the exception of L1 (LED lighting), demonstrated medium to high discounted payback periods. Despite this, these payback periods are still within the expected lifespan of the equipment. With proper maintenance, systems such as solar thermal panels and heat pumps can last significantly longer than their discounted payback periods, ensuring that the initial investment is recouped before the equipment reaches the end of its useful life. This makes these measures both financially viable and effective in the long run, as they continue to provide energy savings and emission reductions even after the payback period has passed.

In contrast, L1 (LED lighting) achieved a short discounted payback period, primarily due to the low cost of LED technology. LEDs are relatively inexpensive to install and operate, which allows for faster financial returns. Moreover, LED lights are known for their long lifespans, often lasting up to 50,000 h of use, which translates to several years of continuous service. This combination of low cost, high efficiency, and long life makes LED lighting an essential component of any decarbonization strategy, offering both immediate financial savings and long-term emission reduction.

Finally, behavioral measures showed the lowest discounted payback periods, largely due to the immediate energy savings that result from changes in daily habits. These measures require minimal upfront investment, as the primary costs are associated with education and awareness campaigns, while the energy savings are realized quickly. Actions such as reducing unnecessary energy consumption, like optimizing appliance use or conserving water, lead to significant reductions in household energy demand, translating into rapid monetary savings.

Because behavioral shifts can be implemented without costly infrastructure changes, the financial return is seen almost immediately. This makes them highly cost-effective in the short term while also contributing to emission reductions. These measures underscore the importance of engaging residents in decarbonization efforts, showing that even small adjustments in energy use habits can have a notable impact on overall energy consumption and emissions. Behavioral measures, therefore, play a vital role in achieving immediate gains while laying the foundation for broader decarbonization strategies.

The final analysis aimed to assess whether the abatement cost of a specific measure remained consistent across different municipalities or regions.

At the municipal level, the measure evaluated was W2, which has the highest abatement potential, and it involved the replacement of gas water heaters with heat pumps. The municipal-level abatement cost and emissions savings for measure W2 are presented in Figure 15.

As shown in the previous figure, Figure 14, the national-level abatement cost for measure W2 was found to be slightly negative. Now, at the municipal level, Figure 15 reveals that the abatement cost is not uniform across municipalities, showing significant variability. In some cases, the negative abatement cost observed at the national level shifts to a positive abatement cost locally, indicating that the measure does not perform as efficiently in all municipalities.

A potential factor influencing the variability in abatement costs could be the number of residents per household. Larger households typically use more hot water, increasing energy consumption; as a result, upgrading to more efficient systems, such as heat pumps, can lead to substantial cost savings and emission reductions. On the other hand, smaller households may have lower energy demands, making the transition less impactful and, therefore, less cost-effective.

For the regions NUTS II, Norte and Azores abatement cost curves were created to assess the decarbonization measures that are applicable to each region from the set analyzed, and these curves are presented in Figures 16 and 17.

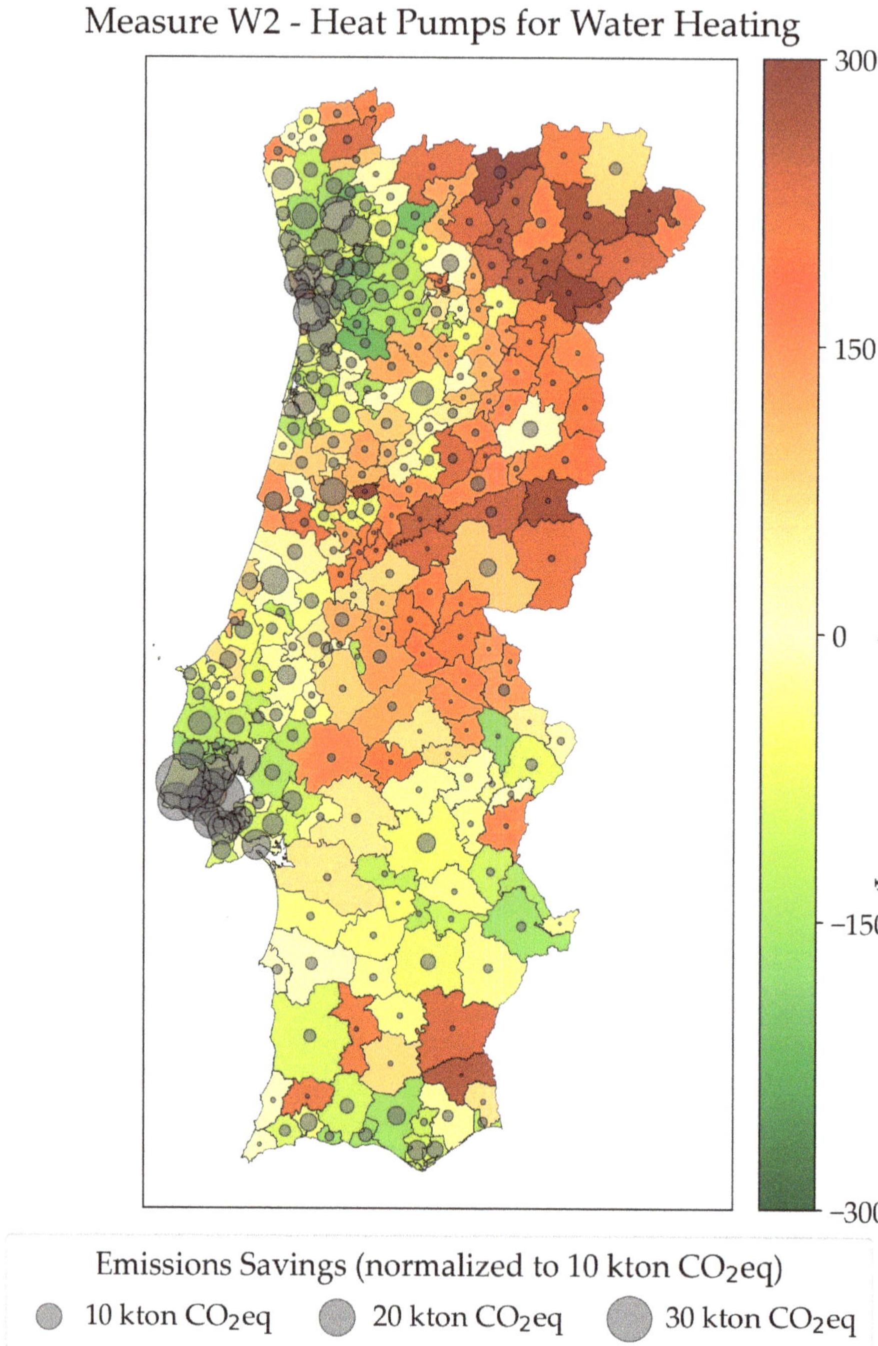

Figure 15. Municipal−level abatement cost and emission reductions for measure W2.

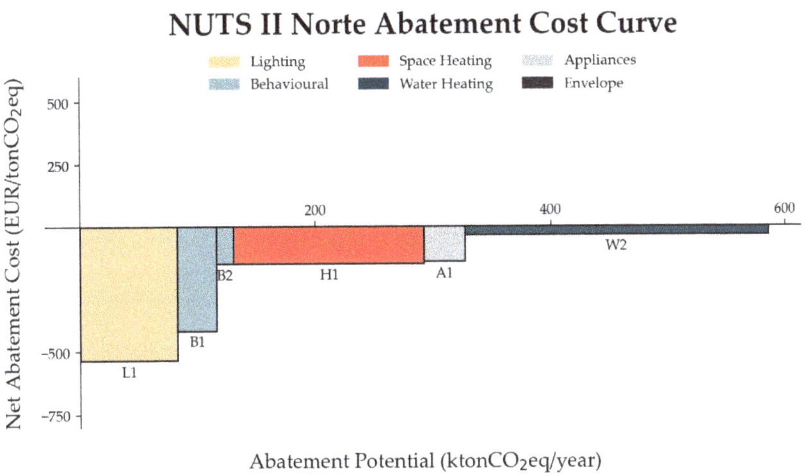

Figure 16. NUTS II Norte abatement cost curve.

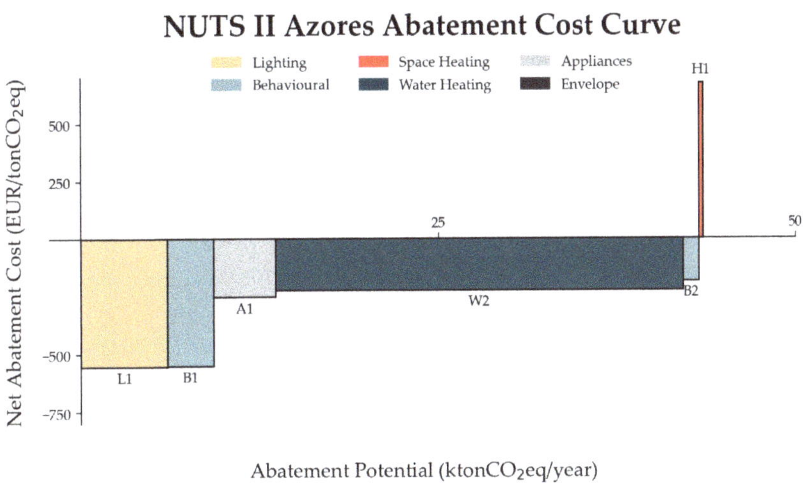

Figure 17. NUTS II Azores abatement cost curve.

As depicted in Figures 16 and 17, the abatement cost curves for NUTS II, Norte and Azores, illustrate notable differences in the performance of the decarbonization measures across these two regions. The most striking contrast is with measure H1, which demonstrates a negative abatement cost in Norte, indicating that the investment is recovered within the equipment's lifetime due to the region's harsher winter climate. In contrast, the same measure in the Azores has a positive abatement cost, suggesting that the milder climate makes this measure less economically viable. Furthermore, the abatement potential of H1 also differs substantially between the two regions, highlighting the influence of local conditions on the effectiveness of decarbonization measures.

Given this variability, it is important to explore how different municipal-level factors drive abatement costs and measure performance. The variability in the penetration rates of technologies, building characteristics, and other local conditions shows that a uniform approach may not be effective across all municipalities; therefore, future work should focus on conducting sensitivity analyses to understand how different parameters and variables affect the abatement cost of decarbonization measures.

Sensitivity analyses are valuable because they give decision-makers a clearer idea of how a measure's abatement cost might change under different conditions. By understanding local factors like building stock, population density, and the use of fossil fuel-based equipment's, municipalities can prioritize which measures to focus on without running extensive simulations for each one. Identifying the most influential variables helps authorities prioritize measures, simplifying decision-making and optimizing resource use. This approach allows municipalities to adopt more effective, scalable strategies and make smart, cost-efficient decisions that fit their specific local needs.

6. Conclusions

A computationally efficient regional residential energy model was developed and validated with a full-scale demonstration in Portugal. The model offers insights into both economic and environmental impacts, helping design decarbonization strategies by analyzing abatement costs, emission reduction potential, and discounted payback periods.

While the model offers distinct advantages, such as computational efficiency, adaptability, and the inclusion of a cost-effectiveness analysis, it also faces certain limitations. These limitations include the necessity of simplifications, assumptions, and the use of archetypes. Despite these limitations, the model demonstrated good performance at the national level, achieving a mean electricity and natural gas consumption estimation below 1%.

The novelty of this model lies in enabling a nation-wide design of cost-effective building solutions with regional disaggregation, as a consequence of its computational efficiency. This is advantageous when compared to traditional UBEMs, since they lack the computational efficiency necessary for national-scale deployment without the support of high-performance infrastructure.

The results highlight that active measures like LED lighting and heat pumps offer quicker payback periods and larger emission reductions compared to passive measures, such as insulation upgrades, which require higher upfront investments and result in lower energy savings. Specifically, LED lighting achieves a discounted payback period of 2 years, while heat pumps range from 10 to 15 years, depending on the measure. In contrast, thermal insulation has a much longer payback period of 41 years. Additionally, the role of behavioral changes demonstrates that even small adjustments in daily energy use can significantly impact energy savings and emission reductions, achieving annual savings of 147 tonnes of CO_2eq and 653 MWh.

The insights provided via this model can contribute meaningfully to a coordinated national decarbonization strategy. By identifying the most effective interventions across diverse municipalities, the model can guide policymakers in designing efficient, regionally tailored approaches to emissions reduction.

Author Contributions: Conceptualization, P.F., P.B., R.G. and M.J.; methodology, P.B., R.G. and M.J.; software, M.J.; validation, M.J.; formal analysis, M.J.; investigation, M.J.; data curation, M.J.; writing—original draft preparation, M.J.; writing—review and editing, P.F., P.B., R.G. and M.J.; visualization, M.J.; supervision, P.F., P.B. and R.G.; project administration, R.G.; funding acquisition, P.F. All authors have read and agreed to the published version of the manuscript.

Funding: This research was funded by Fundação para a Ciência e Tecnologia through the following programs: IN+ Strategic Project (1801P.00962.1.01—IN+ UIDP/EEA/50009/2020—IST-ID) and grant UI/BD/154783/2023.

Data Availability Statement: Data that support the findings of this study are available from the corresponding author, P.F., upon reasonable request.

Acknowledgments: The authors would like to acknowledge the Project BE.Neutral—Agenda de Mobilidade para a neutralidade carbónica nas cidades, contract number 35, funded by the Resilience and Recovery Plan (PRR) through the European Union under the Next Generation EU, and Fundo Ambiental, through Protocol 49—"Plataforma de Apoio à Rede de Cidades Portuguesas climaticamente neutras e inteligentes 2030".

Conflicts of Interest: The authors declare no conflicts of interest.

Abbreviations

The following abbreviations are used in this manuscript:

ACH	Air changes per hour
DGEG	Direção-Geral de Energia e Geologia
EU	European Union
GAP	Glazing area percentage
GHGs	Greenhouse gases
HDDs	Heating degree days
IEA	International Energy Agency
INE	Instituto Nacional de Estatística
LED	Light-emitting diode
LPG	Liquefied petroleum gas
NECP	National Energy and Climate Plan
RCCTE	Portuguese Regulation for Energy Performance of Buildings
REH	Regulation on the Energy Performance of Residential Buildings
UBEM	Urban building energy model

References

1. United Nations. Paris Agreement. 2015. Available online: https://unfccc.int/sites/default/files/english_paris_agreement.pdf (accessed on 15 October 2024).
2. Firth, S.K.; Lomas, K.J.; Wright, A.J. Targeting household energy-efficiency measures using sensitivity analysis. *Build. Res. Inf.* **2010**, *38*, 25–41. [CrossRef]
3. European Commission. Focus on Energy Efficiency in Buildings. 2020. Available online: https://commission.europa.eu/news/focus-energy-efficiency-buildings-2020-02-17_en (accessed on 15 October 2024).
4. Eurostat. Energy, Transport and Environment Statistics. 2020. Available online: https://ec.europa.eu/eurostat/web/products-statistical-books/-/ks-dk-20-001 (accessed on 15 October 2024).
5. Kong, D.; Cheshmehzangi, A.; Zhang, Z.; Ardakani, S.P.; Gu, T. Urban building energy modeling (UBEM): A systematic review of challenges and opportunities. *Energy Effic.* **2023**, *16*, 69. [CrossRef]
6. Ferrando, M.; Causone, F.; Hong, T.; Chen, Y. Urban building energy modeling (UBEM) tools: A state-of-the-art review of bottom-up physics-based approaches. *Sustain. Cities Soc.* **2020**, *62*, 102408. [CrossRef]
7. Kamel, E. A Systematic Literature Review of Physics-Based Urban Building Energy Modeling (UBEM) Tools, Data Sources, and Challenges for Energy Conservation. *Energies* **2022**, *15*, 8649. [CrossRef]
8. Hong, T.; Chen, Y.; Lee, S.H.; Piette, M. CityBES: A Web-Based Platform to Support City-Scale Building Energy Efficiency. 2016. Available online: https://www.researchgate.net/publication/304824985_CityBES_A_Web-based_Platform_to_Support_City-Scale_Building_Energy_Efficiency (accessed on 15 October 2024).
9. Li, H.; Hong, T.; Zhang, W. Modeling and Simulation of District Heating and Cooling Systems Using CityBES. In Proceedings of the 17th Conference of the International Building Performance Simulation Association (BS 2021), Bruges, Belgium, 1–3 September 2021. Available online: https://publications.ibpsa.org/proceedings/bs/2021/papers/bs2021_30155.pdf (accessed on 15 October 2024).
10. Fonseca, J.A.; Schlueter, A. Integrated model for characterization of spatiotemporal building energy consumption patterns in neighborhoods and city districts. *Appl. Energy* **2015**, *142*, 247–265. [CrossRef]
11. Johari, F.; Peronato, G.; Sadeghian, P.; Zhao, X.; Widén, J. Urban building energy modeling: State of the art and future prospects. *Renew. Sustain. Energy Rev.* **2020**, *128*, 109902. [CrossRef]
12. Ali, U.; Shamsi, M.H.; Hoare, C.; Mangina, E.; O'Donnell, J. Review of urban building energy modeling (UBEM) approaches, methods and tools using qualitative and quantitative analysis. *Energy Build.* **2021**, *246*, 111073. [CrossRef]
13. Sola, A.; Corchero, C.; Salom, J.; Sanmarti, M. Multi-domain urban-scale energy modelling tools: A review. *Sustain. Cities Soc.* **2020**, *54*, 101872. [CrossRef]
14. Chen, Y.; Hong, T.; Piette, M.A. Automatic generation and simulation of urban building energy models based on city datasets for city-scale building retrofit analysis. *Appl. Energy* **2017**, *205*, 323–335. [CrossRef]
15. Beltrán-Velamazán, C.; Monzón-Chavarrías, M.; López-Mesa, B. A new approach for national-scale Building Energy Models based on Energy Performance Certificates in European countries: The case of Spain. *Heliyon* **2024**, *10*, e25473. [CrossRef]
16. Pezzutto, S.; Croce, S.; Zambotti, S.; Kranzl, L.; Novelli, A.; Zambelli, P. Assessment of the Space Heating and Domestic Hot Water Market in Europe—Open Data and Results. *Energies* **2019**, *12*, 1760. [CrossRef]

17. International Energy Agency. Heating. Available online: https://www.iea.org/energy-system/buildings/heating (accessed on 15 October 2024).
18. Instituto Nacional de Estatística, I.P.; Direção-Geral de Energia e Geologia. Inquérito ao Consumo de Energia no Sector Doméstico—2020. 2021. Available online: https://www.dgeg.gov.pt/media/jvfgkejh/dgeg-aou-icesd-2020.pdf (accessed on 15 October 2024).
19. Decreto-Lei n.º 118/2013: Regulamento de Desempenho Energético dos Edifícios Habitação (REH). *Diário da República*, 20 August 2013; 1st serie, N.º 159. Available online: https://diariodarepublica.pt/dr/detalhe/decreto-lei/118-2013-499237 (accessed on 15 October 2024).
20. Durmayaz, A.; Kadıoğlu, M.; Şen, Z. An application of the degree-hours method to estimate the residential heating energy requirement and fuel consumption in Istanbul. *Energy* **2000**, *25*, 1245–1256. [CrossRef]
21. Ximenes, R.F. Energy Demand Model to Support the Definition of Sustainable Energy Systems. Master's Thesis, Instituto Superior Técnico, Universidade de Lisboa, Lisbon, Portugal, 2015. Available online: https://scholar.tecnico.ulisboa.pt/api/records/r46rO1tHBuxQm6bS3rR-UYUKqRB3hU2F1ys7/file/44b7ccca63770763a22921198b4f2a25820767eb99e53462867b8d83c49e1c6e.pdf (accessed on 15 October 2024).
22. Ounis, S.; Aste, N.; Butera, F.M.; Pero, C.D.; Leonforte, F.; Adhikari, R.S. Optimal Balance between Heating, Cooling and Environmental Impacts: A Method for Appropriate Assessment of Building Envelope's U-Value. *Energies* **2022**, *15*, 3570. [CrossRef]
23. Giama, E.; Kyriaki, E.; Papadopoulos, A. U-Value: A Key Role Parameter for Sustainable Buildings 2020. Available online: https://www.researchgate.net/publication/340528428_U-Value_A_key_role_parameter_for_sustainable_buildings (accessed on 15 October 2024).
24. Portuguese Government. *Regulamento das Características de Comportamento Térmico de Edifícios (RCCTE)*; Official Gazette of the Portuguese Republic; Portuguese Government: Lisbon, Portugal, 2006.
25. De Souza, G.B.H. A Multi-Objective Decision Support Methodology for Developing National Energy Efficiency Plans. Ph.D. Thesis, Universidade do Porto, Porto, Portugal, 2011. Available online: https://repositorio-aberto.up.pt/bitstream/10216/64682/1/000151875.pdf (accessed on 15 October 2024).
26. Daioglou, V.; van Ruijven, B.J.; van Vuuren, D.P. Model projections for household energy use in developing countries. *Energy* **2012**, *37*, 601–615. [CrossRef]
27. Bréchet, T.; Jouvet, P.A. Why environmental management may yield no-regret pollution abatement options. *Ecol. Econ.* **2009**, *68*, 1770–1777. [CrossRef]
28. Madadizadeh, A.; Siddiqui, K.; Aliabadi, A.A. Review: The Economics Landscape for Building Decarbonization. *Sustainability* **2024**, *16*, 6214. [CrossRef]
29. Gbadeyan, O.J.; Muthivhi, J.; Linganiso, L.Z.; Deenadayalu, N. Decoupling Economic Growth from Carbon Emissions: A Transition toward Low-Carbon Energy Systems—A Critical Review. *Clean Technol.* **2024**, *6*, 1076–1113. [CrossRef]
30. European Commission. Economic Appraisal Vademecum 2021–2027—General Principles and Sector Applications. European Union. 2021. Available online: https://ec.europa.eu/regional_policy/sources/guides/vademecum_2127/vademecum_2127_en.pdf (accessed on 15 October 2024).
31. Instituto Nacional de Estatística. Census 2021. 2021. Available online: https://censos.ine.pt/xportal/xmain?xpgid=censos21_main&xpid=CENSOS21&xlang=pt (accessed on 15 October 2024).
32. IPMA, Instituto Português do Mar e da Atmosfera. Área educativa: Clima Portugal Continental. 2023. Available online: https://www.ipma.pt/pt/educativa/tempo.clima/index.jsp?page=clima.pt.xml (accessed on 15 October 2024).
33. Almeida, L.; Silva, A.S.; Veiga, M.d.R.; Vieira, M.; Mirão, J. Physical and Mechanical Properties of Reinforced Concrete from 20th-Century Architecture Award-Winning Buildings in Lisbon (Portugal): A Contribution to the Knowledge of Their Evolution and Durability. *Constr. Mater.* **2022**, *2*, 127–147. [CrossRef]
34. Coutinho, R.N. Relação Entre o Regulamento de Desempenho energéTico de Edifícios de Habitação com o Conceito de Edifício com Necessidades Quase Nulas de Energia. Master's Thesis, Universidade do Minho, Minho, Portugal, 2014. Available online: https://repositorium.sdum.uminho.pt/handle/1822/29022 (accessed on 15 October 2024).
35. Appolloni, L.; D'Alessandro, D. Housing Spaces in Nine European Countries: A Comparison of Dimensional Requirements. *Int. J. Environ. Res. Public Health* **2021**, *18*, 4278. [CrossRef]
36. RELIABLE: Real-Time Risk Warning System for Building Occupants During Extreme Weather Events. Available online: http://reliable.tecnico.ulisboa.pt/ (accessed on 15 October 2024).
37. Sousa, J.; Bragança, L.; Almeida, M.; Silva, P. Research on the Portuguese Building Stock and Its Impacts on Energy Consumption—An Average U-Value Approach. *Arch. Civ. Eng.* **2013**, *59*, 29. [CrossRef]
38. Quercus—Associação Nacional de Conservação da Natureza. Projecto EcoFamílias—Relatório Final. 2008. Available online: https://www.slideshare.net/slideshow/programa-ecofamlias225/65252138#1 (accessed on 15 October 2024).
39. European Environment Agency. 2022. Greenhouse Gas Emission Intensity of Electricity Generation in Europe. Available online: https://www.eea.europa.eu/ims/greenhouse-gas-emission-intensity-of-1 (accessed on 15 October 2024).
40. Prussi, M.; Yugo, M.; De Prada, L.; Padella, M.; Edwards, R.; Lonza, L. JEC Well-to-Tank Report v5. In *Scientific Analysis or Review, Policy Assessment, Technical Guidance KJ-NA-30269-EN-N*; JEC: Luxembourg, 2020. [CrossRef]

41. Entidade Reguladora dos Serviços Energéticos. Tarifas e Preços Para a Energia Elétrica de Julho a Dezembro de 2022. Available online: https://www.erse.pt/media/jfphbwng/tarifas-e-pre%C3%A7os-se-2sem2022.pdf (accessed on 12 November 2024).
42. Direcção Geral de Energia e Geologia (DGEG). Energia. Available online: https://www.dgeg.gov.pt/pt/estatistica/energia/ (accessed on 12 November 2024).
43. Direção-Geral de Energia e Geologia (DGEG). Preço dos Combustíveis. Available online: https://precoscombustiveis.dgeg.gov.pt/ (accessed on 12 November 2024).
44. Cherubini, F.; Peters, G.P.; Berntsen, T.; Strømman, A.H.; Hertwich, E. CO_2 emissions from biomass combustion for bioenergy: Atmospheric decay and contribution to global warming. *Glob. Chang. Biol. Bioenergy* **2011**, *3*, 413–426. [CrossRef]
45. Instituto Nacional de Estatística, I.P.; Direção-Geral de Energia e Geologia. Inquérito ao Consumo de Energia no Sector Doméstico—2010. 2011. Available online: https://www.dgeg.gov.pt/media/g5ppb25r/i009949.pdf (accessed on 12 November 2024).
46. Meta-Explorer Database. Available online: http://metaexplorer.tecnico.ulisboa.pt/ (accessed on 12 November 2024).
47. Reinhart, C.F.; Cerezo Davila, C. Urban building energy modeling—A review of a nascent field. *Build. Environ.* **2016**, *97*, 196–202. [CrossRef]
48. Konstantinou, T.; Hoces, A.P. Environmental Design Principles for the Building Envelope and More: Passive and Active Measures. In *Energy—Resources and Building Performance*; Reviews of Sustainability and Resilience of the Built Environment for Education, Research and Design; TU Delft OPEN Publishing: Delft, The Netherlands, 2018; pp. 147–180.
49. Januário, M. Cost Analysis of National Suppliers for Energy Efficiency Measures. 2024. Compiled Based on National Supplier Searches Conducted in 2024.
50. POSEUR & DGEG. Custos-Padrão Máximos por Tecnologia Definidos Pela DGEG. 2020. Available online: https://poseur.portugal2020.pt/media/39337/anexo-ii-custos-padr%C3%A3o-m%C3%A1ximos-por-tecnologia-definidos-pela-dgeg.pdf (accessed on 12 November 2024).
51. Mazur-Stommen, S.; Farley, K. ACEEE Field Guide to Utility-Run Behavior Programs. 2013. Report Number B132. Available online: https://www.aceee.org/sites/default/files/publications/researchreports/b132.pdf (accessed on 12 November 2024).

energies

Review

Power Quality Control Using Superconducting Magnetic Energy Storage in Power Systems with High Penetration of Renewables: A Review of Systems and Applications

António J. Arsénio Costa and Hugo Morais *

INESC-ID—Instituto de Engenharia de Sistemas e Computadores-Investigação e Desenvolvimento, Instituto Superior Técnico (IST), Universidade de Lisboa (UL), Rua Alves Redol, 9, 1000-029 Lisbon, Portugal; antoniojcosta@tecnico.ulisboa.pt

* Correspondence: hugo.morais@tecnico.ulisboa.pt

Abstract: The increasing deployment of decentralized power generation based on intermittent renewable resources to reach environmental targets creates new challenges for power systems stability. Several technologies and approaches have been proposed in recent years including the use of superconducting magnetic energy storage. This study focuses on the review of existing superconducting magnetic energy storage systems for power quality control purposes. Such systems can supply and absorb the rated power level within seconds, promoting fast power quality regulation. Systems for power quality services such as frequency regulation, power oscillation damping, power fluctuation suppression, and active power filtering are identified and described. First, the physical characterization of superconducting magnets concerning geometries, materials, associated inductances, and nominal magnetic energy storage capacities is conducted. Then, the functional description of several current conversion circuits and systems used as interfaces for superconducting magnets is performed. The existing methodologies and systems to perform the control of current converters for different power control services and applications are also identified and described. Finally, the results regarding the number of different systems identified for each power quality control service are presented, and their applicability is discussed based on the adopted control approach. Challenges concerning the development of new systems to improve the power quality on grids with high penetration of decentralized energy resources from intermittent renewables are also identified.

Keywords: superconducting magnetic energy storage; power quality control; frequency regulation; power oscillation damping; power fluctuation suppression; active harmonics filtering

1. Introduction

With the large integration of renewable intermittent energy resources and adaptive technologies to comply with the environmental targets, the power system storage capacity should increase to reserve energy for periods when there is no power generation. The existing electrical energy storage technologies were classified and mapped in terms of storage capacity and charging or discharging time [1]. A detailed description of the working principles of several electrical energy storage technologies was performed [2]. The stacking of energy storage services (ESS) to improve the power systems' economic potential was performed [1].

The intermittency of renewable resources causes instability in the grid power quality. To perform a fast regulation of the power quality, high levels of power should be exchanged with the grid in a short time. Superconducting magnetic energy storage (SMES) systems have the technical advantage of being able to charge or discharge the rated power within seconds, performing a fast re-establishment of the grid power balance to address instabilities. The most powerful SMES systems reach peak powers of almost 100 MW [3] and several SMES systems can reach 10 MW, with high-temperature superconducting (HTS) toroidal magnets [4], HTS solenoidal magnets [5], and low-temperature superconducting (LTS) four-pole magnets [6]. Due to their ability to supply or absorb high power in a short time, their use is recommended for rapid grid power regulation to address imbalances between production and consumption, thereby ensuring the quality of supply.

The use of battery energy storage systems (BESS) for power quality control was studied [7]. Lithium-ion BESS are characterized by charging and discharging times in the range of tens of minutes [1,2]. Ultracapacitors are also able to charge and discharge within seconds. The maximum rated power with ultracapacitors is about 1 MW [8]. Once there are SMES systems with 100 MW, the maximum possible rated powers with SMES systems are about one hundred times higher than with ultracapacitors. The energy density of ultracapacitors is within the range of 2.5–15 kWh/kg and that of SMES systems is within the range of 0.5–5 kWh/kg [9]. SMES systems present round-trip efficiencies within 80–97% while ultracapacitors within 75–95% [10]. Although ultracapacitors reach energy densities three times higher than SMES systems, the efficiency of SMES is slightly higher.

The advantages of using SMES in power quality services, due to SMES's high power density, high energy storage efficiency, and low discharging time of the rated power, were highlighted in [11]. The application of SMES to control the power quality in grids with several types of renewable power generation sources [12] and unbalanced loads was identified in [13]. The use of SMES to regulate the power quality from wind double-fed induction generators (DFIG) was highlighted [14]. Improving the performance of a PV/wind microgrid was studied [15]. Hybrid storage using SMES to cooperate with BESS in mitigating photovoltaic (PV) power fluctuations was proposed [16]. The adaptative control of a hybrid SMES/BESS to smooth power fluctuation and enlarge battery lifetime was studied [17]. A hybrid SMES/BESS to help increase the power density in electric vehicle (EV) applications was proposed [9]. The application of SMES to enable high power density in naval applications was identified [18].

Several reviews were published referring to the use application of SMES. The general classification of energy storage technologies including SMES, mapping of their storage capacities and discharging times, possible storage services, and optimization of their stacking was reviewed [1]. A detailed technical description of the working principles of several electrical energy storage (EES) technologies including the SMES and different existing BESS technologies was performed [2]. This included their classification in terms of rated capacity, discharging time duration, specific power and energy capacity, cycle efficiency, capital, and operation/maintenance costs. The review of SMES devices and identification of potential application scenarios in grids with distributed generation systems and integrated with multiple EES technologies was performed [19]. A review of power system frequency control solutions and new challenges looking forward to the high penetration of renewable energy generation in power grids was performed [20]. A comprehensive review of the new trends and future directions in load and frequency control in a flexible power system using several EES technologies and with multiple power generation sources was presented [21]. An overview of the cryocooler-based cooling systems for SMES was performed [22]. The existing reviews are focused on the identification of ESS and the identification of applications and services for the ESS including SMES. Nevertheless, there is no comprehensive synthesis that thoroughly explores the performance and control approaches of SMES to ensure the overall power quality of the system.

Considering the contributions and gaps of the previous review papers, the novelty of this research work consists of:

- Reviewing the different SMES systems that perform power quality control and regulation services for several power applications. The following main power quality control and regulation services are addressed: Frequency regulation; Power oscillation damping; Power fluctuation suppression; Voltage sags and flickering compensation; and Active harmonics filtering.

- For each type of service that SMES can provide, the main gaps and research trends are identified. This analysis is crucial for guiding future developments in the field.
- An overview of the literature is quantified allowing a visual perspective of the works published about the use of SMES in power quality control and regulation services.

After the present introductory section, Section 2 presents the materials and methodology adopted in the current research work. Section 3 presents an introductory description of the several components of SMES systems, their characteristics, functionalities, and reference costs. Sections 4–7 are dedicated to the identification, description, and classification of existing SMES systems designed to perform, respectively, frequency regulation, power oscillation damping, power fluctuation suppression, and active harmonics filtering. The applicability of each SMES system depending on the adopted type of control is also discussed. Research and development challenges concerning the evolution of the SMES systems for each power quality control service are also proposed. Section 8 is dedicated to the presentation of results concerning the number of different SMES systems identified for each power quality control service.

2. Review Methodology

The survey of existing documentation related to the subject in the study was based on a systematic review methodology. The search for references was performed using the IEEE\Xplore, ScienceDirect, and MDPI database platforms. The diagram from Figure 1 shows performed queries, the number of searches, and final references from the adopted systemic review methodology. The used search terms are between quotation marks. As a primary search, queries were made looking for the existence of two different terms in the document titles.

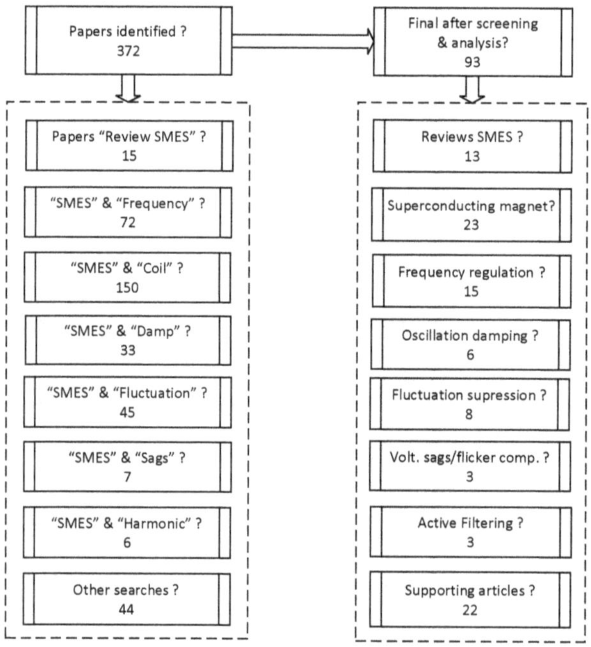

Figure 1. Searches and final references from the adopted systematic review methodology.

The total number of papers initially identified from the queries performed was 372. After the screening of their abstracts and detailed analysis of the body text, a total of 93 references were considered.

3. Superconducting Magnetic Energy Storage System Components

This section begins with the physical characterization of superconducting magnets in terms of geometries, materials, inductances, and nominal magnetic energy storage capacities. Then, a functional description of several current conversion circuits and systems is performed. At final an overview of SMES reference costs is presented.

3.1. Superconducting Magnets

The magnetic energy W_m stored in a superconducting magnet is given by (1), where I_{SC} is the current through the coil and L_{SC} a superconducting magnet inductance.

$$W_m = L_{SC}\,\frac{{I_{SC}}^2}{2} \tag{1}$$

From the manipulation in (2), a variation $\Delta\, I_{SC}$ in the superconducting magnet current results in a variation $\Delta\, W_m$ in the stored magnetic energy given by (3).

$$W_m + \Delta\, W_m = L_{SC}\,\frac{(\,I_{SC} + \Delta\, I_{SC})^2}{2} = L_{SC}\,\frac{{I_{SC}}^2}{2} + L_{SC}\frac{2\;I_{SC}\,\Delta\, I_{SC} + \Delta\, {I_{SC}}^2}{2} \tag{2}$$

$$\Delta\, W_m = L_{SC}\left(I_{SC}\,\Delta\, I_{SC} + \frac{\Delta\, {I_{SC}}^2}{2}\right) \tag{3}$$

From (3), the stored magnetic energy variation depends on the initial magnet current I_{SC}.

Several superconducting magnets have been designed and fabricated for SMES applications. Many magnets use HTS materials. A hybrid energy storage system with a 60 kJ solenoid magnet with double pancake coils (DPC) of yttrium barium copper oxide (YBCO) tapes and lithium-ion batteries to smooth the fluctuated power from wave energy converters was proposed [23]. The design of a 1 MJ/100 kW magnet using DPC of YBCO tapes operating at 22 K was described [24]. The design, coil fabrication and 30 kJ magnet with DPC of bismuth strontium calcium copper oxide (BSCCO) tapes were presented [25]. The conceptual design of a 100 MJ magnet with magnesium diboride (MgB2) coils was described [26]. Induction current losses and magnetization losses were analyzed for a 2.5 MJ toroidal magnet with DPC of YBCO tapes [27]. The operational losses of a real manufactured 10 kJ class toroidal magnet with DPC of BSCCO tapes were analyzed [28]. Its design and manufacturing for a real-time power quality enhancement simulator were described [29]. The conceptual design of a 200 kJ solenoidal magnet with DPC of YBCO tapes was performed [30]. A toroidal magnet with six helicoidal coils of YBCO tapes was studied [31]. The development of a prototype machine using such a magnet was described [32]. A 10 kJ class solenoidal magnet with DPC of YBCO tapes for a real-time grid connection study was designed [33]. The design and fabrication of another 10 kJ solenoidal magnet with DPC of BSCCO tapes were presented [34]. A kW class solenoidal magnet with single coils of gadolinium barium copper oxide GdBCO wires for a hybrid storage dynamic voltage restorer application was designed and evaluated [35]. A toroidal magnet with DPC of BSCCO tapes was studied [36]. The preliminary tests of the pancakes from a 12 T insulated solenoidal magnet with rare-eart barium copper oxide (ReBCO) tapes were presented [37].

Other magnets use LTS materials. A 3 MJ/750 kVA solenoidal magnet with 64-layer coils of niobium-titanium (NbTi) wires for improving power quality was described [38]. Experimental results of the toroidal magnet with 3 helical NbTi coils were presented [39]. The quench properties of this magnet were studied [40]. A solenoid magnet with 16-layer coils of NbTi wires was described [41].

Table 1 states the characteristics of superconducting magnets in the literature, including the: geometry, coil materials, operating temperature T_o, inductance L, operating current I_o and energy capacity E_c.

Table 1. Characteristics of superconducting magnets: geometry, coil materials, operating temperature, inductance, operating current, and energy capacity.

Magnet	Superconductor	T_o [K]	L [H]	I_o [A]	E_c [kJ]	Ref.
Solenoid, 6 double pancake coils	YBCO 4 mm tape	65	24.5	70	60	[23]
Solenoid, 7 double pancake coils	YBCO 12 mm tape	22	11.34	450	1148	[24]
Solenoid, 14 double pancake coils	BSCCO tape	20	2.5	155	30	[25]
Toroid, 18 single coils	MGB2 cable	20	12.6	4×10^3	100×10^3	[26]
Toroid, 28 double pancake coils	YBCO 12 mm tape	14	5.42	960	2.5×10^3	[27]
Toroid, 30 double pancake coils	BSCCO tape	6	0.229	300	10.3	[28,29]
Solenoid, 31 double pancake coils	YBCO 12 mm tape	65	174.7	53.8	253	[30]
Toroid, 52-layer coils	YBCO 12 mm tape	50	2.1	3×10^3	10×10^3	[4]
Toroid, 6 helicoidal coils	YBCO 5 mm tape	4.2	2.39×10^{-3}	1×10^3	1.2	[31,32]
Solenoid, 7 double pancake coils	YBCO 4 mm tape	33	1.02	142	10.3	[33]
Solenoid, 13 double pancake coils	BSCCO tape	20	4.68	86	17.4	[34]
Solenoid, 6 single coils	GDBCO wire	77	3.25×10^{-3}	250	0.1	[35]
Toroid, 16 double pancake coils	BSCCO tape	77	20.24	500	2.53×10^3	[36]
Solenoid, 21 double pancake coils	ReBCO 12 mm tape	4.2	2.79	850	1×10^3	[37]
Solenoid, 13×26 coils	YBCO 4 mm tape	20	144.1×10^{-3}	12×10^3	10.1×10^3	[5]
Solenoid, 64-layer coils	NbTi wires cable	4.2	6	1×10^3	3×10^3	[38]
Solenoid, 44 double pancake coils	NbTi/Cu strand cable	4.2	10.8	4.3×10^3	100×10^3	[3]
4 poles coil configuration	NbTi wires cable	4.2	21.1	1.35×10^3	19.2×10^3	[6]
Toroid, 3 helical coils	NbTi/Cu strand cable	4.2	1.8	552	274	[39,40]
Solenoid, 16 layers coil	NbTi wires cable	4.2	0.13	6.7×10^3	2.9×10^3	[41]

The geometries of superconducting magnets are solenoidal and toroidal. The use of double pancake coils was proposed for most of these two geometries. There are a few toroidal geometries composed of helicoidal coils. Many magnets using HTS materials are cooled at temperatures much lower than the liquid nitrogen (LN_2) boiling temperature of 77 K to increase the critical current I_c and the magnetic energy storage capacity.

3.2. Current Conversion Circuits and Systems

To perform the charge and discharge of active power to and from an SMES, the current converters used should be bidirectional. Three different types of bidirectional current converters can be used to interface the SMES coils with the AC grid [42]. These are namely: (1) set with a voltage source converter (VSC) and a DC chopper; (2) current source controller (CSC); and (3) silicon-controlled rectifier (SCR) "Thyristor".

The voltage source controllers (VSCs) are AC-DC voltage converters, with six insulated gate bipolar transistors (IGBT) and six diodes, that allow a bidirectional flow of power from the AC grid to a DC link and vice versa [43]. The six diodes act as a voltage rectifier from the AC grid to the DC-Link. The six IGBT constitute a voltage inverter that is controlled with space vector pulse-width modulation (SVPWM).

The DC choppers control the flow of charged and discharged power to a load that in this case is the superconducting coil (SC) of an SMES. The application of a DC chopper for dynamic DC voltage restorer was presented [44]. Figure 2 shows the set with a VSC and DC-chopper used to connect the SMES to the AC grid controlling the amount of active power charged and discharged to and from the SC. An example of using this set for power fluctuation suppression and active harmonic filtering is presented [45]. To convert the level of AC voltage, this set should be connected via a transformer to the AC grid point of common coupling (PCC).

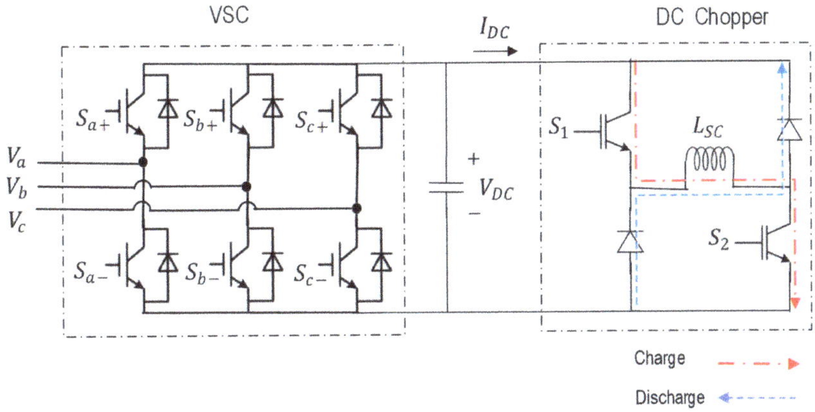

Figure 2. Set with a VSC and DC chopper to control the active power flow to and from an SMES.

The SMES could also be connected via an individual DC chopper to the DC-link of double-fed induction generators (DFIG) or the DC-links between inverters and permanent magnet synchronous generators (PMSG) or photovoltaic (PV)-arrays.

Figure 3 schematizes at a high level the state input variables and reference values in the digital signal processors (DSP) that control the VSC and DC chopper [46].

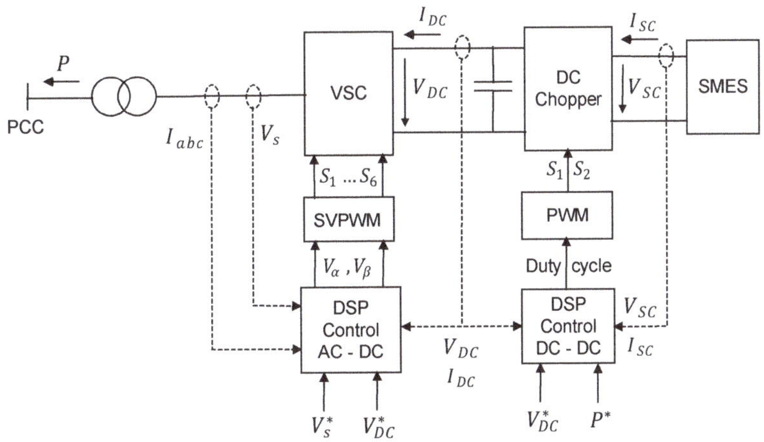

Figure 3. State input variables and reference values in the DSP control of the VSC and DC chopper.

The six gates of the S_{abc+} and S_{abc-} are controlled using space vector pulse-width modulation (SVPWM) [47]. The DSP that controls the VSC reads the AC line currents I_{abc}, line voltage V_s, DC-bus current I_{DC} and voltage V_{DC}. For the control, we consider reference evolutions for the line voltage V_s^* and DC-bus voltage V_{DC}^*. The Clarke space vector components V_α and V_β that generate SVPWM for controlling the VSC are computed by the system presented in Figure 4. This system has four PI regulators that process the current and voltage Park components dq0. A phase-locked loop (PLL) circuit is used to determine the line voltage phase θ to compute the Park and Clarke transformations [48].

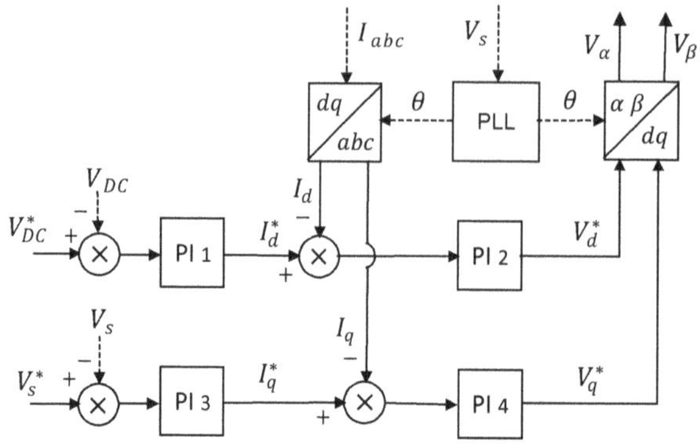

Figure 4. System to compute components V_α and V_β that generate the SVPWM to control the VSC.

The DC chopper [44] governs the charging and discharging of active power between the DC-link and the SC. When S_1 and S_2 are simultaneously "on" (closed), the current I_{DC} is positive flowing from the grid to the SC (red dashed-dotted line). When S_1 and S_2 are simultaneously "off" (open), there could be a discharge of the SC when the current I_{DC} becomes negative flowing toward the grid direction (blue dashed line) and the two diodes start conducting. If the duty cycle D, defined by the percentage of time that S_1 and S_2 are simultaneously on, is greater than 50% there is a charging effect of the SC. On the contrary, the SC is discharged. When the SC is charged, the memorized current and magnetic field increase. These memorized quantities tend to decrease when the SC is discharged.

Figure 5 shows the system to compute the duty cycle D. This system has two PI regulators. The value of D is computed by comparing the measured active power P and the desired active power P^*. The DC-link voltage V_{DC} is regulated to follow its reference V^*_{DC} and the DC-link current I_{DC} adjusted to reach the desired active power. The PI output is compared with a sawtooth signal to generate the duty cycle switching.

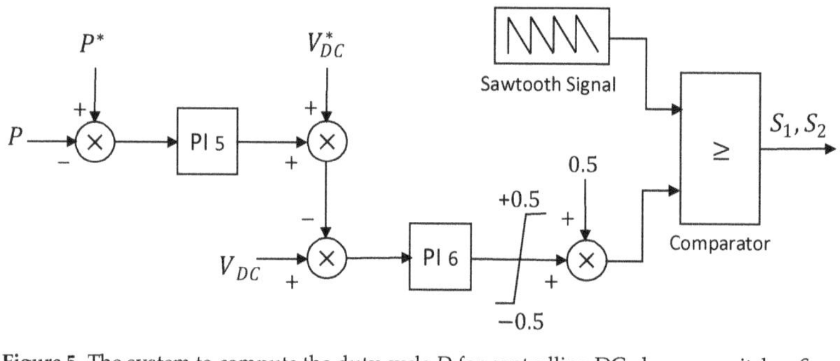

Figure 5. The system to compute the duty cycle D for controlling DC-chopper switches S_1 and S_2.

The circuit scheme for a CSC is shown in Figure 6. This circuit performs as a rectifier towards the SMES coil and an inverter towards the AC grid. The six IGBT gates S_{abc+}, and S_{abc-} are controlled by the outputs of a pulse-width modulator (PWM). The duty cycles D to control the inverter's three phases are generated by comparing the phase voltages V_{abc} with a sawtooth signal. When the phase voltages are higher than $E_d/2$, the plus IGBT are directly polarized and the minus IGBT are inversely polarized. Otherwise, when the phase voltages are lower than $-E_d/2$, the plus IGBT are inversely polarized and the minus IGBT are directly polarized. The IGBT directly polarized and with gate signal "On" contribute to the increase in the current through the SC and increase in the stored magnetic energy. When there is significant energy consumption by loads on the AC grid side, the power flows from the SC to the grid functioning the CSC as an inverter.

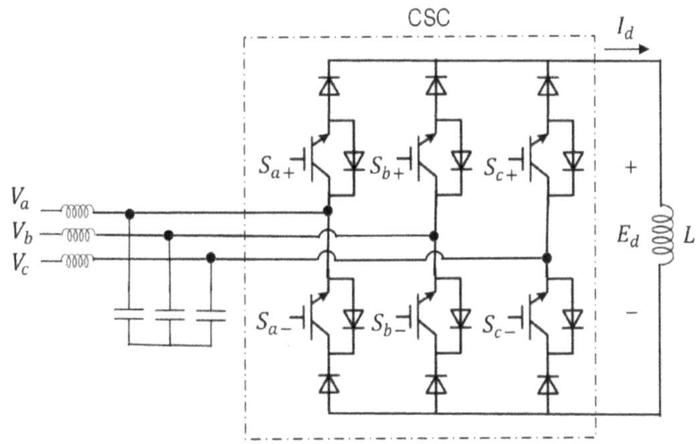

Figure 6. Bidirectional CSC used to interface the SMES coil with the AC grid.

Figure 7 shows an alternative bidirectional AC-DC converter composed of a twelve-pulse silicon-controlled rectifier (SCR) bridge and bypass SCR gate. The charge or discharge of the coil is performed when the firing angle α is less or higher than 90°.

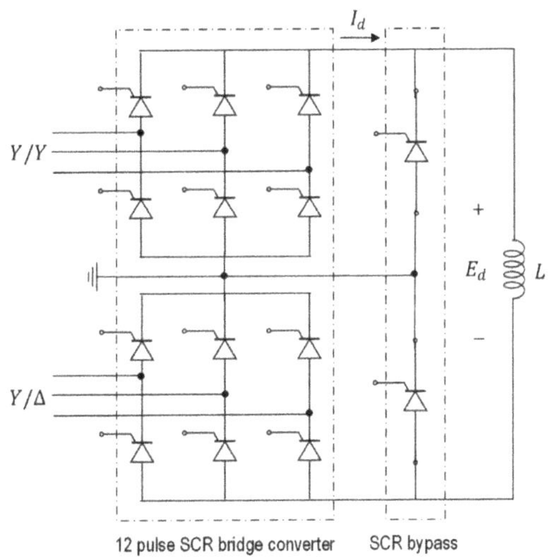

Figure 7. Alternative bidirectional converter with twelve-pulse SCR bridge and SCR bypass gate.

This time converter was proposed for SMES systems used in the automation of an interconnected hydrothermal generation system [49], optimized tilt fractional order co-operative controllers for preserving frequency stability [50], load frequency control of deregulated power system [51], synthetic inertia system control to enhance frequency dynamic performance in microgrids with high renewable penetration [52] for frequency regulation of power systems with dynamic participation from DFIG based wind farm [53].

With the interconnection of various renewable energy resources and adaptive technologies, the voltage quality and frequency stability of modern power systems are becoming unstable. The dynamic characteristic of SMES allows for rapid exchange of electrical power with the grid during small and large disturbances to address those instabilities.

3.3. Performance and Investment Indicators for Real Systems

A real SMES system with 100 MJ storage capacity was installed at the center for advanced power systems (CAPS) at Florida State University (FSU), and connected to a 4.16 kV bus from a 5 MVA generator and a 5 MW motor [3]. An SMES system with 20 MJ storage capacity and a power of 10 MVA, to compensate for active load fluctuation compensation was implemented and connected to a real 11 kV power grid in Nikko City, Tochigi Prefecture, Japan [6]. Field test results have shown that the efficiency of this SMES system converter was about 98% or more.

Typical SMES-specific power costs are within 200–350 $/kW and specific energy costs within 1k–10k $/kWh [10]. The ranges 130–515 $/kW and 700–10k $/kWh were also stated [54].

The SMES capital expenditures (CaPex) marginal costs are within: 21–25% for the superconducting magnet, 15–30%, for the power current converters, 8–12% for the cryostats, being the reminder for other costs such as buildings, installations, grid connections, and licensing [55].

Studies on return on investment (ROI) were performed for four SMES, two belonging to the San Diego gas and electricity and the other two to the West Coast utilities [55]. ROI values of 17%, 24%, 46%, and 71% were determined, meaning that all four cases showed a positive net present value (NPV).

Although SMES presented positive NPV on some applications, SMES technology requires broader adoption to enhance its maturity and reduce the investment risks typically associated with emerging technologies. To address this challenge, a new policy framework is needed to incentivize the adoption of SMES, similar to what was implemented for certain renewable technologies. Additionally, with the expected proliferation of renewables and the consequent reduction in system inertia, power quality control will become increasingly challenging. In this context, SMES can emerge as a key technology to mitigate these challenges.

The main challenge of the technology is to reduce the upfront cost, as well as address operation and maintenance, mainly concerning the provision and replenishment of cryogen. However, it should be highlighted that its high efficiency could compensate for maintenance costs and, in time, upfront costs.

4. Frequency Regulation

In a single isolated grid with power from one driving synchronous machine, the frequency varies depending on the non-balance between the active power generation by the synchronous machine P_M and active power consumption by resistive loads P_L according to expression (4),

$$P_M - P_L = \frac{dW_c}{dt} \quad ; \quad W_c = \frac{I\,\omega^2}{2} \tag{4}$$

where W_c the kinetic energy, I the moment of Inertia, and ω the angular velocity from the synchronous machine rotor mass. So, the grid frequency tends to increase or decrease when the difference $P_M - P_L$ is, respectively, positive or negative. Grid areas could be interconnected with other grid areas. Each area could be supplied by one or more central power generation systems and several decentralized power generation systems. When the difference between the active power injected by all generation systems and the active power demanded by all loads is positive the grid frequency tends to increase. Otherwise, when the difference is negative the grid frequency tends to decrease. Frequency regulation can be made by injecting or absorbing active power to compensate for the active power missing or in excess.

In Continental Europe, the standard frequency range should be within ± 0.05 Hz with a maximum instantaneous deviation of ± 0.8 Hz and a maximum steady-state deviation of ± 0.2 Hz [56].

4.1. Global Frequency Regulation System Models and Parameters

The global frequency regulation system for a specific grid area *i* connected to another grid area *j*, is shown in Figure 8. The grid area *i* integrates auxiliary SMES systems that perform active power compensation contributing to the grid frequency regulation. The difference between the active power resulting from all synchronous machines and the active power to all loads generates a specific frequency variation Δf_i in grid area *i*. The speed of the turbines in the central generation system is regulated by a speed governor that controls the valves of the fluid flow to the turbines.

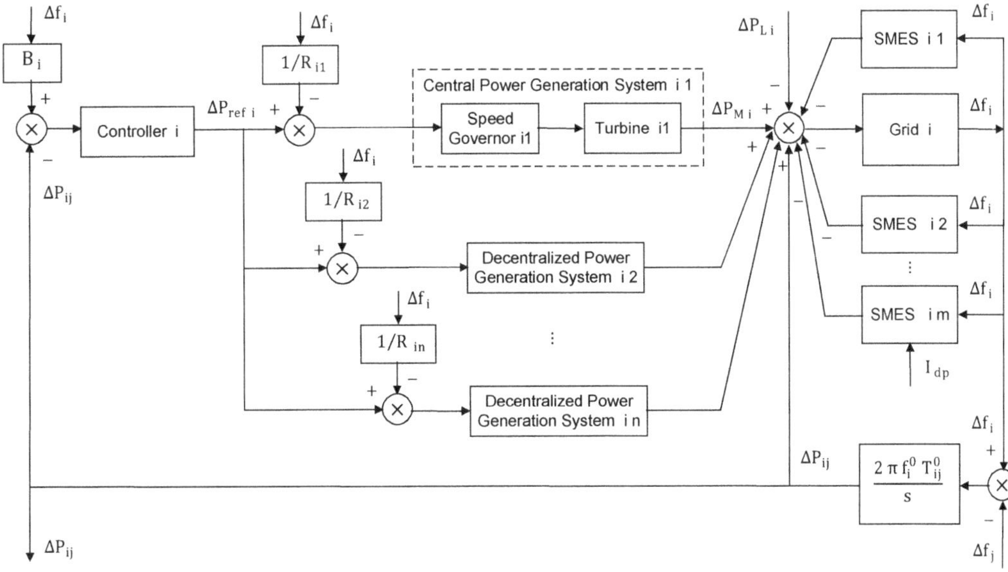

Figure 8. Frequency regulation system using SMES, for a grid area *i* connected to other areas *j*.

Each grid area *i* is defined by a specific frequency bias constant B_i that defines the ratio between grid area active power variation ΔP_i and the frequency variation Δf_i. Each central *n* in a specific area *i* is defined by a specific governor speed regulation parameter R_{in} that defines the ratio between the frequency variation and the power to regulate the fluid flow valve. The tie-line delay coefficient T_{ij} defines the frequency synchronization time between areas *i* and *j*.

Thermal and hydro turbines are generally used in central power generation systems. The transfer function for the set of speed governor and thermal turbine without reheat is given by (5),

$$\frac{1}{1+s\,T_g}\,\frac{1}{1+s\,T_t} \tag{5}$$

where T_g is the speed governor time constant and T_t the thermal turbine time constant. The transfer function for the set of speed governor and thermal turbine with reheat is instead given by (6),

$$\frac{1}{1+s\,T_g}\,\frac{1+s\,K_r T_r}{(1+s\,T_t)(1+s\,T_r)} \tag{6}$$

where K_r and T_r are, respectively, the gain and time constant to account for the reheating effect.

The transfer function for the set of speed governor and hydro turbine is given by (7),

$$\frac{1}{1+s\,T_1}\,\frac{1+s\,T_R}{1+s\,T_2}\,\frac{1+s\,T_w}{1+0.5\,s\,T_w} \tag{7}$$

where T_1 is the time constant of the hydraulic governor, T_2 the time constant of the hydraulic droop transient, T_R the reset time and T_w the water starting time of the hydro turbine.

The transfer function for the grid, having as input the difference between powers from generator machines and the powers to the loads and as output the resulting frequency variation, is given by (8),

$$\frac{K_p}{1+s\,T_p} \tag{8}$$

where K_p and T_p are the grid gain and time constant.

Several frequency regulation systems using SMES were defined. An optimized hybrid fractional order controller was proposed to model SMES in the frequency regulation of multi-area power systems [57]. The modeling of magnetic energy storage in frequency stabilization [58], frequency control of multi-source multi-area hydrothermal systems [59], load frequency control of power systems with renewable integration [60], and frequency control of multi-area power systems with hybrid energy storage devices including SMES was studied [61]. Also, the use of SMES in the predictive load frequency control of hybrid multi-interconnected plants comprising renewable energy was discussed [62], as well as frequency regulation of solar and wind-integrated multi-area systems [63], the optimized fractional-order PI-based SMES to perform frequency control of a PV/Wind-Integrated system [64], and the automatic generation control with SMES in power systems [65]. An optimal design frequency fluctuation control system with SMES unit was presented [66]. Table 2 states typical values for the several model parameters in the global frequency regulation system.

Table 2. Typical values for the several model parameters in the global frequency regulation system.

Parameter	Values	References
B_i [pu MW/Hz]	0.425	[49–51,57–62,65]
R_i [Hz/pu Mw]	$2 \le R_i \le 2.7$	[49–51,57–62,65]
T_{ij} [s]	$0.054 \le T_{ij} \le 0.087$	[49–51,57–62,65]
T_g [s]	$0.07 \le T_g \le 0.1$	[49–52,57–62,65]
T_t [s]	$0.3 \le T_t \le 0.4$	[49–52,57–62,65]
K_r [Hz/pu MW]	$0.33 \le K_r \le 0.5$	[51,62,65]
T_r [s]	10	[51,62,65]
T_1 [s]	$41.6 \le T_1 \le 48.7$	[49–51,57,59]
T_2 [s]	0.513	[49–51,57,59]
T_R [s]	5	[49–51,57,59]
T_w [s]	1	[49–51,57,59]
K_p [Hz/pu MW]	$100 \le K_p \le 120$	[49,51,58,59,61,62,65]
T_p [s]	$15 \le T_p \le 25$	[49,51,58,59,61,62,65]

The transfer functions for wind, PV [21], and diesel [62] decentralized power generation systems are expressed, respectively, by (9), (10) and (11),

$$G_w = \frac{K_{pw1}\left(1+s\,T_{pw1}\right)}{1+s}\,\frac{K_{pw2}}{1+s\,T_{pw2}}\,\frac{K_{pw3}}{1+s} \tag{9}$$

where K_{pw1}, K_{pw2}, and K_{pw3} are the gains, T_{pw1}, T_{pw2}, and T_{pw3} are the time constants of the wind plant system.

$$G_{PV} = \frac{K_{PV1}\,s + K_{PV2}}{s^2 + T_{PV1}\,s + T_{PV2}} \tag{10}$$

where K_{PV1}, and K_{PV2}, are the gains T_{pw1}, and T_{pw2} are the time constants of the PV system with maximum power point tracking (MPPT).

$$G_D = \frac{K_{dies}\,(1+s)}{0.025\,s^2+s} \tag{11}$$

where K_{dies} is the gain of the diesel generator. Typical values for the gains and time constants in expressions (9)–(11), are stated in Table 3 [62].

Table 3. Typical parameter values for wind, PV, and diesel decentralized generation systems.

Wind					PV				Diesel
K_{pw1}	K_{pw2}	K_{pw3}	T_{pw1} [s]	T_{pw2} [s]	K_{PV1}	K_{PV2}	T_{PV1} [s]	T_{PV1} [s]	K_{dies}
1.25	1.4	-	6	0.041	−18	900	100	50	16.5

Several types of control such as proportional-integral (PI) proportional integral derivative (PID), fractional order proportional integral derivative (FOPID), and tilt integral derivative (TID) were tested for the secondary controller to calculate ΔP_{ref} [57].

4.2. Superconducting Magnetic Energy Storage System Models

Frequency regulation using SMES is performed by storing or restoring active power from or into the grid to maintain the system frequency. Figure 9 shows the two-stage lead-lag system model to characterize the SMES in frequency stabilization. The gain and time constants of the SMES coil are given by K_{SMES} and T_{SMES}. The time constants of the two-stage lead-lag compensator are defined by T_1, T_2, T_3, and T_4. Table 4 states typical values for these parameters.

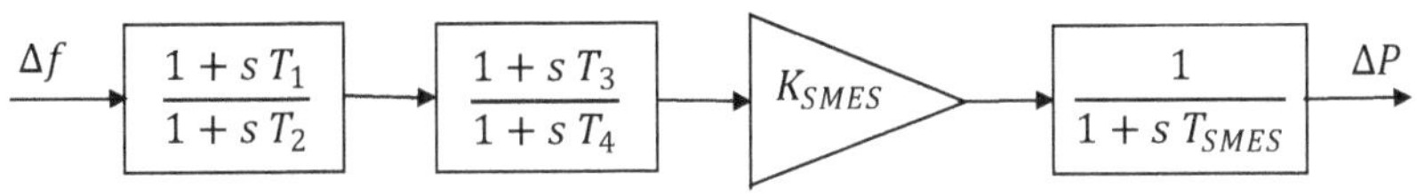

Figure 9. SMES system model with two-stage lead-lag compensator.

Table 4. Typical parameter values for the SMES system model with a two-stage lead-lad compensator.

K_{SMES} [1]	T_{SMES} [s]	T_1 [s]	T_2 [s]	T_3 [s]	T_4 [s]	Reference
2.500	0.046	0.279	0.020	0.900	0.360	[53]
0.297	0.03	0.121	0.8	0.011	0.148	[51,59]
0.2035	0.03	0.2333	0.016	0.7087	0.2481	[21,61–63]

Figure 10 shows the SMES system model with fractional-order proportional integral (FOPI). The converter gain and time delay are given by K_{SMES} and T_{DC}. Quantities ΔE_d and ΔI_d are the incremental change of the coil DC voltage and current. The DC bias to the SMES coil is given by I_{d0}. The control loop gain is given by K_{id}.

The input $\Delta Error$, corresponding to the area control error (ACE), is given by expression (12), where B is the area frequency bias constant Δf the frequency variation and ΔP_{tie} the sum of the power variations in the tie interconnections to other grid areas.

$$\Delta Error = B\,\Delta f + \Delta P_{tie} \tag{12}$$

Table 5 states typical parameter values for the SMES system model with FOPI.

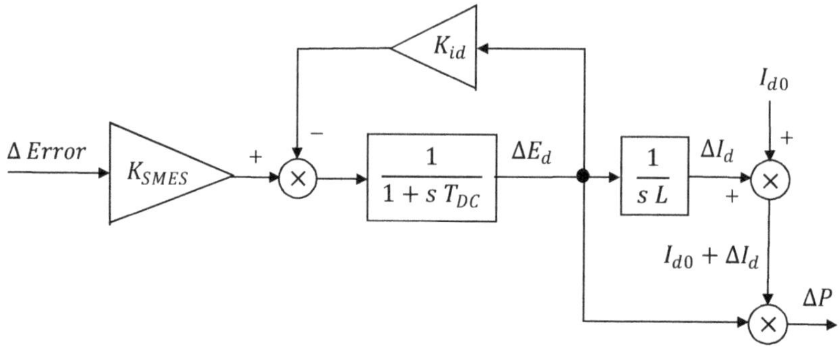

Figure 10. SMES system model with FOPI.

Table 5. Typical parameter values for the SMES system model with FOPI.

K_{SMES} [kV/puMW]	K_{id} [kV/kA]	T_{DC} [s]	I_{d0} [kA]	L [H]	Reference
100	0.2	0.03	4.5	2.65	[49,51,64]
100	0.2	0.03	4.5	0.03	[50]
1	0.01	0.03	1.732	3	[52]
100	0.5	0.026	4	1	[58]
10	0.4	0.003	0.3	2.65	[65]

The modeling of an SMES system to regulate the frequency with an equivalent PI control was proposed and studied [66]. The optimal proportional and integral gains, K_P and K_I, of the SMES equivalent PI control were computed, first using the Ziegler–Nichols method, and then using a fuzzy logic controller with rules that look at the frequency variation ΔF and the output power variation ΔP.

4.3. Systems Overview and Development Challenges in Frequency Control

Several studies on frequency regulation with SMES have been conducted. Some proposed that SMES system models are based on PI lead-lag control and others on FOPI. Table 6 classifies the studies on frequency regulation with SMES systems in terms of: (i) the number of grid areas considered; (ii) types of centralized and decentralized generation resources by area; (iii) system control level (primary or secondary); (iv) type of SMES current conversion interface; and (v) type of SMES system control.

Most of the scenarios considered systems of two grid areas. There was one study on a system with four grid areas [61] and another on a system with six grid areas [62]. Decentralized renewable power generation was considered in half of the frequency regulation studies with SMES. The twelve-pulse SCR bridge was considered as an SMES current converter interface for most of the frequency regulation studies. Concerning the SMES system control, the FOPI and PI lead-lag are the most common systems used. There was one study using PID [57] and another study using a PI-FLC [66].

In terms of research challenges, studies are necessary on the:

(1) Optimization of the SMES energy storage capacity looking for the renewable resources potential and expected load demanding profile.
(2) Use of predictive control for anticipating the possibility that the maximum frequency deviation is exceeded based on the prediction of renewable resources evolution.
(3) Study and assessment of SMES system control techniques for frequency regulation on grids with significant renewable power generation.

Table 6. Overview of scenarios for studies on frequency regulation with SMES systems.

Ref.	# Areas	Centralized Generation	Decentralized Generation	System Control	SMES Interface	SMES System Control
[21]	1	1: Thermal	1: PV + Wind	Secondary	Not specified	PI Lead-Lag
[49]	2	1: Thermal 2: Hydro	-	Secondary	12 pulse SCR bridge	FOPI
[50]	2	1: Thermal 2: Hydro	1: Wind; 2: PV	Secondary	Not specified	FOPI
[51]	2	1: Thermal 2: Hydro	-	Secondary	12 pulse SCR bridge	FOPI
[52]	1	1: Thermal	1: PV + Wind	Secondary	12 pulse SCR bridge	FOPI
[53]	2	1: Steam reheat 2: Hydro	1: Wind	Secondary	12 pulse SCR bridge	PI Lead-Lag
[57]	2	1: Thermal 2: Hydro	1: Wind; 2: PV	Secondary	Not specified	PID
[58]	2	1: Thermal reheat 2: Thermal reheat	-	Secondary	12 pulse SCR bridge	FOPI
[59]	2	1: Hydro + Thermal 2: Hydro + Thermal	-	Secondary	12 pulse SCR bridge	PI Lead-Lag
[60]	2	1: Thermal 2: Thermal reheat	-	Secondary	12 pulse SCR bridge	PI Lead-Lag
[61]	4	1, 2, 3, 4: Thermal	-	Secondary	Not specified	PI Lead-Lag
[62]	6	1: Thermal; 2: Hydro 3: Thermal; 4: Hydro 5: Thermal; 6: Hydro	1: PV; 2: Wind 3: Diesel; 4: PV 5: Wind; 6: Diesel	Secondary	12 pulse SCR bridge	PI Lead-Lag
[63]	2	1: Thermal + Hydro 2: Thermal + Hydro	1: Wind; 2: Diesel	Secondary	Not specified	PI Lead-Lag
[64]	2	1: Thermal 2: Hydro	1: PV; 2: Wind 1: PV; 2: Wind	Secondary	12 pulse SCR bridge	FOPI
[65]	2	1: Thermal reheat 2: Thermal reheat	-	Primary	VSC + DC Chopper	FOPI
[66]	1	1: Thermal	-	Primary	Not specified	PI-FLC

5. Power Oscillation Damping

Power oscillation results from the variation of the synchronous generators' rotor speed mainly after the connection of synchronous generators (SG) or connection/disconnection of significant loads or grid sections. Also, during the connection/disconnection of decentralized renewable power generation plants including DC-AC converters, characterized by different dynamics.

5.1. Power Oscillation Damping Systems Using Superconducting Magnetic Energy Storage

The use of an SMES system with a current source controller (CSC) to damp power oscillations in the connection between an SG and the grid utility was studied in [67]. Figure 11a shows the general unifilar connection scheme with a current source-controlled SMES connected to the point of common coupling (PCC) in between the SG and the connection to the grid utility. When the SG is connected to the power grid there is a transient period when there is an oscillation on the generated active power ΔP traduced by an oscillation on the phase-angle $\Delta\delta$ around the target phase-angle δ_0. Consequently, there is oscillation of the angular velocity $\Delta\omega_r$ around the synchronous rotor velocity ω_{ro}. There is transmission of active power P towards the power grid, if the phase-angle δ_o corresponding to the difference between the SG electromotive force angle and the PCC voltage angle is positive. Also, the difference β between the PCC voltage **V** and grid voltage **U** should be positive.

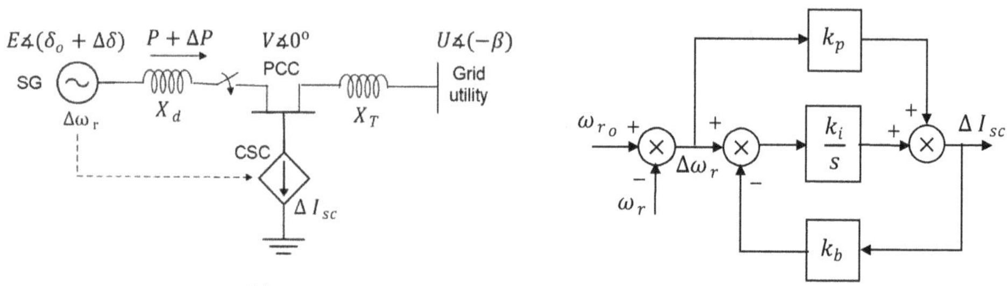

Figure 11. (**a**) General unifilar scheme of a current source controlled SMES in the connection from an SG to the power grid; (**b**) system diagram of the PI-IA in the CSC.

A proportional Integral controller with Integral alleviation (PI-IA) is proposed for the CSC to control the SMES direct current variation ΔI_{sc} depending on the verified SG rotor speed variation $\Delta\omega_r$. The PI-IA transfer function is expressed by (13). According to [67], optimal values for the proportional, integral, and back loop gains are, respectively, $k_p = 470$, $k_i = 10$, and $k_b = 0.056$. A deep reinforcement learning (DRL)-based agent was proposed to obtain the optimal parameters. [68]. Figure 11b schematizes the system diagram of the PI-IA in the CSC.

$$G_{CSC} = \frac{s\,k_p + k_i}{s + k_i\,k_b} \tag{13}$$

A simple conventional PI control could cause the deep SMES charging or discharging not to effectively dampen the power oscillation.

The use of SMES with an adaptive learning controller by using a fuzzy-based goal representation heuristic dynamic programming (Fuzzy-GrHDP) algorithm to perform power oscillation damping on a grid with several generators looking for other SG states is proposed in [69]. The Fuzzy-GrHDP algorithm belongs to the adaptative critic design (ACD) family. The differences between the rotor angles of different generators $\Delta\omega_{ij}$, that is, their synchronous deviations, are considered as input entries. Also, the delays τ_{ij} that account for the response time due to transient changes on the state of different SG are considered as input entries. The state input vector $\overline{x}$ is expressed as (14), where N_{SG} are the number of synchrounous generators throughout the grid.

$$\overline{x} = \left\{\Delta\omega_{12}, \ldots, \Delta\omega_{ij}, \tau_{12} \ldots, \tau_{ij}\right\} \quad ; \quad i \neq j \quad ; \quad i,j = 1, I, N_{SG} \tag{14}$$

The fuzzy logic looks at the deviation $\overline{\Delta x}$ between the actual state input vector $\overline{x}$ and the desired reference state input vector $\overline{x}_{ref}$ and generates the control vector $\overline{u}$ with the set of current increments ΔI_{sci} to be set on the different SMES throughout the grid nodes as expressed by (15), where N is the number of SMES in the power grid.

$$\overline{u} = \{\Delta I_{sc1}, \ldots, \Delta I_{scN}\} \tag{15}$$

Then, an output vector $\overline{y}$ is defined by the set of increments of injected active power ΔP_i by the different SMES in the power grid, as expressed by (16).

$$\overline{y} = \{\Delta P_1, \ldots, \Delta P_N\} \tag{16}$$

The set of increments of injected active power $\overline{y}$ creates a new state of synchronous deviations $\overline{x}$ between the different SG in the real critical power grid. Also, the values of the output vector $\overline{y}$ are used as inputs into a virtual goal power grid to compute optimal values of synchronous deviations $\overline{x}^*$. The difference vector $\Delta x^* = \overline{x} - \overline{x}^*$ is then compared with the objective vector $\overline{J}$ generating an error vector $\overline{e}$ that is used as input to update the vector $\overline{r}$ of fuzzification and de-fuzzification rules. The increment in the rules $\overline{\Delta r}$ is given by a variation in the set weight functions $\overline{\mu}$ calculated according to (17).

$$\overline{\Delta r} = \overline{\mu} \cdot \left(\frac{\partial \overline{e}}{\partial \overline{r}} \right) \tag{17}$$

The new vector of synchronous deviations $\overline{x}$ is then compared again with $\overline{x}_{ref}$ generating new input $\overline{\Delta x}$ for the Fuzzy-GrHDP inference system. Figure 12 schematizes the distributed control using Fuzzy-GrHDP of multiple SMES looking for the synchronous deviations between different SG in a power grid.

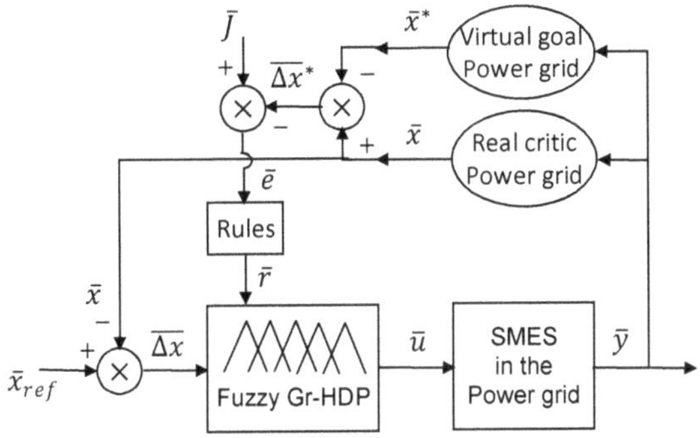

Figure 12. Distributed control using Fuzzy-GrHDP of multiple SMES looking for the synchronous deviations between different SG in a power grid.

Discrete adaptive self-optimizing pole-shifting controls were proposed in [70] to mitigate the power oscillations at the connections between different areas supplied by different synchronous generators. An adaptive pole-shift control technique was proposed [71]. The dynamics from renewable power generation (like Wind and PV) inverters are different than from SG. Hence, the connection/disconnection of renewable power generation inverters imposes different power oscillation dynamics. This discrete adaptive self-optimizing pole-shifting is a flexible technique for use in SMES control to damp power oscillations when the produced power is not only from SG but also from inverters of renewable power generation plants. Adaptative control enables damping from different power oscillation dynamics, not only in the connection of SG but also in the connection of inverters from renewable resources.

Figure 13a shows the principle of the discrete adaptative control system. Figure 13b shows the corresponding z-transform system diagram. The control input $u(t)$ is the current variation ΔI_{sc}, and the system output $y(t)$ the active power variation ΔP.

The coefficients of the numerator $\overline{b} = \{b_1, \ldots, b_{Nb}\}$ and of the denominator $\overline{a} = \{a_1, \ldots, a_{Na}\}$ from the open loop transfer function are estimated using an adaptative constrained recursive least square (C-RLS) algorithm. From the open loop poles are calculated the closed loop poles using a self-optimized pole-shifting, so that the closed-loop system becomes stable and its damping is maximized (poles with a negative real part of maximum amplitude). The coefficients of the numerator $\overline{g} = \left\{g_1, \ldots, g_{Ng}\right\}$ and denominator $\overline{f} = \left\{f_1, \ldots, f_{Nf}\right\}$ of the back loop transfer function are then estimated based on the optimized pole-shifting and trying to match the gain between the open loop system inputs $u(t)$ and outputs $y_{sys}(t)$, according to Equation (18).

$$\frac{u(t)}{y_{sys}(t)} = -\frac{G(z^{-1})}{F(z^{-1})} \tag{18}$$

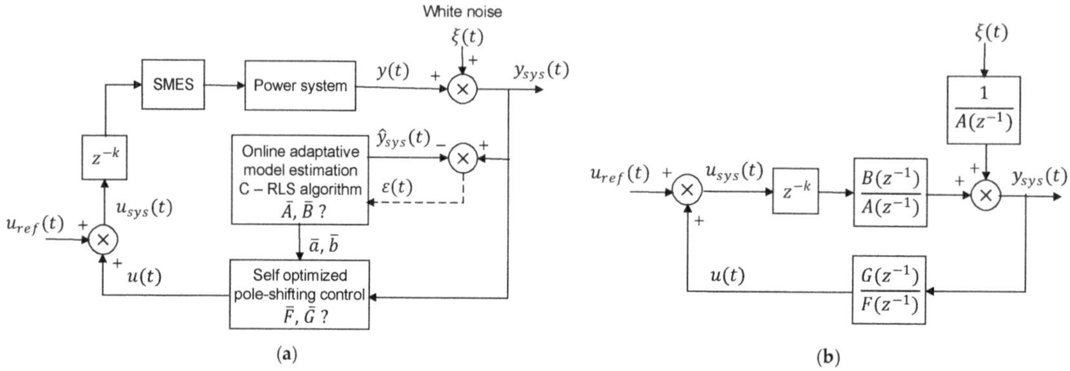

Figure 13. Principle (**a**), and Z-transform diagram (**b**) of the discrete adaptative control system.

5.2. Systems Overview and Development Challenges in Oscillation Damping

Several SMES systems for power oscillation damping have been proposed and studied. A system with CSC using PI-IA to damp power oscillations in the connection between an SG and the grid utility was studied [58,59]. Another system uses a distributed FLC with GrHDP to damp power oscillations based on the measurements of synchronous deviations between rotor angles of different SG in the grid [69]. An adaptative control was proposed to damp power oscillations from the dynamics of WTG [70]. Another was based on the adaptative pole-shifting technique [71]. Finally, an SMES system using multi-loop feedback control to cancel the resonant peak of the LCL filter in the link between an SG and the PCC was proposed and studied [72]. Table 7 classifies the different SMES systems identified for power oscillation damping in terms of: (i) point of connection to the grid; (ii) type of current converter interface; and (iii) adopted SMES control strategy.

Table 7. Different SMES systems that were identified for power oscillation damping.

References	Quality Control	SMES System Connection	SMES Interface	SMES System Control Strategy
[67,68]	Power oscillation damping	PCC	CSC	PI-IA
[69]	Power oscillation damping	PCC	CSC	Distributed FLC Fuzzy GrHDP
[70,71]	Power oscillation damping	WTG AC or DC bus	Adaptable	Adaptative

The SMES systems proposed in [67–69] are applicable for grids with power generation from SG. The SMES systems based on adaptative control [70,71], also apply to grids with power generation from intermittent renewable energy resources characterized by dynamics different from SG.

In terms of research challenges, studies are necessary on the:

(1) Optimization of the SMES energy storage capacity for a specific damping coefficient.
(2) Analysis of the damping gain from using the distributed Fuzzy-GrHDP on a grid with several SG, looking for the synchronous deviations between different SG, instead of performing the power oscillation damping based only on the local generator phase angle.

(3) Use of distributed adaptative control SMES systems, with adaptative algorithms based on the measurements of power oscillations also from other decentralized renewable power generation plants in the same distribution grid.

6. Power Fluctuation Suppression

The integration of even more renewable power resources, such as photovoltaic (PV) and wind turbine generators (WTG), causes power fluctuation due to the inconstancy of the power injected into the grid. A predictive control model to manage power flow on a hybrid wind-PV and diesel microgeneration power plant with additional storage capacity was proposed [73]. The wind forecast at medium voltage distribution networks was performed [74]. Also, with the increase in the number of electrical vehicles (EVs), there are peaks of consumption during specific periods when the EV batteries are charged [75]. With the increase in the imbalance between production and consumption, there is an increase in power fluctuation [76].

6.1. Power Fluctuation Suppression Systems Using Superconducting Magnetic Energy Storage

This subsection presents a review of SMES systems to mitigate power fluctuation, thus increasing the stability of power systems. SMES systems may connect to the PCC of the AC power grid utility using a VSC and DC chopper. SMES may also connect to the DC-link of DFIG or DC μ-grids between inverters and PMSG or PV arrays, using only a DC chopper.

An SMES with YBCO and BSCCO coils connected to the grid using a VSC and DC chopper controlled by a dual processor with a digital signal processor (DSP) and a micro-programmed control unit (MCU) was implemented and tested to compensate for the power fluctuation [46]. The VSC is commanded by SVPWM and the DC Chopper by PWM with a scheme like the one presented in Figure 3.

A fuzzy logic-based and hysteresis current system is proposed in [77] to control the energy exchange between an SMES coil and the DFIG of wind turbine generators (WTG) to reduce fluctuations of the generated power, shaft speed, electromagnetic torque, and the DC-link voltage. The SMES is connected to the DC-link of the DFIG via a DC Chopper, which, with an appropriate fuzzy logic controller (FLC), manages the energy exchange between the SMES and the power system. The set of fuzzy logic rules is developed based on the deviations of the DFIG-generated power $\Delta P_G = P_G - P_G^*$ and of the SMES current $\Delta I_{SC} = I_{SC} - I_{SC}^*$, with the objective of balancing the SMES current I_{SC} with the generated power P_G. The desired SMES current increment ΔI_{SC} is obtained by setting a specific PWM duty cycle to the DC chopper control.

The hysteresis current controller (HCC) generates gate pulses $S_1 \ldots S_6$ for VSC switching based on the comparison between the reference line current I_{abc}^* and the measured line current I_{abc} to minimize voltage flicking and power fluctuation. The reference line's current I_{abc}^* determined from the reference direct current I_d^* and reference quadrature current I_q^* with the Park's inverse transformation. These last two current quantities are obtained using proportional-integral (PI) transformation from the line voltage deviation $\Delta V_s = V_s^* - V_s$ and the DC-link voltage deviation $\Delta V_{DC} = V_{DC}^* - V_{DC}$, respectively. Figure 14 schematizes this compensation system with fuzzy logic and hysteresis current-based control.

A similar fuzzy logic basic system was proposed in [78]. In this last one, the FLC that controls the DC chopper looks at the fuzzy logic rules developed based on the deviations of the wind speed $\Delta V_{wind} = V_{wind} - V_{wind}^*$ and of the SMES current $\Delta I_{SC} = I_{SC} - I_{SC}^*$, to balance the SMES current I_{SC} with the DFIG power curve. The FLC rules help to decide on the appropriate duty cycle to command the DC chopper PWM switching. The adopted VSC switching control is based on the scheme of Figure 3.

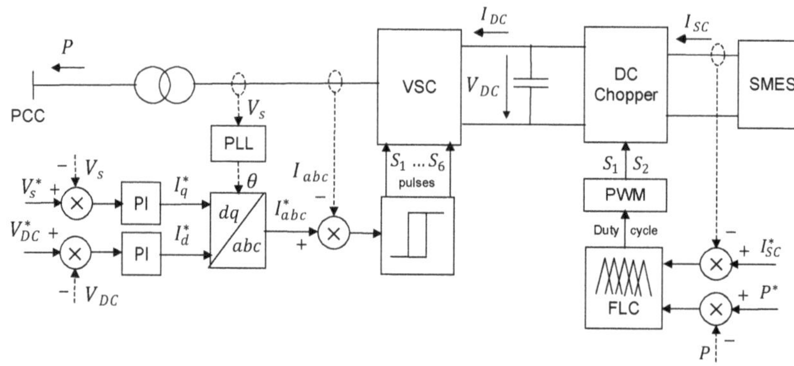

Figure 14. Compensation system with fuzzy logic and HCC to suppress power fluctuation.

6.1.1. Wind Power Generation

The doubly fed induction generator (DFIG) wind turbine presents the problems of generating power fluctuation due to the power generation inconstancy and low voltage ride-through (LVRT) performance. The second consists of the ability to continue connected to the grid during periods when the grid voltage is under the regulated range during short-circuit fault periods, which is not desired. This ability of DFIG is also known as fault ride-through (FRT).

The simultaneous use of SMES with the DC chopper and VSC connected to the AC bus to mitigate power fluctuations and a separate cooperative superconducting fault current limiter (SFCL) that protects the connection to the grid during short-circuit faults was proposed and studied in [79]. Figure 15 illustrates the unifilar scheme of such a system, with simultaneous use of SMES connected with the VSC and DC chopper to the AC bus and cooperative SCFL. When a fault occurs, the SFCL is used to limit the fault current, alleviate the terminal voltage drop, and transient power fluctuation, and avoid the DFIG continuing operating on LVRT conditions. Subsequently, the remaining power fluctuation is suppressed by the SMES. The resistive value of the SFCL as well as the inductance of the SMES coil are simultaneously optimized to minimize the sudden increase in the DFIG kinetic energy during faults, the SFCL energy loss, and the output power fluctuation.

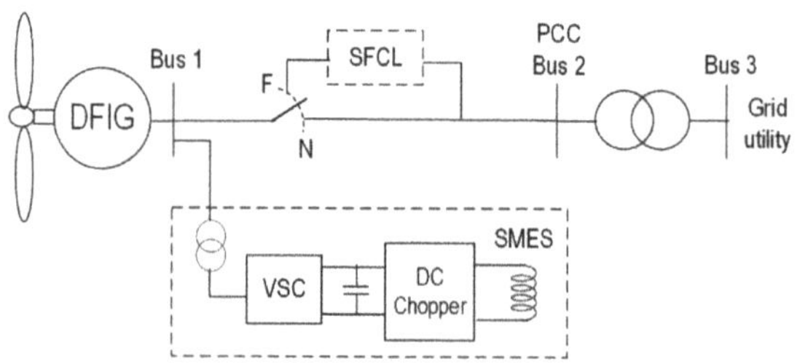

Figure 15. SMES connected to the AC-bus with VSC and DC chopper and cooperative SCFL.

A circuit configuration using superconducting magnetic energy storage with fault current limiting function (SMES-FCL) in the DC microgrid of a DFIG was presented in [80]. Figure 16 shows the unifilar scheme of such a circuit configuration. The SMES-FCL circuit mainly consists of two DC choppers with a common superconducting coil (SC). During normal operation, the SMES-FCL acts as the SMES unit to suppress the power fluctuation of DFIG. When severe faults occur in the system, the SC is automatically used as the fault current limiter. Consequently, the combined SMES-FCL system compensates for power fluctuations and performs protection during short-circuit faults. The energy function method is used to formulate the optimization problem of SC inductance, initial stored energy, and the proportional-integral control parameters of DC-DC choppers. The SMES-FCL in the DC μ-grid of DFIG enhances the LVRT capability and regulates power fluctuation.

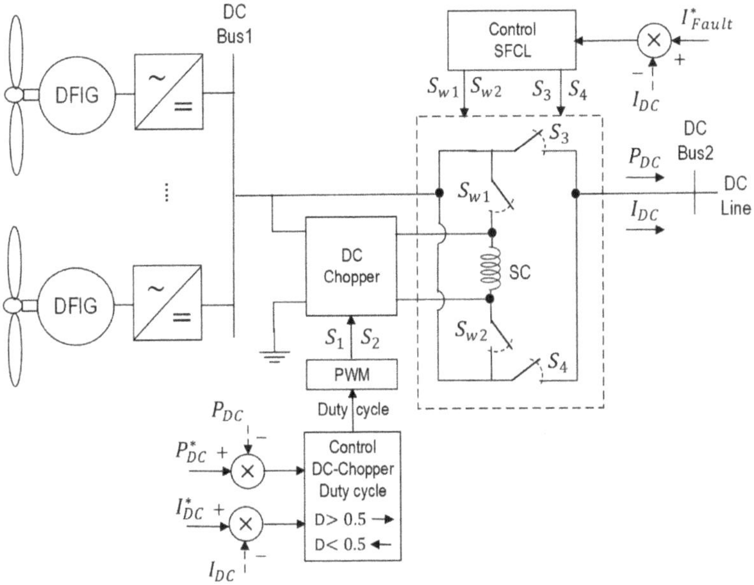

Figure 16. Unifilar scheme of a SMES-FCL in the DC μ-grid of DFIG.

The logic switching scheme for the SFCL switches S_{w1}, S_{w2}, S_3, and S_4 is schematized in Figure 17. Switches S_{w1} and S_{w2} present simultaneously the same logic state. This is also true for switches S_3 and S_4. These two sets with two switches each present simultaneously opposite states. When S_{w1} and S_{w2} are both "on" (closed), S_3 and S_4 are both "off" (open) and the SC is working as SFCL. In the other case, the SC is connected to the DC Chopper and works as an SMES. In this last case with the SC serving as SMES, when switches S_1 and S_2 of the DC chopper are both "on" and the SMES is in the charging mode. Otherwise, when S_1 and S_2 are both "off", the SMES is in the discharging mode.

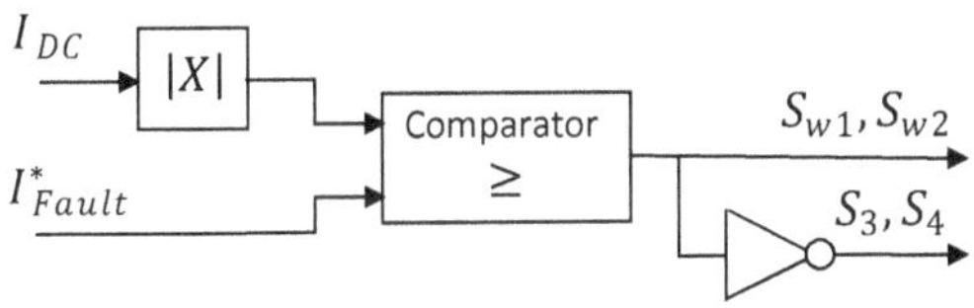

Figure 17. Logic switching scheme for the SFCL switches S_{w1}, S_{w2}, S_3 and S_4.

The control that determines the duty cycle D of the PWM that commands S_{w1}, the DC chopper switches S_1 and S_2 is based on the logic scheme presented in Figure 4.

Figure 18 shows the scheme of a system with the optimal superconducting coil (SC) integrated into the DC-link of the DFIG wind turbine, to avoid operation with FRT and suppress output power fluctuation, that was proposed and studied in [81]. The SC is connected using an AC-DC converter from the DFIG stator terminal to the DC μ-grid between the rotor side converter (RSC) and the grid side converter (GSC). A DC chopper controls the exchanged energy to and from the SC. During normal operation, the SC acts as an energy storage device to exchange energy with the system so that the power fluctuation of the DFIG wind turbine can be alleviated. When the switches S_1 and S_2 of the DC chopper are both "on" (closed) the SC is charged. Otherwise, the SC is discharged. When severe short-circuit faults occur in the system, the SC is used as the current limiting inductor to suppress both overcurrent in the rotor and stator, and overvoltage in the DC link of the DFIG.

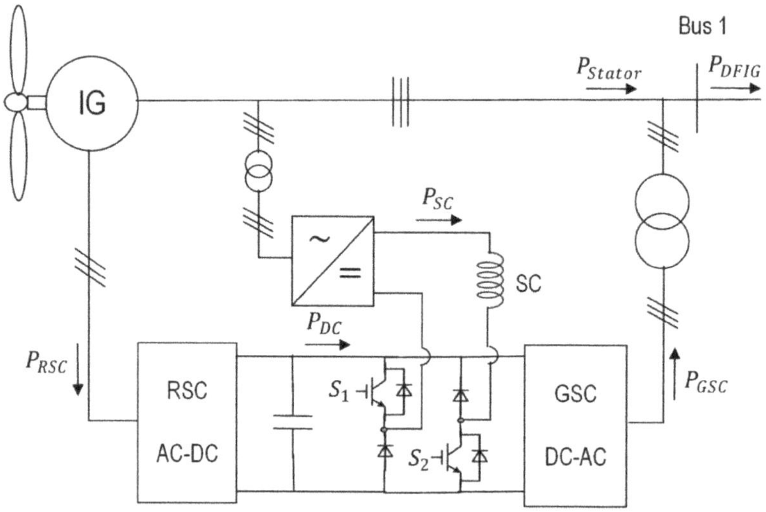

Figure 18. A system with the optimal SC integrated into the DC link of the DFIG wind turbine.

A particle swarm optimization (PSO) was performed to optimize the inductance of the SC, the initial current, the initial necessary stored energy in the SC, and the proportional-integral (PI) parameters of the GSC and RSC, to achieve the following objectives: (1) alleviation of the output power fluctuation; (2) reduction in the DC-link overvoltage; and (3) suppression the overcurrent in both stator and rotor. The DFIG wind turbine with the optimal SC maximizes the power suppression effect and enhances the FRT capability. Although the cost of integrating the optimized SC into the DFIG, this solution has the benefits of smoothing power fluctuations, avoiding the operation with FRT, limiting fault current, and improving the power system's transient stability.

An interline AC-DC unified power quality conditioner (UPQC) protection device with SMES was proposed and designed to avoid the operation with low voltage ride through (LVRT) and smooth output power for doubly fed induction generator (DFIG)/permanent magnet synchronous generator (PMSG) hybrid wind energy conversion system (WECS) [82]. The AC side of the proposed UPQC is connected in series with the terminal of DFIG as a dynamic voltage restorer (DVR). The DC side of the UPQC is connected in parallel with the DC bus of PMSG for DC bus voltage maintenance and output power adjustment.

6.1.2. Photovoltaic Power Generation

Connecting intermittent sources to the grid introduces challenges in various technical aspects such as power quality, protection, generation dispatch control, and reliability. In this context, leveling intermittent source output is necessary to maintain the grid's stability. A comprehensive literature review on problems when intermittent PV is connected to the grid and the methods of smoothing the output power fluctuation from PV was performed [83].

An SMES with the dual functions of active power storing for smoothing power fluctuations and shunt active power filter (SAPF) for constraining harmonic and unbalanced currents in a photovoltaic (PV) microgrid was proposed and studied [45]. As shown in Figure 19, the SMES is interfaced with a DC chopper and VSC to the PCC. A control scheme like the one shown in Figure 14, using PI-based and hysteresis current control (HCC) of the VSC and FLC of the DC chopper was proposed for this system.

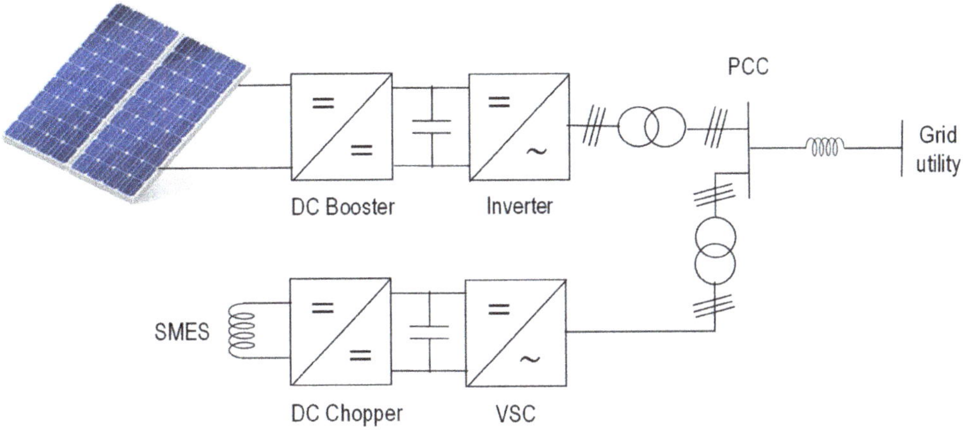

Figure 19. PV System with SMES interfaced with a DC chopper and VSC to the PCC.

Figure 20 shows another system with SMES connected with a DC-chopper to the DC-link between the DC booster and inverter of a PV system to perform power fluctuation suppression and active current filtering control that was proposed in [84]. An improved sliding mode controller (SMC) was developed for the DC chopper to stabilize the DC-link voltage. The stability of the proposed SMC controller was demonstrated based on the Lyapunov stability theory.

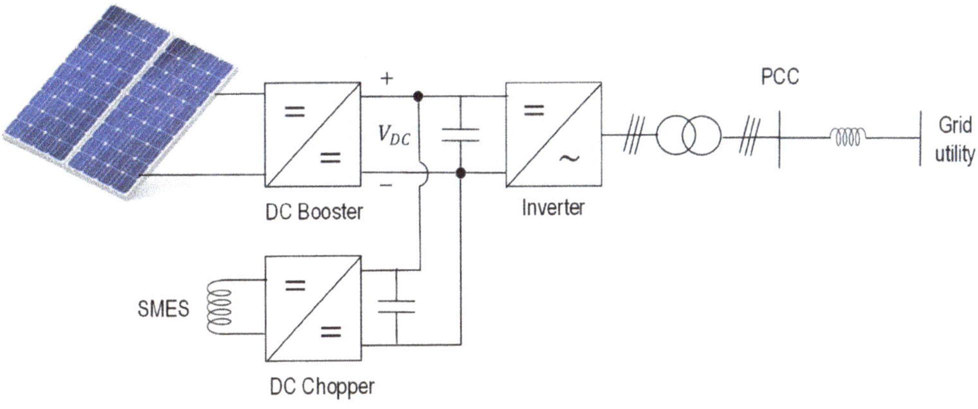

Figure 20. PV System using SMES connected with a DC chopper to the DC link between the DC Booster and inverter.

6.2. Systems Overview and Development Challenges in Fluctuation Suppression

Several SMES systems identified to mitigate the power fluctuations caused by the intermittency of renewable resources such as PV and wind are stated in Table 8. For the ones that connect to the AC grid, the VSC was controlled by a PI following the grid line reference voltage V_S^*. In some of the VSCs, switches were commanded using SVPWM. In other VSCs, switches were commanded using HCC. The DC chopper switches were commanded using PWM controlled by a PI following the DC link reference voltage V_{DC}^* or the DC link reference current I_{DC}^*. Some of the DC chopper switches were commanded using PWM controlled by FLC. The use of SMC based on the Lyapunov stability theory was proposed for controlling the DC chopper to stabilize the DC-link voltage.

Table 8. Overview of SMES systems to mitigate power fluctuations.

References	Quality Control	SMES System Connection	SMES Interface	SMES System Control Strategy
[46]	Fluctuation suppression	PCC	VSC + DC chopper	PI (V_S^*)→SVPWM + PI (V_{DC}^*) →PWM
[77]	Fluctuation suppression	PCC	VSC + DC chopper	PI (V_S^*)→HCC + FLC→PWM
[79]	Fluctuation suppression + Fault limiter	DFIG Stator terminal	VSC + DC chopper + SFCL	PI (V_S^*)→SVPWM + PI (V_{DC}^*) →PWM + I_{Fault} comparator
[80]	Fluctuation suppression + Fault limiter	WTG DC μ−grid	DC chopper + SFCL	PI (I_{DC}^*) →PWM+ I_{Fault} comparator
[81]	Fluctuation suppression + Fault limiter	DFIG DC link RSC-GSC	AC-DC converter DC Chopper	PI (V_S^*)→SVPWM + I_{Fault} comparator
[45,83]	Fluctuation suppression + Harmonics filter	PV PCC	VSC + DC chopper	PI (V_S^*)→HCC + FLC→PWM + Active filtering
[84]	Fluctuation suppression + Harmonics filter	PV DC link Boost-Inverter	DC Chopper	SMC→PWM Lyapunov stability Active filtering

In some DFIG, the SMES coil becomes a SFCL when the line current exceeds a maximum reference value I_{Fault} to disconnect the DFIG during short-circuit fault periods because of its LVRT capability [79–81].

In terms of research challenges, the following are necessary:

(1) Studies comparing the efficiency, time-response and fluctuation suppression precision of SMES systems connected via VSC and DC chopper to the AC grid bus or via only a DC-Chopper to the DC-link of DFIG or to the DC μ-grid of wind or PV generation plants.
(2) Studies of the transient effects when a superconducting magnet with both SMES and SFCL functionalities, changes from a specific stored magnetic energy to support a short-circuit fault current.
(3) Studies comparing the precision of SVPWM and HCC modulations on the control of VSC in an SMES system for power suppression.
(4) The proposal and study of adaptative control techniques that continuously adapt to the system response and improve the power fluctuation suppression.

7. Voltage Sags and Flickering Compensation

Voltage sags and flickers are caused by reactive power decompensation. In Continental Europe, the standard voltage range should be within ±5% [85].

An SMES voltage sag compensation system with 0.3 MJ NbTi magnet able to absorb or inject reactive power that compensates for the phase angle deviation created by reactive loads, was designed and implemented [86]. The performance of an SMES System with a DC-chopper and VSC with hysteretic voltage control (HVC) to compensate for the steady voltage sag caused by phase angle deviation was evaluated [87]. An SMES system with a 3 MJ superconducting magnet, a DC-chopper with FLC, and a VSC with HVC, to reduce the sag and flicker was studied [88].

8. Active Harmonics Filtering

With the increasing deployment of power electronic converters, current harmonics are injected into the grid affecting the quality of the supplied energy. Thus, it is even more necessary to perform current harmonics filtering to not affect sensitive consumer loads. Table 9 resumes the maximum voltage total harmonic distortion (THD) depending on the voltage level according to [89].

Table 9. Maximum voltage total harmonic distortion (THD) depends on the voltage level.

Bus Voltage V [kV]	Maximum THD [%]
$V \leq 1$ kV	8.0
1 kV $< V \leq$ 69 kV	5.0
69 kV $< V \leq$ 161 kV	2.5
$V >$ 161 kV	1.5

Passive or active filters can be used to remove current harmonics from the grid. Passive filters are based on the creation of a lower impedance absorption path for higher frequency current harmonics. Active filters are based on the cancellation of the harmonic content by subtraction of symmetric amplitude harmonics.

8.1. Active Harmonics Filtering Systems Using Superconducting Magnetic Energy Storage

Optimized active control of voltage source controllers (VSCs) connecting SMES systems to the grid utility via LCL filter is studied in [72]. This active control method to perform the damping of the LCL resonant peak uses a multi-loop feedback control. A negative band pass filter (NBPF) on one feedback looks to suppress the resonant peak generated by the LCL mesh and another feedback looks to control the generated line current deviation. A quasi-proportional resonant (QPR) regulator is adopted for the inner loop to realize the sinusoidal current without static error. Figure 21a shows the diagram for active control of VSC with SMES connected to the grid via LCL mesh. Figure 21b shows the block diagram of VSC feedback control based on LCL virtual impedance and NBPF.

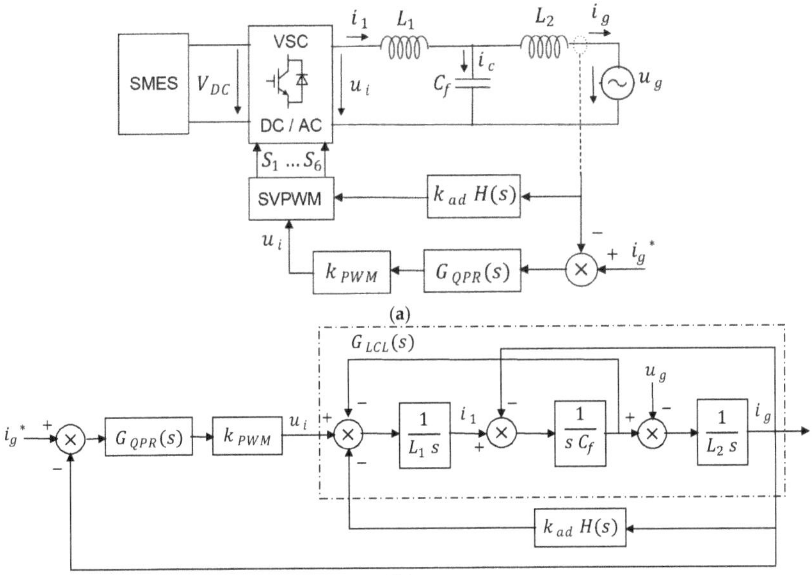

Figure 21. (**a**) Diagram for active control of VSC with SMES connected via LCL mesh to the power grid; (**b**) block diagram of VSC feedback control based on LCL virtual impedance and NBPF.

The transfer function of the NBPF is expressed by Equation (19), where Q is the quality factor and ω_{res} the central resonance angular frequency.

$$H(s) = \frac{s\ \omega_{res}/Q}{s^2 + s\ \omega_{res}/Q + \omega_{res}^2} \tag{19}$$

The transfer function of the QPR regulator is expressed in Equation (20), where K_p, K_R, and ω_c are constants that enable to control of the poles' location maximizing the damping.

$$G_{QPR}(s) = K_p + \frac{2\ K_R\ \omega_c\ s}{s^2 + 2\ \omega_c\ s + \omega_0^2} \tag{20}$$

To obtain system stability, the damping feedback coefficient $k = K_{PWM}\ k_{ad}$ and quality factor Q should satisfy the relationship (21).

$$0 < k < (L_1 + L_2)\ Q\ \omega_{res} \tag{21}$$

The synchronous reference frame (SRF) theory [90,91], is used to perform the determination of the harmonic content. Figure 22 shows the basic system that determines the harmonic content based on the SRF theory.

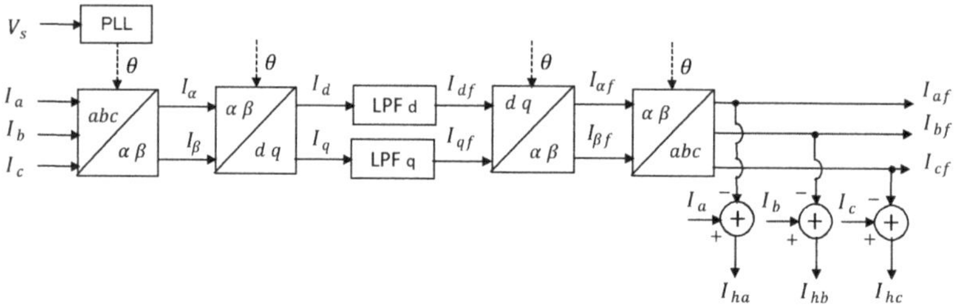

Figure 22. A system that determines the harmonic content based on the SRF theory.

The three-phase current components I_{abc} from the grid are converted into quadrature components $I_{\alpha\beta}$ using Clarke's transformation and then into I_{dq} using Park's transformation. The quadrature components I_{dq} are then filtered by a low pass filter (LPF) to remove higher harmonic content. The resulting filtered quadrature components (identified with f in the index) are again converted by the inverse Park and Clarke transformations into final three-phase current components without harmonic content. These last ones are subtracted from the initial three-phase current components to find the harmonic content in each phase. The obtained harmonic content is used for the cancellation of harmonic by active filters.

An active harmonics filter by appropriate control of a VSC and DC-chopper connecting an SMES to the grid was analyzed and simulated [92]. A PI with HCC was adopted for the VSC control to follow the reference line voltage and current. The two switches of the DC chopper are controlled by a comparator of the DC-link voltage with the reference voltage value.

As described in the previous Section 6.1.2, the two SMES systems that execute power fluctuation suppression in PV generation perform also active current harmonics filtering.

8.2. Active Harmonics Filtering Challenges

With respect to the active harmonics filtering challenges will be on the:

(1) Determination of the minimum optimum SMES energy storage capacity that is necessary to generate cancellation currents equal in magnitude but opposite in phase to the detected harmonics.
(2) Development of VSC control systems based simultaneously on the SRF theory presented in Figure 21 to perform active harmonics filtering [90,91], and on the multiloop

VSC feedback control with LCL virtual impedance and NBPF shown in Figure 21b to prevent LCL filter resonance [72].

9. Quantitative Overview

Each SMES system is defined by current converters and types of control used. Some SMES systems to suppress fluctuations from DFIG act also as fault current limiters to avoid the LVRT effect during periods when the grid voltage reduces significantly during the occurrence of short-circuit faults. Also, some SMES systems to suppress fluctuations from PV power generation also perform active harmonic filtering.

Figure 23 presents a graphic of the number of identified systems for each class of power quality regulation service. The systems for only fluctuation suppression and the ones with fault current limiter or active current harmonics were considered in different classes.

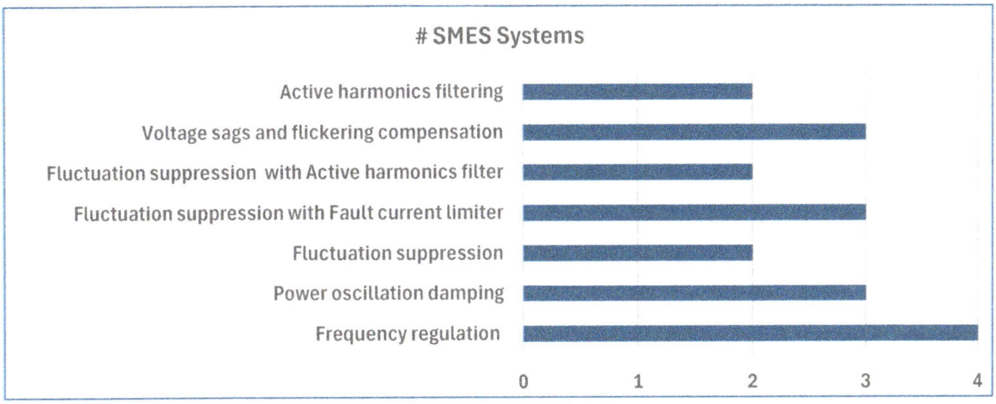

Figure 23. Graphic with the number of identified systems for each class of power quality regulation service.

Figure 24 presents a pie chart with the percentage of identified systems for each class of power quality regulation service.

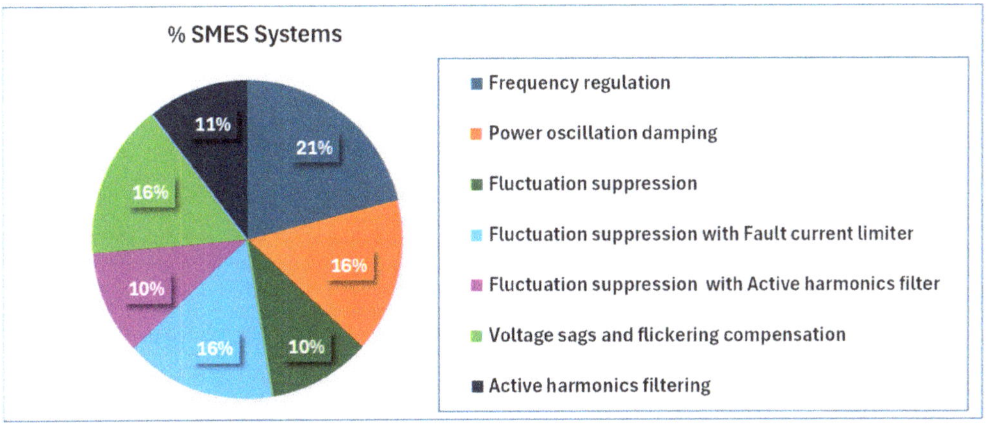

Figure 24. Pie chart with the percentage identified systems for each class of power quality regulation service.

The percentage of identified SMES systems that perform power fluctuation suppression is about 50%, power oscillation damping 19%, and frequency regulation 25%. Special attention has been paid to suppressing the power fluctuations created by the intermittency of decentralized energy renewable resources, especially from wind and photovoltaic generation.

10. Conclusions

Superconducting magnetic energy storage (SMES) systems can charge, present high-rated powers, high efficiencies, and very short charge and discharge periods of a few seconds. Because of their ability to perform a fast high active power compensation, their applicability is more suited to power quality control and regulation purposes. Studies have also shown the potential economic viability of SMES systems for grid applications.

As a result of this research, several SMES systems for power quality control and regulation services were identified and classified. This classification was performed concerning the type of power quality control and regulation service, point of connection, power current converters used, and the control strategies adopted.

Different models of SMES systems for frequency regulation purposes were identified to enable modeling the contribution of these systems in an overall frequency regulation system. These include models of SMES systems as a second-order lead-lag compensator or as a fractional order proportional integral (FOPI) controller.

The connection or disconnection of decentralized renewable power generation plants with inverters, to or from the grid, is characterized by a different dynamic from that of traditional synchronous generators (SG). Adaptative SMES systems have also been proposed to track the different power oscillation dynamics that result from the connection or disconnection of different power generation systems. Distributed control SMES systems using fuzzy logic control to account for the influence of other non-local generators on power oscillations were also proposed.

Power generation from distributed energy resources (DER) based on renewable resources is intermittent, creating power fluctuation in the grid. Special attention has been paid recently to proposing SMES systems for power fluctuation suppression. Several SMES systems have been proposed to suppress the power fluctuation from wind energy conversion systems (WECS) and photovoltaic (PV) systems. Some SMES fluctuation suppression systems were designed specifically for double-fed induction generators (DFIG). These include an extra fault current limiting capability to avoid DFIG in continuing injecting power during short-circuit faults when the grid voltage is low, due to their low voltage ride-through (LVRT) capability. Others include active harmonic filtering to mitigate the total harmonic distortion from inverters.

This document identifies SMES system solutions for power quality control and regulation services within grids with increasing instability due to the even greater integration of power generation from DER based on renewable resources.

Author Contributions: Conceptualization, A.J.A.C. and H.M.; methodology, A.J.A.C. and H.M.; software, A.J.A.C. and H.M.; validation, A.J.A.C. and H.M.; formal analysis, A.J.A.C. and H.M.; investigation, A.J.A.C. and H.M.; resources, H.M.; data curation, A.J.A.C. and H.M.; writing—original draft preparation, A.J.A.C.; writing—review and editing, A.J.A.C. and H.M.; visualization, A.J.A.C. and H.M.; supervision, H.M.; project administration, H.M.; funding acquisition, H.M. All authors have read and agreed to the published version of the manuscript.

Funding: This research was funded by project C644914747-00000023/56—"Aliança para a Transição Energética", financed by the Recovery and Resilience Plan (RRP)—Component 5: "Agendas Mobilizadoras para a Inovação Empresarial", included in the Next Generation EU funding program.

Data Availability Statement: The original contributions presented in the study are included in the article, further inquiries can be directed to the corresponding author.

Acknowledgments: The authors acknowledge the support given by the laboratories from the electrical engineering department of IST-UL.

Conflicts of Interest: The authors declare no conflicts of interest.

References

1. Hjalmarsson, J.; Thomas, K.; Boström, C. Service stacking using energy storage systems for grid applications—A review. *J. Energy Storage* **2023**, *60*, 106639. [CrossRef]
2. Luo, X.; Wang, J.; Dooner, M.; Clarke, J. Overview of current development in electrical energy storage technologies and the application potential in power system operation. *Appl. Energy* **2015**, *137*, 511–536. [CrossRef]
3. Luongo, C.A.; Baldwin, T.; Ribeiro, P.; Weber, C.M. A 100 MJ SMES demonstration at FSU-CAPS. *IEEE Trans. Appl. Supercond.* **2003**, *13*, 1800–1805. [CrossRef]
4. Sparacio, S.; Napolitano, A.; Savoldi, L.; Viarengo, S.; Laviano, F. Design of a Module for a 10 MJ Toroidal YBCO Superconducting Magnetic Energy Storage. *IEEE Trans. Appl. Supercond.* **2022**, *32*, 5700105. [CrossRef]
5. Zhao, Z.; Wang, Y.; Gao, Y.; Yang, Z.; Li, Z.; Pi, W. Mechanical characterization of a 10-MJ HTS SMES magnet wound by quasi-isotropic strands and directly stacked tape conductors. *Superconductivity* **2023**, *5*, 100042. [CrossRef]
6. Katagiri, T.; Nakabayashi, H.; Nijo, Y.; Tamada, T.; Noda, T.; Hirano, N.; Nagata, T.; Nagaya, S.; Yamane, M.; Ishii, Y.; et al. Field Test Result of 10MVA/20MJ SMES for Load Fluctuation Compensation. *IEEE Trans. Appl. Supercond.* **2009**, *19*, 1993–1998. [CrossRef]
7. Singh, H.; Vadhera, S. Comparative analysis between SMES and BESS in application of VSC-HVDC system. In Proceedings of the 2015 Annual IEEE India Conference (INDICON), New Delhi, India, 17–20 December 2015; pp. 1–6.
8. Adetokun, B.B.; Oghorada, O.; Abubakar, S.J. Superconducting magnetic energy storage systems: Prospects and challenges for renewable energy applications. *J. Energy Storage* **2022**, *55*, 105663. [CrossRef]
9. Molina-Ibáñez, E.-L.; Rosales-Asensio, E.; Pérez-Molina, C.; Pérez, F.M.; Colmenar-Santos, A. Analysis on the electric vehicle with a hybrid storage system and the use of Superconducting magnetic energy storage (SMES). *Energy Rep.* **2021**, *7*, 854–873. [CrossRef]
10. Kebede, A.A.; Kalogiannis, T.; Van Mierlo, J.; Berecibar, M. A comprehensive review of stationary energy storage devices for large scale renewable energy sources grid integration. *Renew. Sustain. Energy Rev.* **2022**, *159*, 112213. [CrossRef]
11. Huang, Y.; Ru, Y.; Shen, Y.; Zeng, Z. Characteristics and Applications of Superconducting Magnetic Energy Storage. *J. Phys. Conf. Ser.* **2021**, *2108*, 012038. [CrossRef]
12. Jin, J.X.; Chen, X.Y. Study on the SMES Application Solutions for Smart Grid. *Phys. Procedia* **2012**, *36*, 902–907. [CrossRef]
13. Alkhafaji, A.S.; Trabelsi, H. Uses of Superconducting Magnetic Energy Storage Systems in Microgrids under Unbalanced Inductive Loads and Partial Shading Conditions. *Energies* **2022**, *15*, 8597. [CrossRef]
14. Yunus, A.M.S.; Saini, M. Overview of SMES units application on smart grid systems. In Proceedings of the 2016 International Seminar on Intelligent Technology and Its Applications (ISITIA), Lombok, Indonesia, 28–30 July 2016; pp. 465–470.
15. Jacques, B.J.; Armel, T.K.F.; Astrid, M.N.B.; René, T. Contribution to the improvement of the performances of PV/wind microgrids integrating a hybrid SMES/battery energy storage unit. *J. Energy Storage* **2024**, *85*, 111011. [CrossRef]
16. Wang, P.; Lu, J.; Wu, Y.; Xu, L.; Li, J.; Mao, H.; Tian, Y.; Xu, B.; Huang, J. A Novel Cooperative Control for SMES/Battery Hybrid Energy Storage in PV Grid-Connected System. *IEEE Trans. Appl. Supercond.* **2024**, *34*, 5401605. [CrossRef]
17. Papageorgiou, P.G.; Papafilippou, P.T.; Oureilidis, K.O.; Christoforidis, G.C. An Adaptive Controller of a Hybrid Storage System for Power Smoothing with Enlarged Battery Lifetime. *IEEE Trans. Sustain. Energy* **2024**, *15*, 2567–2580. [CrossRef]
18. Hernando, C.; Munilla, J.; García-Tabarés, L.; Pedraz, G. Optimization of High Power SMES for Naval Applications. *IEEE Trans. Appl. Supercond.* **2023**, *33*, 5700205. [CrossRef]
19. Jin, J.X.; Xu, W.; Chen, X.Y.; Zhou, X.; Zhang, J.Y.; Gong, W.Z.; Ren, A.L.; Xin, Y. Developments of SMES devices and potential applications in smart grids. In Proceedings of the IEEE PES Innovative Smart Grid Technologies, Washington, DC, USA, 16–20 January 2012; pp. 1–6.
20. Bevrani, H.; Golpîra, H.; Messina, A.R.; Hatziargyriou, N.; Milano, F.; Ise, T. Power system frequency control: An updated review of current solutions and new challenges. *Electr. Power Syst. Res.* **2021**, *194*, 107114. [CrossRef]
21. Khan, I.A.; Mokhlis, H.; Mansor, N.N.; Illias, H.A.; Awalin, L.J.; Wang, L. New trends and future directions in load frequency control and flexible power system: A comprehensive review. *Alex. Eng. J.* **2023**, *71*, 263–308. [CrossRef]
22. Filippidis, S.; Bouhouras, A.; Poulakis, N.; Christoforidis, G. A Review of the Cryocooler-Based Cooling Systems for SMES. *IEEE Trans. Appl. Supercond.* **2021**, *31*, 4902713. [CrossRef]
23. Zhang, H.; Nie, Z.; Xiao, X.; Aggarwal, R.; Kang, Q.; Ainslie, M.; Zhu, J.; Coombs, T.; Yuan, W. Design and Simulation of SMES System Using YBCO Tapes for Direct Drive Wave Energy Converters. *IEEE Trans. Appl. Supercond.* **2013**, *23*, 5700704. [CrossRef]
24. Zimmermann, A.W.; Sharkh, S.M. Design of a 1 MJ/100 kW high temperature superconducting magnet for energy storage. *Energy Rep.* **2020**, *6*, 180–188. [CrossRef]
25. Wang, Q.; Dai, Y.; Song, S.; Wen, H.; Bai, Y.; Yan, L.; Kim, K. A 30 kJ Bi2223 High Temperature Superconducting Magnet for SMES with Solid-Nitrogen Protection. *IEEE Trans. Appl. Supercond.* **2008**, *18*, 754–757. [CrossRef]
26. Atomura, N.; Takahashi, T.; Amata, H.; Iwasaki, T.; Son, K.; Miyagi, D.; Tsuda, M.; Hamajima, T.; Shintomi, T.; Makida, Y.; et al. Conceptual design of MgB2 coil for the 100 MJ SMES of advanced superconducting power conditioning system (ASPCS). *Phys. Procedia* **2012**, *27*, 400–403. [CrossRef]
27. Lee, S.; Park, S.-H.; Kim, W.-S.; Lee, J.-K.; Lee, S.; Park, C.; Bae, J.-H.; Kim, S.-H.; Seong, K.-C.; Choi, K.; et al. Analysis of Eddy Current Losses and Magnetization Losses in Toroidal Magnets for a 2.5 MJ HTS SMES. *IEEE Trans. Appl. Supercond.* **2011**, *21*, 1354–1357. [CrossRef]

28. Kim, K.; Kim, A.-R.; Kim, J.-G.; Park, M.; Yu, I.-K.; Sohn, M.-H.; Eom, B.-Y.; Sim, K.; Kim, S.; Kim, H.-J.; et al. Analysis of Operational Loss Characteristics of 10 kJ Class Toroid-Type SMES. *IEEE Trans. Appl. Supercond.* **2011**, *21*, 1340–1343. [CrossRef]
29. Kim, A.-R.; Kim, G.-H.; Kim, K.-M.; Kim, J.-G.; Kim, D.-W.; Park, M.; Yu, I.-K.; Kim, S.-H.; Sim, K.; Seong, K.-C. Design and Manufacturing of a SMES Model Coil for Real Time Digital Simulator Based Power Quality Enhancement Simulation. *IEEE Trans. Appl. Supercond.* **2010**, *20*, 1339–1343. [CrossRef]
30. Trillaud, F.; Cruz, L.S. Conceptual Design of a 200-kJ 2G-HTS Solenoidal μ-SMES. *IEEE Trans. Appl. Supercond.* **2014**, *24*, 5700205. [CrossRef]
31. Kamada, H.; Ninomiya, A.; Nomura, S.; Yagai, T.; Nakamura, T.; Chikaraishi, H. Development of 1-T Class Force-Balanced Helical Coils Using REBCO Tapes. *IEEE Trans. Appl. Supercond.* **2020**, *30*, 4600905. [CrossRef]
32. Kimura, Y.; Kamada, H.; Nomura, S.; Yagai, T.; Nakamura, T.; Tsutsui, H.; Chikaraishi, H.; Yanagi, N.; Imagawa, S. Development of a Prototype Winding Machine for Helical Coils Using High-Temperature Superconducting Tapes. *IEEE Trans. Appl. Supercond.* **2016**, *26*, 4202104. [CrossRef]
33. Kim, A.-R.; Jung, H.-Y.; Kim, J.-H.; Park, M.; Yu, I.-K.; Kim, S.-H.; Sim, K.; Kim, H.-J.; Seong, K.-C. Design of a 10 kJ Class SMES Model Coil for Real Time Digital Simulator Based Grid Connection Study. *IEEE Trans. Appl. Supercond.* **2009**, *19*, 2023–2027. [CrossRef]
34. Wang, Q.; Song, S.; Lei, Y.; Dai, Y.; Zhang, B.; Wang, C.; Lee, S.; Kim, K. Design and Fabrication of a Conduction-Cooled High Temperature Superconducting Magnet for 10 kJ Superconducting Magnetic Energy Storage System. *IEEE Trans. Appl. Supercond.* **2006**, *16*, 570–573. [CrossRef]
35. Zheng, Z.-X.; Chen, X.-Y.; Xiao, X.-Y.; Huang, C.-J. Design and Evaluation of a Mini-Size SMES Magnet for Hybrid Energy Storage Application in a kW-Class Dynamic Voltage Restorer. *IEEE Trans. Appl. Supercond.* **2017**, *27*, 5700911. [CrossRef]
36. Jubleanu, R.; Cazacu, D. Design and Numerical Study of Magnetic Energy Storage in Toroidal Superconducting Magnets Made of YBCO and BSCCO. *Magnetochemistry* **2023**, *9*, 216. [CrossRef]
37. Vialle, J.; Badel, A.; Tixador, P.; Ciceron, J.; Forest, F.; Pasquet, R. Preliminary Tests of Pancakes From a 12 T REBCO Insulated Solenoid Magnet. *IEEE Trans. Appl. Supercond.* **2021**, *31*, 4600805. [CrossRef]
38. Kim, H.J.; Seong, K.C.; Cho, J.W.; Bae, J.H.; Sim, K.D.; Kim, S.; Lee, E.Y.; Ryu, K.; Kim, S.H. 3 MJ/750 kVA SMES System for Improving Power Quality. *IEEE Trans. Appl. Supercond.* **2006**, *16*, 574–577. [CrossRef]
39. Nomura, S.; Kasuya, K.; Tanaka, N.; Tsuboi, K.; Tsutsui, H.; Tsuji-Iio, S.; Shimada, R. Experimental Results of a 7-T Force-Balanced Helical Coil for Large-Scale SMES. *IEEE Trans. Appl. Supercond.* **2008**, *18*, 701–704. [CrossRef]
40. Nomura, S.; Kasuya, K.; Tanaka, N.; Tsuboi, K.; Tsutsui, H.; Tsuji-Iio, S.; Shimada, R.; Arai, K.; Ninomiya, A.; Ishigohka, T. Quench Properties of a 7-T Force-Balanced Helical Coil for Large-Scale SMES. *IEEE Trans. Appl. Supercond.* **2009**, *19*, 2004–2007. [CrossRef]
41. Hayashi, H.; Hatabe, Y.; Nagafuchi, T.; Taguchi, A.; Terazono, K.; Ishii, T.; Taniguchi, S. Test Results of Power System Control by Experimental SMES. *IEEE Trans. Appl. Supercond.* **2006**, *16*, 598–601. [CrossRef]
42. Ali, M.H.; Wu, B.; Dougal, R.A. An Overview of SMES Applications in Power and Energy Systems. *IEEE Trans. Sustain. Energy* **2010**, *1*, 38–47. [CrossRef]
43. Sun, Q.; Lv, H.; Gao, S.; Wei, K.; Mauersberger, M. Optimized Control of Reversible VSC with Stability Mechanism Study in SMES Based V2G System. *IEEE Trans. Appl. Supercond.* **2019**, *29*, 5401106. [CrossRef]
44. Chen, X.; Xie, Q.; Bian, X.; Shen, B. Energy-saving superconducting magnetic energy storage (SMES) based interline DC dynamic voltage restorer. *J. Power Energy Syst.* **2022**, *8*, 238–248. [CrossRef]
45. Jin, J.X.; Wang, J.; Yang, R.H.; Zhang, T.L.; Mu, S.; Fan, Y.J.; Xing, Y.Q. A superconducting magnetic energy storage with dual functions of active filtering and power fluctuation suppression for photovoltaic microgrid. *J. Energy Storage* **2021**, *38*, 102508. [CrossRef]
46. Zhu, J.; Bao, X.; Yang, B.; Chen, P.; Yang, Y.; Qiu, M. Dynamic Simulation Test Research on Power Fluctuation Compensation Using Hybrid SMES of YBCO and BSCCO Tapes. *IEEE Trans. Appl. Supercond.* **2012**, *22*, 5700404. [CrossRef]
47. Abdalrahman, G.; Abdalla, O.H.; Jibril, M.I.; Awad, A.A. Space Vector Pulse Width Modulation Technique Applied to Two Level Voltage Source Inverter. *SUST J. Eng. Comput. Sci.* **2018**, *19*, 10–17.
48. Shen, Y.; Zheng, Z.; Wang, Q.; Liu, P.; Yang, X. DC Bus Current Sensed Space Vector Pulsewidth Modulation for Three-Phase Inverter. *IEEE Trans. Transp. Electrif.* **2021**, *7*, 815–824. [CrossRef]
49. Abraham, R.J.; Das, D.; Patra, A. Automatic generation control of an interconnected hydrothermal power system considering superconducting magnetic energy storage. *Int. J. Electr. Power Energy Syst.* **2007**, *29*, 571–579. [CrossRef]
50. Elmelegi, A.; Mohamed, E.A.; Aly, M.; Ahmed, E.M.; Mohamed, A.-A.A.; Elbaksawi, O. Optimized Tilt Fractional Order Cooperative Controllers for Preserving Frequency Stability in Renewable Energy-Based Power Systems. *IEEE Access* **2021**, *9*, 8261–8277. [CrossRef]
51. Pappachen, A.; Fathima, A.P. Load frequency control in deregulated power system integrated with SMES–TCPS combination using ANFIS controller. *Int. J. Electr. Power Energy Syst.* **2016**, *82*, 519–534. [CrossRef]
52. Magdy, G.; Bakeer, A.; Alhasheem, M. Superconducting energy storage technology-based synthetic inertia system control to enhance frequency dynamic performance in microgrids with high renewable penetration. *Prot. Control Mod. Power Syst.* **2021**, *6*, 36. [CrossRef]
53. Bhatt, P.; Ghoshal, S.P.; Roy, R. Coordinated control of TCPS and SMES for frequency regulation of interconnected restructured power systems with dynamic participation from DFIG based wind farm. *Renew. Energy* **2012**, *40*, 40–50. [CrossRef]

54. Colmenar-Santos, A.; Molina-Ibáñez, E.-L.; Rosales-Asensio, E.; Blanes-Peiró, J.-J. Legislative and economic aspects for the inclusion of energy reserve by a superconducting magnetic energy storage: Application to the case of the Spanish electrical system. *Renew. Sustain. Energy Rev.* **2018**, *82*, 2455–2470. [CrossRef]
55. Soman, R.; Ravindra, H.; Huang, X.; Schoder, K.; Steurer, M.; Yuan, W.; Zhang, M.; Venuturumilli, S.; Chen, X. Preliminary Investigation on Economic Aspects of Superconducting Magnetic Energy Storage (SMES) Systems and High-Temperature Superconducting (HTS) Transformers. *IEEE Trans. Appl. Supercond.* **2018**, *28*, 5400805. [CrossRef]
56. European Union. *Commission Regulation (EU) 2017/1485 of 2 August 2017 Establishing a Guideline on Electricity Transmission System Operation*; European Union: Brussels, Belgium, 2017.
57. Mohamed, E.A.; Ahmed, E.M.; Elmelegi, A.; Aly, M.; Elbaksawi, O.; Mohamed, A.-A.A. An Optimized Hybrid Fractional Order Controller for Frequency Regulation in Multi-Area Power Systems. *IEEE Access* **2020**, *8*, 213899–213915. [CrossRef]
58. Banerjee, S.; Chatterjee, J.K.; Tripathy, S.C. Application of magnetic energy storage unit as load-frequency stabilizer. *IEEE Trans. Energy Convers.* **1990**, *5*, 46–51. [CrossRef]
59. Chandrakala, V.; Sukumar, B.; Sankaranarayanan, K. Load Frequency Control of Multi-source Multi-area Hydro Thermal System Using Flexible Alternating Current Transmission System Devices. *Electr. Power Compon. Syst.* **2014**, *42*, 927–934. [CrossRef]
60. Vyshna, K.; Thresia, M.M.; Shahin, M. Design of SMES for Load Frequency Control in Power Systems with Renewable Integration. In Proceedings of the 2023 5th International Conference on Energy, Power and Environment: Towards Flexible Green Energy Technologies (ICEPE), Shillong, India, 15–17 June 2023; pp. 1–6.
61. Khan, I.A.; Mokhlis, H.; Mansor, N.N.; Illias, H.A.; Usama, M.; Daraz, A.; Wang, L.; Awalin, L.J. Load Frequency Control Using Golden Eagle Optimization for Multi-Area Power System Connected Through AC/HVDC Transmission and Supported with Hybrid Energy Storage Devices. *IEEE Access* **2023**, *11*, 44672–44695. [CrossRef]
62. Ali, H.H.; Kassem, A.M.; Al-Dhaifallah, M.; Fathy, A. Multi-Verse Optimizer for Model Predictive Load Frequency Control of Hybrid Multi-Interconnected Plants Comprising Renewable Energy. *IEEE Access* **2020**, *8*, 114623–114642. [CrossRef]
63. Pati, S.S.; Subudhi, U. Frequency regulation of Solar-Wind integrated multi-area system with SMES and SSSC. In Proceedings of the 2021 IEEE International Power and Renewable Energy Conference (IPRECON), Kerala, India, 24–26 September 2021; pp. 1–4.
64. Alam, M.S.; Al-Ismail, F.S.; Abido, M.A. PV/Wind-Integrated Low-Inertia System Frequency Control: PSO-Optimized Fractional-Order PI-Based SMES Approach. *Sustainability* **2021**, *13*, 7622. [CrossRef]
65. Tripathy, S.C.; Juengst, K.P. Sampled data automatic generation control with superconducting magnetic energy storage in power systems. *IEEE Trans. Energy Convers.* **1997**, *12*, 187–192. [CrossRef]
66. Tran, X.K.; Chen, Y.; Phan, G.T.; Huynh, D.H. Optimal Design for Frequency Fluctuation Control System with SMES Unit. In Proceedings of the 2020 4th International Conference on Robotics and Automation Sciences (ICRAS), Chengdu, China, 14–16 June 2020; pp. 50–54.
67. Li, T.; Xie, J.; Gong, X.; Chen, M.; Zeng, Q.; Chen, X. Mechanism Analysis and Controller Design of SMES-Based Power Oscillation Damping. *IEEE Trans. Appl. Supercond.* **2021**, *31*, 5403704. [CrossRef]
68. Li, T.; Hu, W.; Zhang, B.; Zhang, G.; Zhang, Z.; Chen, Z. SMES Damping Controller Design and Real-Time Parameters Tuning for Low-Frequency Oscillation. *IEEE Trans. Appl. Supercond.* **2021**, *31*, 3800604. [CrossRef]
69. Tang, Y.; Mu, C.; He, H. SMES-Based Damping Controller Design Using Fuzzy-GrHDP Considering Transmission Delay. *IEEE Trans. Appl. Supercond.* **2016**, *26*, 5701206. [CrossRef]
70. Zhou, X.; Sun, H.; Jiang, L.; Zeng, P. A SMES-based adaptive damping controller design in wind power system. In Proceedings of the 2016 IEEE 8th International Power Electronics and Motion Control Conference (IPEMC-ECCE Asia), Hefei, China, 22–26 May 2016; pp. 212–216.
71. Rai, D.; Gokaraju, R.; Faried, S.O. Adaptive Control Using Constrained RLS and Dynamic Pole-Shift Technique for TCSCs. *IEEE Trans. Power Deliv.* **2014**, *29*, 224–234. [CrossRef]
72. Sun, Q.; Miao, Y.; Zhao, L.; Xu, J.; Gao, S.; Wei, K.; Wang, L. Study on LCL-Type Grid-Connected VSC Based on Optimized Active Damping with NBPF for SMES System. *IEEE Trans. Appl. Supercond.* **2024**, *34*, 5701505. [CrossRef]
73. dos Santos Costa, A.J.A.; Valério, D.; da Costa Branco, P.J. Predictive control model to manage power flow on a hybrid wind-photovoltaic and diesel microgeneration power plant with additional storage capacity. *IET Cyber-Phys. Syst. Theory Appl.* **2018**, *3*, 206–211. [CrossRef]
74. Amezquita, H.; Carvalho, P.M.S.; Morais, H. Wind Forecast at Medium Voltage Distribution Networks. *Energies* **2023**, *16*, 2887. [CrossRef]
75. Amezquita, H.; Guzman, C.P.; Morais, H. Forecasting Electric Vehicles' Charging Behavior at Charging Stations: A Data Science-Based Approach. *Energies* **2024**, *17*, 3396. [CrossRef]
76. Zhang, W.; Zhang, D.; Mu, B.; Wang, L.Y.; Bao, Y.; Jiang, J.; Morais, H. Decentralized Electric Vehicle Charging Strategies for Reduced Load Variation and Guaranteed Charge Completion in Regional Distribution Grids. *Energies* **2017**, *10*, 147. [CrossRef]
77. Yunus, A.M.S.; Abu-Siada, A.; Mosaad, M.I.; Albalawi, H.; Aljohani, M.; Jin, J.X. Application of SMES Technology in Improving the Performance of a DFIG-WECS Connected to a Weak Grid. *IEEE Access* **2021**, *9*, 124541–124548. [CrossRef]
78. Salama, H.S.; Vokony, I. Voltage and Frequency Control of Balanced/Unbalanced Distribution System Using the SMES System in the Presence of Wind Energy. *Electricity* **2021**, *2*, 205–224. [CrossRef]
79. Ngamroo, I.; Karaipoom, T. Cooperative Control of SFCL and SMES for Enhancing Fault Ride Through Capability and Smoothing Power Fluctuation of DFIG Wind Farm. *IEEE Trans. Appl. Supercond.* **2014**, *24*, 5400304. [CrossRef]

80. Ngamroo, I.; Karaipoom, T. Improving Low-Voltage Ride-Through Performance and Alleviating Power Fluctuation of DFIG Wind Turbine in DC Microgrid by Optimal SMES with Fault Current Limiting Function. *IEEE Trans. Appl. Supercond.* **2014**, *24*, 5700805. [CrossRef]
81. Karaipoom, T.; Ngamroo, I. Optimal Superconducting Coil Integrated into DFIG Wind Turbine for Fault Ride Through Capability Enhancement and Output Power Fluctuation Suppression. *IEEE Trans. Sustain. Energy* **2015**, *6*, 28–42. [CrossRef]
82. Jin, J.X.; Yang, R.H.; Zhang, R.T.; Fan, Y.J.; Xie, Q.; Chen, X.Y. Combined low voltage ride through and power smoothing control for DFIG/PMSG hybrid wind energy conversion system employing a SMES-based AC-DC unified power quality conditioner. *Int. J. Electr. Power Energy Syst.* **2021**, *128*, 106733. [CrossRef]
83. Shivashankar, S.; Mekhilef, S.; Mokhlis, H.; Karimi, M. Mitigating methods of power fluctuation of photovoltaic (PV) sources—A review. *Renew. Sustain. Energy Rev.* **2016**, *59*, 1170–1184. [CrossRef]
84. Wang, J.; Mu, S.; Zhou, Q.; Li, Y.J.; Zhang, T.L. Design and Performance Analysis of PV-SMES-Based Microgrid Integrated with Power Fluctuation Suppression and Active Filtering Control. *IEEE Trans. Appl. Supercond.* **2021**, *31*, 5402705. [CrossRef]
85. Masetti, C. Revision of European Standard EN 50160 on power quality: Reasons and solutions. In Proceedings of the 14th International Conference on Harmonics and Quality of Power—ICHQP 2010, Bergamo, Italy, 26–29 September 2010; pp. 1–7.
86. Jiang, X.; Zhu, X.; Cheng, Z.; Ren, X.; He, Y. A 150 kVA/0.3 MJ SMES voltage sag compensation system. *IEEE Trans. Appl. Supercond.* **2005**, *15*, 1903–1906. [CrossRef]
87. Zheng, Z.X.; Xiao, X.Y.; Li, C.S.; Chen, Z.; Zhang, Y. Performance Evaluation of SMES System for Initial and Steady Voltage Sag Compensations. *IEEE Trans. Appl. Supercond.* **2016**, *26*, 5701105. [CrossRef]
88. Yunus, A.M.S.; Habriansyah, I.; Abu-Siada, A.; Masoum, M.A.S. Sag and Flicker Reduction Using Hysteresis-Fuzzy Control-Based SMES Unit. *Can. J. Electr. Comput. Eng.* **2019**, *42*, 52–57. [CrossRef]
89. *IEEE Std 519-2014*; IEEE Recommended Practice and Requirements for Harmonic Control in Electric Power Systems. IEEE: Piscataway, NJ, USA, 2014. [CrossRef]
90. Tong, L.; Qian, Z.; Kuang, N.; Lingxiao, X.; Peng, F.Z. Optimal Design of Synchronous Reference Frame Harmonic Detection Method. In Proceedings of the 2007 IEEE Power Electronics Specialists Conference, Orlando, FL, USA, 17–21 June 2007; pp. 1071–1076.
91. Wang, X.; Liu, J.; Hu, J.; Meng, Y.; Yuan, C. Frequency characteristics of the d-q synchronous-frame current reference generation methods for active power filter. In Proceedings of the 2007 7th International Conference on Power Electronics, Daegu, Republic of Korea, 22–26 October 2007; pp. 945–950.
92. Simoesõ, F.; Pires, V.F.; Murta-Pina, J. Assessment of Using Superconducting Magnetic Energy Storage for Current Harmonic Compensation. In Proceedings of the 2020 International Young Engineers Forum (YEF-ECE), Costa da Caparica, Portugal, 3 July 2020; pp. 73–77.

energies

MDPI

Review

Assessing Critical Raw Materials and Their Supply Risk in Energy Technologies—A Literature Review

Francesco Montana [1], Maurizio Cellura [1,2], Maria Luisa Di Silvestre [1], Sonia Longo [1,2], Le Quyen Luu [1,3], Eleonora Riva Sanseverino [1,*] and Giuseppe Sciumè [1]

[1] Department of Engineering, University of Palermo, Viale delle Scienze Bld. 9, 90128 Palermo, Italy; francesco.montana@unipa.it (F.M.); maurizio.cellura@unipa.it (M.C.); marialuisa.disilvestre@unipa.it (M.L.D.S.); sonia.longo@unipa.it (S.L.); lequyen.luu@unipa.it (L.Q.L.); giuseppe.sciume01@unipa.it (G.S.)

[2] Centre for Sustainability and Ecological Transition, University of Palermo, Complesso Monumentale dello Steri, Piazza Marina 61, 90133 Palermo, Italy

[3] Institute of Science and Technology for Energy and Environment, Vietnam Academy of Science and Technology, A30 Building, 18 Hoang Quoc Viet, Cau Giay District, Hanoi 10072, Vietnam

* Correspondence: eleonora.rivasanseverino@unipa.it

Abstract: Climate change is leading modern society to seek innovative solutions for sustainable development and a zero-carbon economy. Nevertheless, new technologies strongly rely on precious raw materials and might suffer from supply chain risks. The European Union has identified a set of raw materials deemed to be critical or strategic because they appear essential for energy transition technologies. Consequently, long-term energy system planning must factor in the availability of these critical raw materials when selecting specific technologies, as their supply could be affected by global policies or conflicts. This paper provides a literature review on the assessment of critical raw materials in energy technologies comparing the main approaches on critical raw materials content assessment in technologies, long-term planning studies considering critical raw materials, and the development of indicators for critical raw materials content in energy technologies. The main findings of this review suggest that existing reliable databases with the bill of materials, such as life cycle inventories, should be exploited and that proper indicators to rank the criticality of materials and the importance of a specific technology should be developed. These findings are discussed and organized proposing a method for the optimal planning of an energy technologies mix in regional or national energy systems considering the availability and future supply of critical raw materials.

Keywords: critical raw materials; energy; planning; review; supply chain

1. Introduction

The global transition towards renewable energy is essential for mitigating climate change, reducing greenhouse gas emissions, and promoting energy security. In the last decades, researchers and industries focused their efforts on investigating new technologies or improving the existing ones with the aim of attaining energy efficiency and carbon emissions reduction. One of the most notable examples of this trend is the introduction of energy efficiency labels in the European Union (EU) with the EU Directive 92/75/EC, i.e., labels indicating the yearly energy consumption mandatorily applied on appliances such as refrigerators, washing machines, light bulbs, and televisions [1].

However, the energy transition is highly dependent on the availability and supply of innovative materials, precious metals, or rare raw materials. Among these, those showing the highest supply risk and the highest economic importance are known as critical raw materials (CRMs). Moreover, a subset of materials relevant to the green transition, digital transition, defense, and space applications, are collectively known as strategic raw materials (SRMs) [2]

These materials are selected and updated every three years by the European Union. Within the supply risk feature, the rate of substitution of a CRM with another raw material more easily available and the possibility of recycling the material at the end of the life of the technology [3] are considered.

Both the EU and the USA are heavily dependent on imports of CRMs. The European Commission has highlighted the vulnerability of European industries due to the lack of domestic supply of materials like lithium, cobalt, and rare earth elements [4]. The USA has recognized similar risks, implementing policies to secure supply chains by investing in domestic mining and fostering alliances with resource-rich countries. More in detail, three main supply risk categories can be identified as related to CRMs: (1) geopolitical risk, (2) environmental risk, and (3) social risk.

One of the major risks associated with the supply of CRMs is their geographic concentration in a few countries. For instance, over 74% of the world's cobalt mine production in 2023 came from the Democratic Republic of Congo (DRC) [5], while China dominates the production of rare earth elements, accounting for more than 90% of global output [6]. This concentration creates geopolitical vulnerabilities, as supply disruptions due to political instability, trade disputes, or regulatory changes can lead to shortages, price volatility, and increased costs for renewable energy technologies. The 2010 rare earths embargo by China on Japan, which triggered concerns about the reliability of supply chains, is a notable example of such geopolitical risk [7]. Economic factors exacerbate these geopolitical risks. The volatile nature of commodity markets, particularly for CRMs, leads to boom-and-bust cycles that create price uncertainty. For example, fluctuations in the prices of lithium and cobalt are influenced not only by market demand but also by the long lead times required to develop new mining projects and the high capital investment needed [7].

From the environmental point of view, the extraction and processing of CRMs are associated with significant environmental and social challenges, which pose additional risks to the supply chain. Mining for lithium, cobalt, and rare earth elements has been linked to land degradation, water contamination, and habitat destruction, particularly in developing countries where environmental regulations may be weak [4]. For instance, lithium extraction from salt flats in South America, a key region for global lithium production, has raised concerns about the depletion of freshwater resources, a critical issue in arid regions like Chile's Atacama Desert [8].

Moreover, social risks are also prevalent in CRM supply chains, including human rights abuses and labor exploitation. In the DRC, the mining of cobalt has been associated with child labor and hazardous working conditions. These issues not only pose ethical concerns but also introduce legal and reputational risks for companies that rely on these materials. The European Union's Conflict Minerals Regulation, which came into force in 2021, is an attempt to address these risks by ensuring that companies source CRMs responsibly from conflict-affected and high-risk areas [4].

Nevertheless, it is worth highlighting that not all the authors agree on the considerations here illustrated. For instance, with specific reference to renewable energy technologies, Overland pointed out in 2019 that the CRM supply risk fear started back in 2010 when China imposed an embargo on Japan concerning rare earth exports due to a territorial dispute. The author states that the widespread global diffusion of rare earths should discourage these fears, also because only a few rare earths are employed in a specific, not-so-common wind turbine technology [7].

Focusing on the energy sector, CRMs like lithium, cobalt, rare earth elements, and others play a crucial role in the manufacturing of renewable energy technologies such as wind turbines, solar photovoltaics, and battery storage systems, but the risks associated with CRMs create supply chain bottlenecks and delays, hindering the timely deployment of renewable energy technologies [9]. This aspect becomes of paramount importance in the context of the European Green Deal, a long-term path to reach a climate-neutral society by 2050 in a fair, cost-effective, and competitive way. To achieve this target, the provision of energy from zero-emissions sources is one of the main measures, but supply risks of technologies pose a possible threat to the green transition, and it is thus evident that these aspects should be included in energy systems planning. Nevertheless, as pointed out by Aljabery et al. in their review on multi-carrier energy systems [10], the most common objective functions that are generally used in the literature are costs and emissions reduction, while CRM supply risk is not even mentioned.

This literature review synthesizes the available studies concerning energy technologies containing CRMs. According to the main outcomes of this study, some criteria for optimally planning a power technology mix in regional or national energy systems are proposed. The document is structured as follows. Section 2 provides a recap on the most common CRMs used in the manufacture of energy technologies, describing their main uses and the reasons why these materials were listed among the CRMs. Complementarily, Section 3 lists the energy technologies mostly impacted by the CRM content, describing their main operating principles and their role in the energy transition. Section 4 describes the methodology adopted to carry out this literature review, which is illustrated in Section 5. The results of this review are discussed in Section 6, including a proposal for an optimal planning method for regional or national energy systems considering the availability and future supply of critical raw materials. Lastly, the conclusions of this paper are provided in Section 7.

2. Critical Raw Materials Used in Energy Technologies

CRMs are essential for the manufacturing of various energy technologies. Renewable energy technologies, such as wind turbines or solar panels but also energy storage systems, depend on materials like lithium, cobalt, nickel, rare earth elements, and other metals to function efficiently. Lithium and cobalt are critical for battery storage systems, while neodymium and dysprosium are essential for permanent magnets in wind turbines. In detail, the introduction of CRMs in energy technologies is essential to improve their performance, efficiency, and scalability. Their supply chain is vulnerable due to geopolitical factors, economic constraints, and environmental concerns. In this section, a brief description of the key CRMs used for the manufacture of energy technologies is provided. The ChatGPT GenAI software tool was used to generate a preliminary list of critical raw materials mostly used in the energy sector with enriched content included. Additionally, the map of all CRM mines, gathered from the website of the European Geological Data Infrastructure, is shown in Figure 1 [11].

2.1. Lithium

Lithium is a lightweight metal and has a high electrochemical potential, making it crucial for rechargeable batteries. In detail, due to their high energy density and relatively low self-discharge rate, lithium-ion batteries are widely used in energy storage applications, such as mobile phones, electric vehicles, and facility-scale storage systems. However, their increasing demand has led to concerns about future lithium supply since its extraction is geographically concentrated, primarily in Chile, Argentina, and Australia [6].

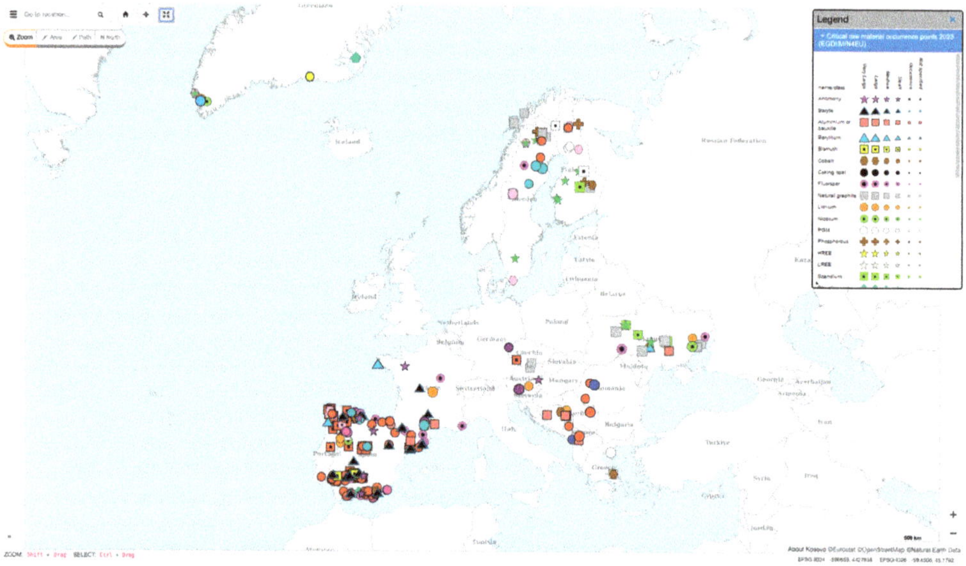

Figure 1. Map of the CRM mines in Europe [11].

2.2. *Rare Earth Elements (REEs)*

Rare earth elements, particularly neodymium, dysprosium, and praseodymium, are vital for high-performance magnets used in wind turbines (especially direct-drive systems), electric motors, as well as some advanced energy storage technologies. Despite the adjective "rare", these elements are actually not scarce but are difficult to extract and refine, often leading to significant environmental degradation. For this reason, China dominates global REE production, controlling over 90% of the share [6].

2.3. *Cobalt*

Due to its stabilizing properties, cobalt is predominantly used in the cathodes of lithium-ion batteries, especially in those designed for long-duration energy storage such as electric vehicles and large storage applications. This raw material is mostly mined in the DRC, where production is frequently associated with ethical concerns like child labor and unsafe working conditions [5,12]. Moreover, the political instability of this region creates geopolitical risks that could impact global supply chains, making cobalt one of the most supply-constrained CRMs.

2.4. *Nickel*

Nickel is used to increase the energy density of batteries, improving their performance. It is an essential component in advanced battery chemistries like nickel–cobalt-aluminum and nickel–cobalt–manganese batteries. Nickel mining is energy intensive and has significant environmental impacts, particularly when extracted from laterite ores. The high demand for batteries and increasing stainless steel production creates high pressure on the supply side [6].

2.5. *Copper*

Copper is widely used as the main component for electrical wires and is thus essential for conducting electricity because its high conductivity makes it a preferred material for electrical infrastructure. Although copper is widely distributed, the extraction process is energy-intensive and subject to environmental concerns. Increased demand from the installation of worldwide power infrastructures related to wide penetration of renewable energy sources and loads electrification is putting pressure on copper supplies [6].

2.6. Graphite

Graphite is used in the anodes of lithium-ion batteries, contributing to their ability to store energy. Synthetic and natural graphite are both critical for battery production. Although synthetic graphite production is technically feasible, it is an energy-intensive process, while natural graphite is primarily mined in China (79% in 2021 [13]). This leads to concerns about the sustainability and availability of this material as demand increases for energy storage solutions.

2.7. Platinum Group Metals

Platinum group metals (PGMs) are a group of five elements. The group comprises platinum, palladium, ruthenium, iridium, osmium, and rhodium. Platinum and palladium are the key elements of this group, and they are considered essential for the catalytic processes in hydrogen fuel cells, which are becoming increasingly important for energy storage and transport in a zero-carbon economy. PGMs are primarily mined in South Africa and Russia, making them vulnerable to supply disruptions due to political instability or trade sanctions [14].

3. Energy Technologies Dependent on CRMs and SRMs

The dependence of energy technologies on CRMs and SRMs varies significantly, influencing their deployment potential at the commercial scale. In this section, a list of key energy technologies that include CRMs and SRMs is provided, along with a description of how these materials are used. In this way, this section complements the view already given in Section 2. The ChatGPT GenAI software tool was used to generate a preliminary list of energy technologies mostly impacted by their critical raw materials content with enriched content included. In addition, a recap of the information provided in Sections 2 and 3 is provided in Table 1 [15].

Table 1. Recap of the main energy technologies with an indication of the relative importance of CRM content (1 is low, 2 is medium, and 3 is high) [15].

	Cobalt	Copper	Lithium	Nickel	PGM	REE	Graphite
BESS	3	3	3	3	1	3	3
Wind	1	3	1	2	1	3	1
Solar PV	1	3	1	1	1	1	1
Hydrogen technologies	1	1	1	3	3	2	1
Superconductors	1	1	1	1	1	3	1
Geothermal	1	1	1	3	1	1	1
Nuclear	1	2	1	2	1	1	1
Electricity networks	1	3	1	1	1	1	1

3.1. Battery Energy Storage Systems (BESSs)

Battery energy storage systems store electrical energy generated in a specific period to release it in another period. These systems are usually employed to store energy from intermittent renewable sources (such as solar or wind) when production is high and the load is low and release it when the load is higher, stabilizing the energy supply. Another main use of batteries is storing the energy used to power electric motors in electric vehicles, which are crucial for decarbonizing the transportation sector.

Lithium-ion batteries, currently the most widely used energy storage system, rely on many CRMs such as lithium, cobalt, nickel, and manganese. In detail, lithium is the primary material in this technology, cobalt enhances the energy density and stability in cathodes, graphite is used as the anode material, and nickel and manganese are used in some selected lithium-based batteries.

Nickel–metal hydride batteries are also commonly used in all-electric and plug-in electric vehicles, while alternative storage technologies such as sodium–sulfur and flow batteries are less CRM-dependent but currently less efficient or more expensive [16].

3.2. Wind Turbines

Wind turbines convert kinetic energy from wind into electrical energy using rotor blades connected to a generator. Direct-drive wind turbines, which eliminate the need for a gearbox, depend heavily on rare earth elements for their high-performance magnets like neodymium and dysprosium, while offshore wind turbines require significant amounts of CRMs due to their higher material intensity [17–19]. In addition, nickel is used in the manufacturing of some wind turbine components, especially in offshore wind turbines due to its corrosion resistance, while copper is essential for electrical wiring [16,19].

3.3. Solar Photovoltaics (PVs)

Solar PV systems convert sunlight directly into electricity using semiconductor materials. Silicon-based PV panels are the dominant technology among PVs. They rely on large quantities of silicon metal, used for cell manufacturing, but also on copper and silver for the electrical contacts. Although silicon is quite abundant, it is considered critical due to its growing economic importance in the electronics industry. Emerging PV technologies such as thin films use gallium as alloys (copper indium gallium diselenide cells) [16,17,19].

3.4. Hydrogen Fuel Cells and Electrolyzers

Hydrogen-based fuel cells generate electricity exploiting the chemical reaction between hydrogen and oxygen to produce water. On the opposite, hydrogen electrolyzers use electricity to break the chemical bond between hydrogen and oxygen in water molecules with the aim of obtaining hydrogen. PGM, mainly platinum and palladium, are used as catalysts in proton-exchange membrane fuel cells, while polymer electrolyte membrane electrolyzers commonly contain platinum in cathodes and iridium in anodes [14,20]. Additionally, nickel is commonly included in electrodes for solid oxide fuel cells [15,16,21].

3.5. Superconductors

Superconductors are used in power grids to transfer electricity with minimal energy loss, which is essential for efficient energy distribution. However, they depend on rare earth elements, mainly yttrium and neodymium, commonly used in high-temperature superconductors to improve their performance [22].

3.6. Electricity Networks (Power Transmission and Distribution Systems)

Power lines transport and distribute electricity from generations to final consumers, involving a complex network of transformers, substations, and lines. As is commonly known, copper is extensively used in wiring and electrical connections due to its excellent conductivity.

4. Methodology

The literature review illustrated in this paper was performed according to the commonly adopted scientific approach described by the following steps [23]:

1. Formulate the questions addressed in this review.
2. Describe the methods to find and select the best of the research relevant to answer those questions.
3. Identify the methods to compare and synthesize the disparate studies found.

4.1. Question Addressed in the Review

Regarding the first step, this review aims to discuss the existing approaches to assess the CRM content in energy technologies. Furthermore, if some research gaps can be clearly identified, the authors aim to highlight these gaps and propose a method to overcome the current barriers.

4.2. Methods to Find and Select the Research

The research relevant to address the aspects discussed in the previous section was performed consulting two main generic databases, Scopus and Google Scholar, and two publisher-specific databases, IEEE Xplore and MDPI. These databases were selected since they are commonly considered among the most comprehensive and reliable in scientific studies. The investigation was performed in the third quarter of 2024 combining the following keywords: critical raw materials, energy planning, energy transition, risk, supply chain, and sustainable development. Furthermore, in order to expand the number of documents collected, their references were also investigated. The analysis focused on any kind of literature studies, including journal articles, conference papers, books or book chapters, and technical reports. Nevertheless, since the specific topic of this review has been addressed only recently by academics, the final set of studies is limited and is actually composed of 13 documents, recapped in Table 2.

Table 2. List of documents analyzed for this review paper with their main details.

Ref.	Year	Study Focus	Technology/Sector	CRMs/Materials Assessed
[24]	2014	CRM consumption estimation for wind energy systems	Wind turbines (onshore and offshore)	13 materials from CRM 2023 list
[25]	2015	Comparison of CRM content in different wind turbine technologies	Gearbox vs. direct-drive wind turbines	Neodymium, dysprosium, copper, strontium
[26]	2017	Material flows for multi-terawatt PV installations	Photovoltaics	Silicon, gallium
[27]	2019	Lithium market assessment in the energy transition	Electric vehicles	Lithium
[28]	2020	Optimal design under fossil fuel scarcity	Multi-energy systems	No direct CRM focus
[29]	2024	Energy hub optimization with hydrogen integration	Multi-energy systems	CRMs mentioned but not assessed
[30]	2024	CRM demand assessment for achieving climate targets	Energy system model (TIMES-VTT)	7 materials from CRM 2023 list
[31]	2022	Economic approach to optimal energy transition under mineral scarcity	General energy transition	No specific CRM mentioned
[4]	2023	Ranking CRMs using Euclidean distance	CRMs	All
[17]	2021	CRM indicators for energy system modeling	Wind and photovoltaic systems	8 materials from CRM 2023 list
[32]	2024	Security of energy systems based on CRM content	General energy systems	4 materials from CRM 2023 list + REE
[33]	2022	Comparative assessment of energy technologies	Renewable and non-renewable systems	30 materials from CRM 2020 list

4.3. Compare and Synthesize the Studies

The studies included in the final set of this bibliometric research may be categorized according to the following criteria:

1. CRM content assessment in specific energy technologies.
2. Optimal long-term planning/design studies on a regional/national scale considering CRMs.

3. Development of indicators for CRM content in energy technologies.

The final set of research studies was analyzed and compared with the others belonging to the same category. The comparison allowed for identifying common trends and research gaps.

5. Literature Review

In this section, the studies subject to review were collected and compared according to their category, highlighting research trends and gaps with respect to the main topic of this literature review.

5.1. CRM Content Assessment in Specific Energy Technologies

In this category of studies, the authors focused on assessing, for specific technologies, the quantity of precious materials, most of them being CRMs. The most notable examples are described here below.

In [24], current and future critical and valuable materials consumption in EU27 for wind energy systems were estimated based on the wind energy production scenarios and roadmap by country. The analysis was based on 13 materials, most of them being in the CRM 2023 list, comparing turbines with different rated sizes and distinguishing between onshore and offshore technologies, according to the bill of materials gathered from the Ecoinvent database.

A similar analysis was performed by Habib and Wenzel in [25], focusing on two different wind turbine technologies (gearbox against direct drive) rather than on the rated size and comparing the content of five raw materials, including two rare earth elements, namely neodymium and dysprosium, and two metals, copper and strontium, all of them being CRMs.

Davidsson and Höök, in [26], investigated the growth rates and material flows required to sustain long-term strategies on multi-terawatt photovoltaics installations, focusing on seven raw materials, including the CRMs silicon and gallium.

Although the studies belonging to this category assess and highlight the relevance of the criticality of some raw materials, they disregard including these considerations in energy systems modeling or planning. Furthermore, the references for the CRM content usually are manufacturer-confidential data, making this information hard to compare with other databases.

5.2. Optimal Long-Term Planning Studies

In this category of studies, the authors developed energy system models to assess final demands or installation of renewables according to national targets. The most notable examples are described here below.

Hache et al., in [27], propose a decision-making tool for assessing the future raw material market in the context of energy transition with particular attention to the electric vehicles market and lithium supply. A Times Integrated Assessment Model (TIAM-IFPEN) for the transportation sector was considered for the modeling, including a projection on the growth of lithium demand in the year 2030 for each vehicle category (it is not explicitly stated how the authors estimated the lithium content). This assessment model is constrained by the lithium supply, based on the extraction data from 2017. The interactions between the widespread of electric vehicles and the lithium supply were analyzed. The lithium sector and a detailed representation of the transportation sector were then implemented into the TIAM-IFPEN processes constituting the global energy system. Energy supply, demand, and market dynamics were modeled in order to represent energy dynamics over a long-term, multi-period time horizon at the local, national, multi-regional, or global level.

In [28], an optimal design under fossil fuel scarcity was considered. The authors formulated the problem considering the uncertainty about fossil fuel provision in a geographical island. Although this study does not deal with CRMs, a similar approach might be used for optimal energy planning considering strategic dependencies and critical raw materials scarcity. The study developed a copula-based interval full-infinite programming method for a multi-energy complementary power generation system under multiple uncertainties. According to the authors, this method has the following advantages with respect to the existing literature: (a) deals with different forms of uncertainty such as probability distribution or crisp interval values; (b) handles the joint distribution of any edge distribution; and (c) model the dynamic features of the system.

Massaro et al. [29] developed a multi-carrier multi-component energy hub optimization model to simulate regional-scale energy systems based on the inclusion of hydrogen as an energy carrier for electricity storage or mobility. The study includes several final demands, such as electricity, high- and low-temperature heating, cooling, freshwater, natural gas, and hydrogen, all of them being assessed under uncertain conditions through a probability distribution function. Furthermore, the inclusion of renewable energy systems was also included in the study, assessing the availability of solar radiation under uncertain conditions. Although CRMs are mentioned in the study, no objective function or constraint was included in the mathematical model.

In [30], the authors describe the use of the TIMES-VTT IAM energy system model to assess the demand for critical minerals and metals (including several CRMs) under the assumption of attaining the 1.5–2.0 °C mitigation targets. Their results show how the availability of cobalt and dysprosium for battery manufacturing might result in a constraint for the energy transition.

Lastly, Pommeret et al. address the problem of optimal energy transition under mineral scarcity from an economic point of view, focusing on climate policies but disregarding explicit references to any specific energy technology [31].

This category of studies is quite heterogeneous, both in terms of objective functions, i.e., the target of the optimization, and in terms of the variables to be evaluated. Nevertheless, the approaches here appear promising for an integrated optimization model for long-term energy systems planning aimed at reducing critical raw materials supply. Regarding the treatment of CRM content, these studies generically base their simulations on projections on technologies installation or on economic optimizations and then use average data from the literature on the CRMs' intensity (kg/MW) for the various technologies resulting from the simulations [15,19].

5.3. Indicators for CRM Content

A part of the literature is also focused on identifying an indicator for comparing technologies based on the content of CRMs. This kind of study starts from the calculation method developed by the European Commission to assess whether a raw material is critical, involving two main indicators named "economic importance" (EI) and "supply risk" (SR), both of them ranging between 0 and 10. According to the current methodology, these indicators are evaluated for several candidate raw materials (54 in 2014, 78 in 2017, 83 in 2020, and 87 in 2023 [34]), and an arbitrary threshold was set for these two indicators. Starting from 2017, a raw material is critical when $EI \geq 2.8$ and $SR \geq 1$ [35]. Nevertheless, according to this method, there is no material that is "more critical" than another, although a quantitative value is assigned to both indicators. This aspect makes it difficult to compare CRMs among them with the aim of selecting a technology with a "low criticality" index.

For these reasons, the authors in [4] combined the two indicators EI and SR using the Euclidean distance method, suggesting that other methods might also be used in future studies. In this way, a ranking of the first five more critical CRMs was presented in the study. Furthermore, the authors noticed that the overall criticality of CRMs increased from 2017 to 2020 and from 2020 to 2023.

Using a different approach, Talens Peiró et al. proposed using a combination of the four indicators easily usable in energy systems modeling as a support to decision-makers [17]. These indicators combine the approach used by the European Commission in terms of supply risk with more classical data from the Life Cycle Assessment (LCA) method. In detail, the set of indicators chosen by the authors is made up of (a) Global Warming Potential, indicating the equivalent carbon emissions over the lifecycle, (b) the Cumulative Energy Demand, quantifying the equivalent use of energy resources over the lifecycle, (c) the supply risk, and (d) the EOL-RIR rate, indicating the percentage rate of recyclability of the raw materials. The study also employs these indicators to compare four wind turbine technologies and four photovoltaic technologies.

Vai et al., in [32], assess the security of energy systems in terms of CRM content and evaluate seven indicators. These indicators calculate the supply risk and the reliability of the energy system both in terms of technology capacity and technology activity. Nevertheless, due to the high number of indicators, the authors are forced to aggregate these indicators in a synthetic index in order to compare the results.

Martin et al. also assessed supply risk together with other indicators for energy technologies, comparing the performance of several renewable and non-renewable electricity production systems [33].

These studies provide significant results in terms of comparison and ranking among CRMs, helping in the development of a methodology for identifying an energy technology mix with a low content of CRMs. As a critical comment, on the one hand, with reference to [17], it is evident that the economic aspects cannot be disregarded in energy systems modeling and planning. On the other hand, with reference to [32], the use of a limited set of indicators is recommended since the use of many indicators causes two well-known issues in the literature on multi-criteria analysis: (a) the risk that each material can outperform the others according to one criterion, thus making all the candidates good and, consequently (b) the aggregation of the indicators in an index, causing loss of precious details and information related to the main indicators.

6. Discussion

According to the main approaches illustrated in the previous section, in order to develop a method to plan a technology mix in regional or national energy systems using an optimization model, the following main criteria should be considered:

1. Regarding the quantification of CRM content in each technology, the bill of materials might be gathered from reliable sources provided by third parties such as IEA reports [15] or life cycle inventory (LCI) databases, as already performed in [33,36]. On this topic, there are several existing LCI databases such as Ecoinvent [37], the Environmental Footprint (EF) database of the European Commission [38], and EXIOBASE [39]. The most common LCI database, Ecoinvent, covers more than 20,000 datasets for various processes. These datasets contain the inputs (and outputs) of the products/processes, including the raw materials used for energy technologies. According to a recent announcement on their website, the next version of Ecoinvent will update several datasets related to energy and power system planning such as the consumption mix of crude oil petroleum, methane emissions from coal mining, new electricity mixes, Li-ion chemistries, Na-ion electrolytes, and new data on metals [40]. While Ecoinvent provides global data, the EF database focuses on EU products and processes, for example, the Product Environmental Footprint (PEF) data for some representative

products and the organizational environmental footprint [41].
The first two databases consider the technologies in detail from a bottom-up approach, i.e., the materials, energy, and other inputs to manufacture the technologies, and the emissions and other outputs being originated from these processes. In contrast, EXIOBASE is an example of an LCI database considering the whole economy from the top-down approach. This database estimates the emissions, materials, and resource extraction (including 662 materials and resources) by industries at the global scale [42]. Although it is not suitable for the optimization of one sector, it is useful to simultaneously examine several energy production and consumption sectors such as transportation, electric power, chemistry, etc. In addition, it will be interesting to combine several databases to obtain the most effective results.

2. Regarding the selection of CRMs, a preliminary ranking among the materials should be introduced in order to focus on the most critical raw materials and technologies.
3. Regarding energy system modeling, a comprehensive multi-carrier, multi-component model should be developed, including reliable methods to assess the uncertainty related to the various aspects of real systems.
4. Regarding the comparison among different technologies, a proper list of indicators should be selected, including economic, energy, and environmental aspects together with a CRM assessment, limiting the list to no more than four or five indicators to avoid the need for aggregation in synthetic indexes. Additionally, since both EI and SR are concerned with macroeconomic aspects, the CRM content might be limited using constraints.
5. Regarding the inclusion of a specific technology, since there exist several options for each category of technology (e.g., silicon-based or copper indium gallium diselenide cells for photovoltaic systems), a preliminary comparison might be introduced.
6. Regarding the size of the energy system, to properly consider macroeconomic aspects deriving from the selection of a specific material/technology, a regional or national scale should be selected. Nevertheless, this should also be accompanied by further analyses such as consequential LCA studies.

The integration of such methodologies into the energy system planning model could greatly increase awareness about critical raw materials provision and supply chain vulnerabilities and guide policies that maximize the compatibility between resource availability and decarbonization goals achievement. The following Figure 2 provides a recap of the main findings discussed in this section.

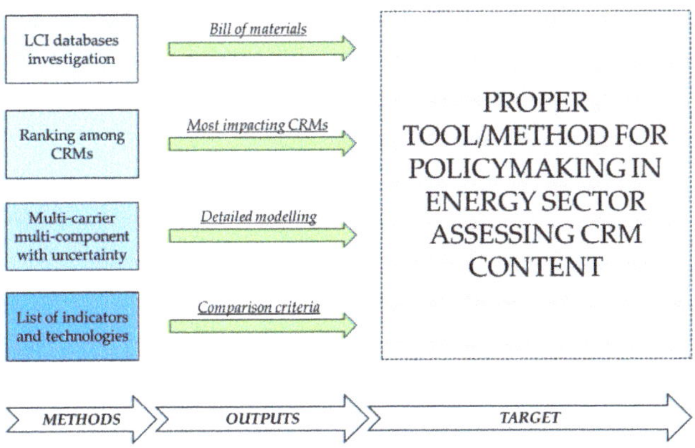

Figure 2. Flow diagram of the proposed methodology.

7. Conclusions

This paper illustrates the most common approaches adopted in the existing scientific literature about the CRM content in energy technologies. First, an overview of the most common CRMs employed in the energy sector was provided, followed by a description of the technologies mostly impacted by the CRMs' supply risks. Subsequently, the main criteria adopted for the present review study were illustrated, and the analysis of the available literature according to categories was presented, identifying main ideas and research gaps. In the discussion section, the authors summed up the considerations emerging from the literature review and proposed an integrated approach for the optimal planning of large-scale energy systems considering the CRM supply risk as a constraint for the optimization problem.

According to the outcomes of this review study, we further recap some recommendations for the policymakers to include a CRM assessment in energy system planning. In detail, assessing CRMs in energy technology policies will support the sustainable development of the energy system and ensure regional and national energy security, with considerations of material shortage, as well as the long-term economic feasibility of the system. At the same time, it is necessary to link the energy system planning of a country to those of its energy or input supply partners. The transboundary energy system will partly tackle the problem of material shortage, especially the limited deposits of critical raw materials. For energy practitioners, it is suggested that the traditional research, development, innovation, and investment in the supply–demand balance of energy systems need to extend to include several factors of environmental aspects such as carbon footprints and critical raw materials. These factors not only impact the environmental profile but also contribute to the economic sustainability of the energy projects through carbon credits and the investment costs of energy technologies.

Future research should focus on developing proper indicators or indexes and collecting data for the comparison of energy technologies in terms of their CRM content without disregarding more common approaches such as cost reduction, primary energy saving, and carbon emissions abatement, useful also to help develop advanced tools for assisting policymakers in making proper aware decisions.

Author Contributions: Conceptualization, M.C., M.L.D.S., F.M., S.L., F.M. and E.R.S.; methodology, all authors; investigation, F.M., L.Q.L. and E.R.S.; writing—original draft preparation, F.M., L.Q.L. and G.S.; writing—review and editing, F.M., L.Q.L. and E.R.S.; supervision, M.C., M.L.D.S., S.L. and E.R.S. All authors have read and agreed to the published version of the manuscript.

Funding: This research received no external funding.

Acknowledgments: ChatGPT GenAI (ChatGPT 16 May 2024 Version, OpenAI) was used in this paper. All the information was checked according to the references included in this paper.

Conflicts of Interest: The authors declare no conflicts of interest.

Abbreviations

Abbreviation	Meaning
BESS	Battery energy storage system
CRM	Critical raw material
DRC	Democratic Republic of Congo
EI	Economic importance
EF	Environmental footprint
EU	European Union

LCA	Life Cycle Assessment
LCI	Life cycle inventory
PEF	Product Environmental Footprint
PGMs	Platinum group metals
REEs	Rare earth elements
SR	Supply risk
SRM	Strategic raw material

References

1. European Council. *Council Directive 92/75/EEC of 22 September 1992 on the Indication by Labelling and Standard Product Information of the Consumption of Energy and Other Resources by Household Appliances*; European Council: Brussels, Belgium, 1992.
2. European Union European Critical Raw Materials Act 2023. Available online: https://single-market-economy.ec.europa.eu/sectors/raw-materials/areas-specific-interest/critical-raw-materials/critical-raw-materials-act_en (accessed on 26 December 2024).
3. European Commission. *Proposal for a Regulation of the European Parliament and of the Council Establishing a Framework for Ensuring a Secure and Sustainable Supply of Critical Raw Materials and Amending Regulations (EU) 168/2013, (EU) 2018/858, 2018/1724 and (EU) 2019/1020*; European Commission: Brussels, Belgium, 2023.
4. Feichtinger, G.; Posch, W. Evaluation of European Critical Raw Material Assessments under Energy Transition Considerations. In Proceedings of the 3rd EURECA-PRO Conference on Responsible Consumption and Production 2023, Chania, Greece, 26–29 September 2023.
5. United Nations Economic Commission for Europe. *Critical Minerals for the Sustainable Energy Transition. A Guidebook to Support Intergenerational Action*; United Nations Economic Commission for Europe: Switzerland, Geneva, 2024.
6. Gielen, D. *Critical Minerals for the Energy Transition*; IEA: Abu Dhabi, United Arab Emirates, 2021.
7. Overland, I. The Geopolitics of Renewable Energy: Debunking Four Emerging Myths. *Energy Res. Soc. Sci.* **2019**, *49*, 36–40. [CrossRef]
8. Liu, W.; Agusdinata, D.B.; Myint, S.W. Spatiotemporal Patterns of Lithium Mining and Environmental Degradation in the Atacama Salt Flat, Chile. *Int. J. Appl. Earth Obs. Geoinf.* **2019**, *80*, 145–156. [CrossRef]
9. Riva Sanseverino, E.; Luu, L.Q. Critical Raw Materials and Supply Chain Disruption in the Energy Transition. *Energies* **2022**, *15*, 5992. [CrossRef]
10. Aljabery, A.A.M.; Mehrjerdi, H.; Mahdavi, S.; Hemmati, R. Multi Carrier Energy Systems and Energy Hubs: Comprehensive Review, Survey and Recommendations. *Int. J. Hydrog. Energy* **2021**, *46*, 23795–23814. [CrossRef]
11. European Geological Data Infrastructure (EGDI) Critical Raw Materials Mines Map in Europe. Available online: https://maps.europe-geology.eu/#baslay=baseMapGEUS&extent=-406679.70556335896,734556.7038816502,9408915.50311318,5463424.18722842&layers=egdi_mineraloccurr_critical_raw_materials_2023&filter_0=crm_2023.multi=&mine_status.multi= (accessed on 3 December 2024).
12. Berthet, E.; Lavalley, J.; Anquetil-Deck, C.; Ballesteros, F.; Stadler, K.; Soytas, U.; Hauschild, M.; Laurent, A. Assessing the Social and Environmental Impacts of Critical Mineral Supply Chains for the Energy Transition in Europe. *Glob. Environ. Change* **2024**, *86*, 102841. [CrossRef]
13. IRENA; NUPI. *Constructing a Ranking of Critical Materials for the Global Energy Transition*; IRENA: Abu Dhabi, United Arab Emirates, 2024.
14. Kamran, M.; Raugei, M.; Hutchinson, A. Critical Elements for a Successful Energy Transition: A Systematic Review. *Renew. Sustain. Energy Transit.* **2023**, *4*, 100068. [CrossRef]
15. IEA. *The Role of Critical Minerals in Clean Energy Transitions*; IEA: Paris, France, 2021.
16. Bobba, S.; Carrara, S.; Huisman, J.; Mathieux, F.; Pavel, C. *Critical Raw Materials for Strategic Technologies and Sectors in the EU—A Foresight Study*; European Commission: Brussels, Belgium, 2020.
17. Talens Peiró, L.; Martin, N.; Villalba Méndez, G.; Madrid-López, C. Integration of Raw Materials Indicators of Energy Technologies into Energy System Models. *Appl. Energy* **2022**, *307*, 118150. [CrossRef]
18. Brumme, A. *Wind Energy Deployment and the Relevance of Rare Earths*; Springer Fachmedien Wiesbaden: Wiesbaden, Germany, 2014; ISBN 978-3-658-04912-6.
19. Alves Dias, P.; Pavel, C.; Plazzotta, B.; Carrara, S. *Raw Materials Demand for Wind and Solar PV Technologies in the Transition towards a Decarbonised Energy System*; Publications Office: Luxembourg, 2020.
20. Minke, C.; Suermann, M.; Bensmann, B.; Hanke-Rauschenbach, R. Is Iridium Demand a Potential Bottleneck in the Realization of Large-Scale PEM Water Electrolysis? *Int. J. Hydrog. Energy* **2021**, *46*, 23581–23590. [CrossRef]

21. Cheng, T.-L.; Lei, Y.; Chen, Y.; Fan, Y.; Abernathy, H.; Song, X.; Wen, Y.-H. Oxidation of Nickel in Solid Oxide Cells during Electrochemical Operation: Experimental Evidence, Theoretical Analysis, and an Alternative Hypothesis on the Nickel Migration. *J. Power Sources* **2023**, *569*, 232991. [CrossRef]
22. Commission of the European Communities (EC). *Tackling the Challenges in Commodity Markets and on Raw Materials*; European Commission: Brussels, Belgium, 2011.
23. Glasziou, P. How to Write a Review. In *How to Write a Paper*; Wiley: Hoboken, NJ, USA, 2012; pp. 89–97.
24. Kim, J.; Guillaume, B.; Chung, J.; Hwang, Y. Critical and Precious Materials Consumption and Requirement in Wind Energy System in the EU 27. *Appl. Energy* **2015**, *139*, 327–334. [CrossRef]
25. Habib, K.; Wenzel, H. Reviewing Resource Criticality Assessment from a Dynamic and Technology Specific Perspective—Using the Case of Direct-Drive Wind Turbines. *J. Clean. Prod.* **2016**, *112*, 3852–3863. [CrossRef]
26. Davidsson, S.; Höök, M. Material Requirements and Availability for Multi-Terawatt Deployment of Photovoltaics. *Energy Policy* **2017**, *108*, 574–582. [CrossRef]
27. Hache, E.; Seck, G.S.; Simoen, M.; Bonnet, C.; Carcanague, S. Critical Raw Materials and Transportation Sector Electrification: A Detailed Bottom-up Analysis in World Transport. *Appl. Energy* **2019**, *240*, 6–25. [CrossRef]
28. Zhu, Y.; Tong, Q.; Yan, X.; Liu, Y.; Zhang, J.; Li, Y.; Huang, G. Optimal Design of Multi-Energy Complementary Power Generation System Considering Fossil Energy Scarcity Coefficient under Uncertainty. *J. Clean Prod.* **2020**, *274*, 122732. [CrossRef]
29. Massaro, F.; Di Silvestre, M.L.; Ferraro, M.; Montana, F.; Riva Sanseverino, E.; Ruffino, S. Energy Hub Model for the Massive Adoption of Hydrogen in Power Systems. *Energies* **2024**, *17*, 4422. [CrossRef]
30. Koljonen, T.; Lehtilä, A.; Kiviranta, K.; Koponen, K.; Similä, L. Modelling of Demands of Selected Minerals and Metals in Clean Energy Transition with 1.5–2.0 °C Mitigation Targets. In *Aligning the Energy Transition with the Sustainable Development Goals: Key Insights from Energy System Modelling*; Springer Nature: Cham, Switzerland, 2024; pp. 225–245.
31. Pommeret, A.; Ricci, F.; Schubert, K. Critical Raw Materials for the Energy Transition. *Eur. Econ. Rev.* **2022**, *141*, 103991. [CrossRef]
32. Vai, A.; Colucci, G.; Nicoli, M.; Savoldi, L. A Comprehensive Metric to Assess the Security of Future Energy Systems Through Energy System Optimization Models. In Proceedings of the 10th Applied Energy Symposium: Low Carbon Cities & Urban Energy Systems (CUE2024), Shenzhen, China, 11–13 May 2024.
33. Martin, N.; Madrid-López, C.; Villalba-Méndez, G.; Talens-Peiró, L. New Techniques for Assessing Critical Raw Material Aspects in Energy and Other Technologies. *Environ. Sci. Technol.* **2022**, *56*, 17236–17245. [CrossRef] [PubMed]
34. European Commission: Directorate-General for Internal Market, Industry, Entrepreneurship and SMEs. *Study on the Critical Raw Materials for the EU 2023—Final Report*; Publications Office of the European Union: Luxembourg, 2023.
35. European Commission: Directorate-General for Internal Market, Industry, Entrepreneurship and SMEs. *Study on the Review of the List of Critical Raw Materials—Final Report*; Publications Office: Luxembourg, 2017.
36. Martin, N.; Talens-Peiró, L.; Villalba-Méndez, G.; Nebot-Medina, R.; Madrid-López, C. An Energy Future beyond Climate Neutrality: Comprehensive Evaluations of Transition Pathways. *Appl. Energy* **2023**, *331*, 120366. [CrossRef]
37. Ecoinvent Ecoinvent Database Web Page. Available online: https://ecoinvent.org/ (accessed on 29 October 2024).
38. European Commission Environmental Footprint Database Web Page. Available online: https://eplca.jrc.ec.europa.eu/LCDN/contactListEF.html (accessed on 29 October 2024).
39. EXIOBASE Consortium EXIOBASE Database Web Page. Available online: https://www.exiobase.eu/ (accessed on 29 October 2024).
40. Ecoinvent Ecoinvent Blog Web Page. Available online: https://ecoinvent.org/blog/coming-soon-ecoinvent-version-3-11/ (accessed on 29 October 2024).
41. European Commission EPLCA—About Us Web Page. Available online: https://eplca.jrc.ec.europa.eu/aboutUs.html (accessed on 29 October 2024).
42. EXIOBASE Consortium EXIOBASE—About Us Web Page. Available online: https://www.exiobase.eu/index.php/about-exiobase (accessed on 29 October 2024).

energies

MDPI

Article

Comparative Analysis of Smart Building Solutions in Europe: Technological Advancements and Market Strategies

Negar Mohtashami *, Nils Sauer, Rita Streblow and Dirk Müller

Institute for Energy Efficient Buildings and Indoor Climate, E.ON Energy Research Center, RWTH Aachen University, 52074 Aachen, Germany; nils.sauer@rwth-aachen.de (N.S.); rstreblow@eonerc.rwth-aachen.de (R.S.); dmueller@eonerc.rwth-aachen.de (D.M.)
* Correspondence: negar.mohtashami@eonerc.rwth-aachen.de

Abstract: This paper provides a comprehensive comparative analysis of smart building solution providers within Europe, emphasizing the technological advancements and market strategies employed by companies selected for the study. As energy efficiency becomes a critical focus due to rising global energy demands and climate change concerns, smart building technologies have emerged as pivotal in optimizing energy use and enhancing occupant comfort. This study examines 19 products from 15 prominent manufacturers, categorized into six product categories: smart thermostats, smart valves, HVAC control, data acquisition and energy management software, smart home ecosystems, and home energy management systems. Using a comparative assessment matrix and SWOT analysis, the paper evaluates these products across five key areas: service impacts, market penetration, investment topics, business models, and value propositions. Findings highlight a strong focus of manufacturers in energy efficiency and comfort services, while identifying opportunities for improvement in energy flexibility and health integration. This analysis aims to guide stakeholders in strategic planning and decision-making, offering insights into the current and future landscape of the smart building solutions market.

Keywords: building energy management systems (BEMSs); artificial intelligence (AI); Internet of Things (IoT); smart technologies

1. Introduction

With global energy demand on the rise and the pressing need to address climate change, the deployment of smart building solutions has emerged as the cornerstone of shaping the future of the built environment toward more sustainable living spaces [1]. The building sector is responsible for a large proportion of greenhouse gas emissions; in the European Union, it constitutes 36% of greenhouse gas emissions [2]. Therefore, joint strategies are being deployed globally to curtail energy utilization within the building sector, notably reflected in the European Climate Law, which prescribes a reduction in net greenhouse gas emissions by 55% until 2030 compared to 1990 levels [3]. The drive towards electrification, reflected by a broader deployment of electric heat pumps in the building sector [4], plays a crucial role in diminishing the demand for fossil fuels, but transitioning to renewable energy sources for electricity supply also requires a more sophisticated and load-optimized management of the increasingly volatile power grids [5]. This demand for grid-stabilizing technologies and efficient Building Energy Management Systems (BEMSs) can be expected to rise in the upcoming years, since the share of electricity in total worldwide final energy consumption is estimated to increase from currently 20% to more than 50% in the Net Zero Emissions projection of the World Energy Outlook by the year 2050 [6]. At the same time, the pursuit of energy optimization must not compromise indoor comfort. Estimates suggest we spend 80–90% of our lifetime indoors, making thermal comfort and air quality significant factors influencing our quality of life and health [7,8].

Smart building solutions can be a way to simultaneously address comfort requirements and the demand for greener, grid-friendly buildings. Researchers in the past decade have discussed how smart buildings and grids can increase energy efficiency without compromising user comfort, using intelligent control techniques to optimize BEMSs [9–11]. Smart building solutions hereby refers to the integration of data-driven and needs-based technologies, oftentimes utilizing real-time data from various sensors and user inputs to optimize building operations and resources according to the specific demands of the building occupants and external conditions [12]. Characterized by the integration of Internet of Things (IoT) devices, advanced energy management algorithms, and the intelligent linkage of building controls to occupancy and gird data, these technologies offer unprecedented opportunities to enhance building efficiency, occupant comfort, and environmental stewardship [13,14]. In Germany, for instance, digital building technologies are estimated to enable a reduction of approximately 30% of the national emission targets within the building sector [15]. The advent of data-driven technologies therefore unlocks opportunities to boost human health and productivity while advancing energy efficiency within the built environment [16].

Multinational technology conglomerates, established industrial manufacturers, renewable energy corporations, artificial intelligence startups, and smart home technology innovators are leading the vanguard of this technological revolution across Europe. Each manufacturer brings a unique set of innovations, strategies, and visions to the table, contributing to an ecosystem that is both diverse and rich in potential. This paper aims to delve into a comparative analysis of leading manufacturers and providers of smart building solutions, unraveling the current stand, latest developments, and future trends of this dynamic market. The core of this paper's analysis lies in the comprehensive evaluation of different product categories, examining their technological offerings, impacts on different service areas (e.g., energy efficiency, comfort, or flexibility), investment topics, business models, and value propositions of their solutions.

The methodology employed in assessing these product categories starts with a selection of prominent manufacturers and products from the Smart Building Solutions market for the subsequent analysis. After a categorization of the selected products into certain product categories, a Comparative Assessment Matrix (CAM) is developed as an evaluation framework, partly leaning on the Smart Readiness Indicator (SRI) methodology [17]. Information gathered through detailed research on manufacturers' websites, technical sheets, online newspaper articles, blog posts, and LinkedIn profiles is analyzed and synthesized into the CAM and enhanced through color coding to visually detect weak spots and saturated areas of the market. The CAM is used to identify the current and future developments and trends concerning business models, investment topics, and technological innovations. Finally, a SWOT analysis of the Smart Building Solutions market is carried out, shedding light on the current strengths and weaknesses of the market, opportunities to elaborate, and possible threats to tackle. This approach aims to offer a practical and accessible perspective on the current landscape of the Smart Building Solutions market, helping stakeholders in their decision-making and strategic planning.

2. Selection of Manufacturers

To analyze the smart building solutions market, a selection of manufacturers and products is first made, which are then rated and compared with each other in terms of their strengths, weaknesses, and functionalities. This section will explain the selection of smart home manufacturers and their products for comparative analysis. After explaining the methodology by which manufacturers are selected, an overview of the selected manufacturers will be given.

2.1. Methodology for Manufacturer Selection

The methodology for the selection of manufacturers and products is illustrated in the top area of Figure 1. Two main criteria served as the requirements for the selection of manufacturers and products. The selected manufacturers should represent prominent manufacturers of the smart building solutions market concerning sales volume and international prominence. Since many manufacturers in the smart home domain offer a variety of products, it was decided to restrict the considered products to those which contribute notably on the energy efficiency and indoor climate comfort of a building; thus, security-related products are not considered in this analysis since they do not have an impact on these areas. The manufacturers and products were then identified by means of an internet search and interviews with experts from the industry.

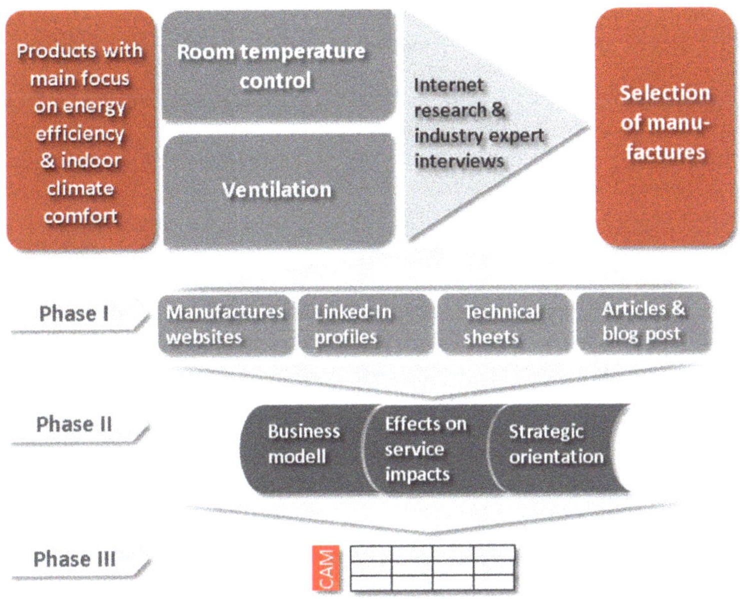

Figure 1. Methodology for selection and analysis of manufacturers and products.

2.2. Overview of Selected Manufacturers and Classification into Product Categories

The selected manufacturers and products are shown in Table 1. In total, 15 manufacturers and 19 different products were selected. Most of the products can be classified into one of three product categories: "smart thermostats", "smart valves", and "HVAC control". Products which cannot be classified in one of these categories are listed in the column "others". The columns "HVAC control" and "others" only list the manufacturers since most of them offer just one product.

Smart thermostats do not get installed directly at the radiator but are rather spatially separated from it, thus being able to operate centrally for multiple rooms as well as for each room individually. These thermostats influence various parameters, such as radiator valve positions, and can provide crucial data for building automation systems. Often, these thermostats are either integrated into such systems or sold in combination with them. Additionally, smart room thermostats can utilize weather data, room data (such as open windows or heating performance), and occupancy data for energy optimization. These data can be used for adjustments of other parameters in the heating system, such as the supply temperature or pump speed. To enhance comfort, smart room thermostats offer various functionalities, such as programmable schedules or self-learning heating behavior that adapts to occupancy patterns. These features contribute not only to increased comfort but also to efficient energy use [1,18].

Table 1. Selected products and manufacturers.

Smart Thermostats		Smart Valves		HVAC Control	Others
Product	**Manufacturer**	**Product**	**Manufacturer**	**Manufacturer**	**Manufacturer**
Smart Thermostat V3+	Tado	Smart Radiator Thermostat V3+	Tado	BrainBox AI	METRON
Smart Modulating Thermostat	Netatmo	Smart Radiator Valves	Netatmo	BeeBryte	Shelly
Room Thermostat II 230 V	Bosch	Radiator thermostat II	Bosch	DABBEL	Bosch (Energiemanager)
Nest Learning Thermostat	Google			R8 Technologies	
EasyControl	Bosch			MeteoViva	
Homix	Enel X			Evogy	
KONO Smart Thermostat	Lux Products				

Smart valves function similarly to smart room thermostats but are located directly at the radiator. Unlike room thermostats, the temperature sensor and actuator (valve) are not spatially separated. The functionalities of smart radiator thermostats often include similar comfort and energy-saving features as their central counterparts, such as programmable heating times and self-learning behavior. These thermostats offer a simple retrofit solution that optimizes energy use without requiring extensive modifications to the heating system [19].

Heating, Ventilation, and Air Conditioning (HVAC) control systems are products designed to regulate air-based systems and can manage components such as air heaters, air coolers, and airflow (fan) volumes. These systems frequently incorporate functionalities that utilize weather data, occupancy data, air quality data, and temperature data to ensure optimal indoor climate conditions. Many of the HVAC systems considered in this analysis make use of AI technologies, enabling self-learning behavior that adapts to the specific needs and habits of users [20]. These AI-driven systems continuously adjust ventilation parameters based on the collected data, achieving energy savings of up to 40% [21–23]. By learning from patterns in environmental and usage data, HVAC systems can predict and respond to changes more effectively, minimizing energy consumption while maintaining a pleasant indoor environment. HVAC control systems are predominantly used in non-residential buildings [24]. While some manufacturers can be classified as developers of data acquisition and management software, some others could be classified as manufacturers of a Smart Home Ecosystem (SHE), since their portfolio can be seen as a construction kit for smart home applications rather than a range of specific smart home products. The sixth category is Home Energy Management Systems (HEMS), which, in this study, is seen as a digital system that monitors and controls energy within a household. In total, the 19 products are categorized into six different product categories, with a strong number in the three categories of smart thermostats, smart valves, and HVAC control (16 of 19 products) and only one product in each of the other three categories of data acquisition and energy management software, SHE, and HEMS.

3. Comparative Assessment Methodology

To provide the CAM framework for evaluation of the smart technologies, products are first researched and gathered in the first phase. In the second phase, they are classified into a product category (see Section 2.2) and, consequently, in the third phase, they are rated and evaluated in comparison to other products from the same product category. Figure 1 shows the steps of the methodology to derive CAM assessments. The CAM covering these five areas is explained in more detail in the subsequent sections.

3.1. *Development of Assessment Matrix*

The evaluation criteria encompass five key areas. The service impacts are assessed across seven impact categories, which are derived from the SRI methodology: energy efficiency, maintenance and fault prediction, comfort, convenience, health and well-being, information to occupants, and energy flexibility and storage. Each product is rated on a scale from zero to four in these categories according to the abilities of each product. A clear definition of what each level of service impact means for each category is available in the SRI assessment sheets.

Categorizing product usage is assessed by building type—residential, commercial, or industrial. Investment topics provide insights into each manufacturer's research, marketing, and investment strategies, allowing for predictions of future market developments. The business model analysis focuses on customer targeting and profit generation methods, while the value proposition section highlights unique marketing aspects that differentiate products from their competitors. The evaluation of each criterion is documented in a table, forming the CAM in Figure 2.

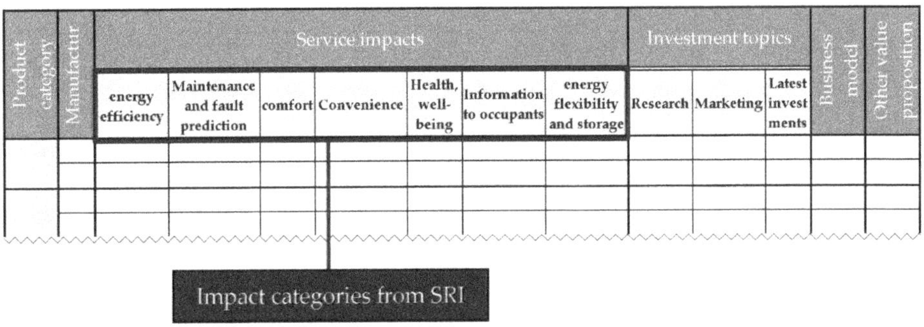

Figure 2. Comparative assessment matrix (CAM).

3.2. *Phase I: Data Collection*

The foundational phase involves a careful gathering of relevant data, essential for conducting a holistic comparative analysis. This phase comprises research on each of the selected manufacturers and their products using manufacturers' websites, technical sheets, online newspaper articles, blog posts, and LinkedIn profiles as the main sources. To ensure the validity and reliability of the data, we prioritized technical sheets and official manufacturer websites as primary sources, given their detailed product specifications and presumed accuracy. Secondary sources, such as online articles, blog posts, and LinkedIn profiles, were used to complement this information but were critically assessed for credibility through cross-verification with more reliable sources. In cases of conflicting information, we relied on data provided directly by manufacturers in their technical sheets and websites, assuming these values were derived through standardized testing processes. However, the inherent risk of relying solely on manufacturer claims is acknowledged as a limitation of the study. The search concentrated on information from which conclusions for the established areas of the CAM could be drawn later in the process, e.g., information regarding the abilities and functionalities of products.

3.3. Phase II: Data Analysis

In the second phase, the collected data are systematically analyzed and filtered to extract relevant information for the key areas. The information was then methodically organized into the CAM, aligning manufacturers and products against the defined criteria. Using the gathered data and analyzing a product's functionalities, the impacts of each product on the seven service impacts were determined.

Market penetration was measured by searching the collected data for the number of countries where a product is available and the various building types where the product is used. This analysis is supposed to provide insights into the geographical distribution and application breadth of the products.

The evaluation of the key area "investment topics" focused on manufacturers' current research themes, marketing campaigns, and recent R&D investments. This included an analysis of the information regarding research into specific technologies and functionalities, joint product developments with other market players, marketing strategies that differentiate products from competitors, and recent investments such as acquisitions and mergers.

To evaluate the business models, the collected data were analyzed for indications of the addressed customer types (e.g., private or business customers) and the sales models (e.g., fixed prices, contracting, subscription models). The examination of the investment topics and business models is intended to provide insights into the current trends and developments in the smart building solutions market.

Information that cannot be assigned to any of the previous key areas but is still regarded as relevant (e.g., unique innovative approaches or unique selling points) is categorized under the key area "other value proposition". This serves to retain potentially valuable information.

3.4. Phase III: Synthesis and Visualization

In the final phase of the analysis, we enhanced the CAM by applying a scoring system to visually articulate the data synthesized from the literature research. Each smart building solution is assigned a score from 0 to 4 in each of the seven service impact areas derived from the SRI (key area 1) according to the effect a product has on it. A score of 0, reflected by red color, means 'No Service Impact', indicating no measures for energy reporting or management are present. A score of 1, highlighted by yellow, 'Reporting on Current Consumption on Building Level', signifies basic reporting features without real-time analytics. Advancing to a score of 2 with bright green color, 'Real-time Feedback or Benchmarking on Building Level', solutions provide current energy data and comparative benchmarks at a building-wide scale. A score of 3, the green color, depicts 'Real-time Feedback or Benchmarking on Appliance Level', offering granular insights into the energy usage of individual appliances. The highest score of 4, which is shown with a dark green color, underpins 'Real-time Feedback or Benchmarking on Appliance Level with Automated Personalized Recommendations', and is attributed to solutions that deliver the most sophisticated level of service, with real-time, appliance-specific data analysis and AI-driven recommendations. The scoring criteria are summarized in Table 2.

Table 2. Color map assignments to individual service impact levels.

No Service Impact	0
reporting on current consumption on building level	1
real-time feedback or benchmarking on building level	2
real-time feedback or benchmarking on appliance level	3
real-time feedback or benchmarking on appliance level with automated personalized recommendations	4

The goal of this visualization is the intuitive visual detection of saturated areas as well as opportunities to elaborate in the smart building solutions market. This step was designed to make the comparative analysis accessible and informative for stakeholders, enabling informed decision-making and strategic planning in the smart building sector.

4. Results

This section will dive into the key results from the CAM. A detailed overview of the scores each product category reached in the seven service impact areas of the first key area "service impacts" will be given in Section 4.1. Subsequently, Section 4.2 summarizes key findings from the other four key areas. Section 4.3 deals with cross-key area results derived from the combination of multiple key area findings.

4.1. Key Area "Service Impacts"

Figure 3 displays the scores of each product category across the seven service impact areas derived from the SRI. In the CAM, high scores in the areas of energy efficiency, comfort, and convenience can be observed, indicating an advanced state of the market concerning these topics. All companies promise substantial reductions in energy consumption, with some manufacturers promising significant efficiency, with an up-to 40% reduction in energy demand [21–23]. By reducing energy consumption and optimizing building operations, these technologies contribute significantly to sustainability goals and offer profound environmental benefits. The CAM elucidates the potential for smart building technologies to play a pivotal role in mitigating climate change, being in line with current estimations and research [15,25,26].

On the contrary, the area of energy flexibility and storage has the lowest scores of all areas, with more than half of the products reaching a score of 0. This gap may be attributed to high costs of energy storage technologies, limited interoperability with existing building systems, and regulatory uncertainties, which collectively discourage widespread adoption. Future research could focus on modular, cost-effective storage solutions, enhanced demand-side management, and standardization frameworks to support broader integration. The area of health, well-being, and accessibility marks the second weak spot of the smart building solutions market, with around 30% of the products providing no service impacts in this area (score 0). Challenges in integrating these functionalities include the complexity of addressing diverse health needs and the absence of standardized metrics to guide development. Key functionalities such as advanced air quality monitoring, circadian lighting, and noise management systems are critical for improving occupant well-being. Future research could explore integrating these features through adaptive smart platforms, aligning with user preferences and health standards. The areas of maintenance and fault prediction and comfort show a high discrepancy between manufacturers, displaying the whole range of scores (0–4).

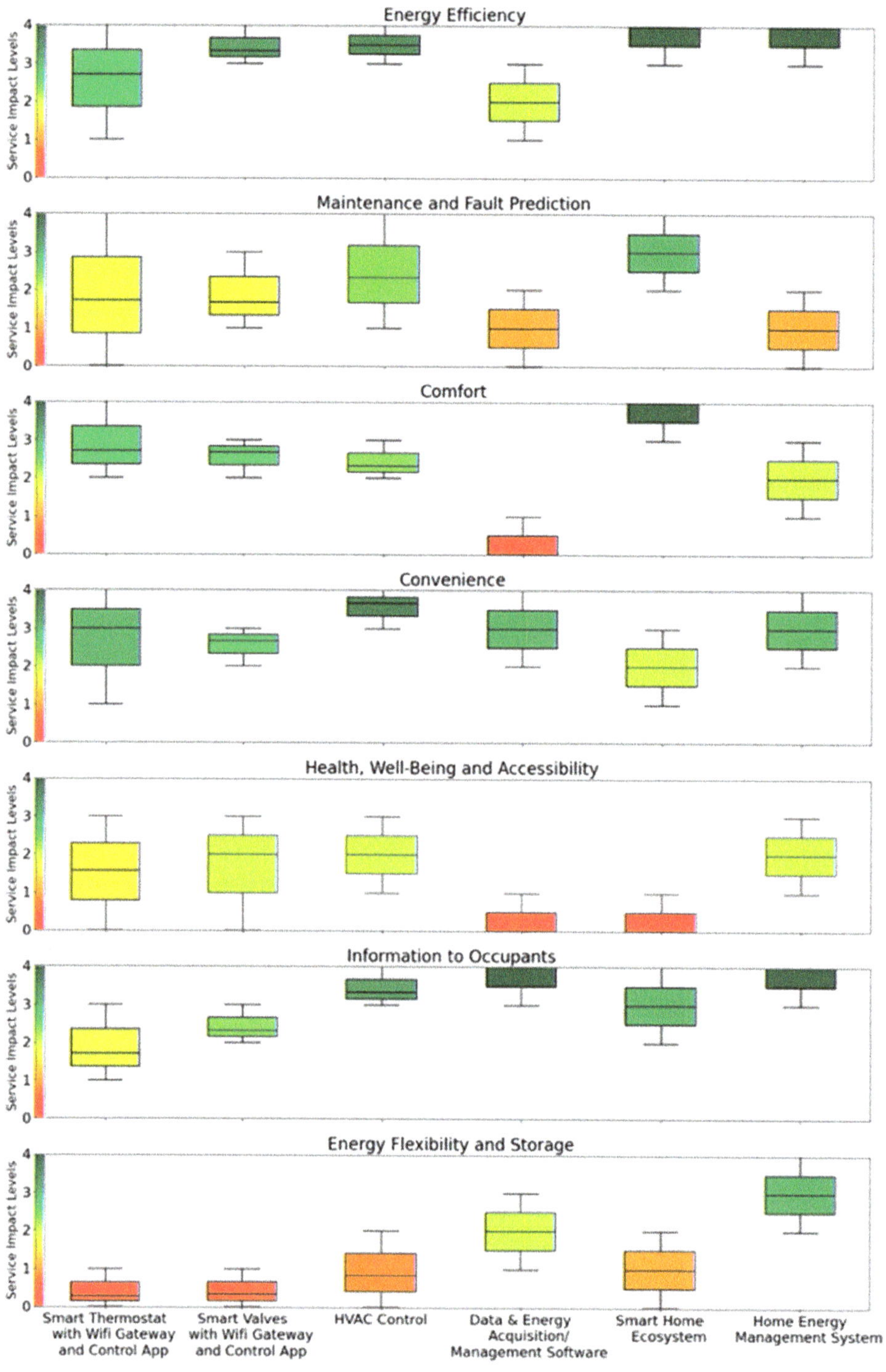

Figure 3. CAM results for key area "service impacts".

4.2. Core Results from Other Key Areas

The CAM has yielded insightful interpretations across various dimensions, including marketing strategies, investment topics, and business models. This section delves into these findings.

4.2.1. Key Area: "Investment Topics"

The analysis highlights varied market strategies among the manufacturers. While companies such as Enel X, Tado°, Netatmo, and Google adopt a user-friendly approach aiming to democratize smart home technologies [27–30], companies such as Bosch emphasize experience through offering wide-spread consulting services to customers [31,32]. Overall, 60% of the evaluated manufacturers are actively investing in building their partner networks. Other common research topics include compatibility, scalability, and IoT integration.

4.2.2. Key Area: "Business Model"

The analysis of the business models of the considered manufacturers reveals a variety of approaches. While in the residential sector sales to private individuals dominate the landscape, the commercial and industrial sector sees an uptake of contracting and subscription models. In some cases, the subscription price is linked to the achieved energy savings, meaning that the price is calculated as a percentage of the achieved savings [21,33,34].

4.3. Cross-Key Area Results

Focus on certain product categories: Most of the products considered in this analysis can be found in the three product categories of smart thermostats, smart valves, and HVAC control systems, indicating a high density and saturation of solutions in these categories. On the contrary, the categories of data acquisition and energy management software, SHE, and HEMS are each comprised of one product only, indicating that they are not as equally addressed as the other categories by the most prominent manufacturers. This does not mean the smart building solutions market hardly offers any solutions for these categories but that the most prominent manufacturers seem to currently focus on the first three product categories predominantly. Here might lie an opportunity for smaller manufacturers and start-ups to fill this gap.

Technological innovation: The development and exploration of AI-based technologies for predictive and real-time adjustments in HVAC systems has reached a ready-for-market status, underscoring a broader industry trend towards harnessing data-driven insights for energy conservation. The trend in the harvesting and processing of building operational data also manifests itself in the rising penetration of data visualization and control apps, allowing building occupants and owners to gain insights into building operation and performance data, underlining the industry's efforts of creating user integration [1].

Holistic approaches: Another result from the CAM is the industry's shift towards user-centric innovations. Manufacturers are increasingly focusing on not just the functional aspects of their products but also on enhancing user comfort and convenience, with nearly all products being capable of voice control integration. While some manufacturers approach convenience by minimalistic and self-installation-friendly products [30], other manufacturers embrace a more luxurious high-end approach [35]. The examined providers exemplify the trend towards creating more intuitive and responsive smart building environments, integrating self-learning and auto-adapting methods like geofencing or malfunctioning detection algorithms. The societal interest in health-consciousness is reflected in the broader deployment of health-related features, such as with Bosch's Room Thermostat II 230V integrating air quality sensors and monitoring functionalities [36]. These trends illustrate a multifaceted focus not only on energy efficiency but also on enhancing indoor environmental quality and enhanced information flows [12]. This is in line with Industry 5.0 movement and an increased focus on the societal impacts of products which can be observed in other economic sectors [37].

5. Roadmap of Smart Building Solutions Market

Interpreting the smart building solutions market through CAM reveals the strategic positioning, internal dynamics, and external pressures within this market. This analysis enables us to distill the strengths, weaknesses, opportunities, and threats (SWOT) for each manufacturer, providing a comprehensive overview of the sector's competitive and operational landscape. Figure 4 encapsulates the SWOT for the smart building solutions market based on the analyzed prominent manufacturers. The identified SWOT is confirmed by previously conducted research on the benefits and challenges of the adoption of smart building technologies [1,38].

	Opportunities	**Threats**
	• rising demand for comfort-increasing products • information to occupants: interfaces, graphical dashboards, real-time monitoring, detailed reports • IoT integration (gateways necessary)	• increased customer expectations • highly competitive market • quickly changing regulations • requirements differing by climate zone and country • installation challenges
Strengths	**S-O strategies**	**S-T strategies**
• highly developed technologies concerning energy efficiency • user-friendly and convenient products • control apps with advanced functionalities	➢ new business models, e.g., subscription/contracting ➢ IoT gateways and ports to anticipate future market developments	➢ customer interaction ➢ strategic partnerships and cooperation ➢ installation support
Weaknesses	**W-O strategies**	**W-T strategies**
• energy flexibility and storage barely addressed • lacking integration of health functionalities into one product • maintenance and fault prediction algorithms • difficult installation of some products	➢ personalized health recommendations ➢ AI-based predictive maintenance algorithms ➢ linking data to cloud ➢ energy management systems for flexibility and storage	➢ expansion to new countries ➢ incentives to build trust

Figure 4. SWOT analysis of smart buildings solution market.

Strengths identified through the CAM underscore the sector's vigorous engagement with technological innovation, particularly the integration of AI and IoT, which has become a defining feature for leading manufacturers. This integration has been extensively applied in various sectors, including smart buildings, to improve efficiency, sustainability, and user experience [39,40]. The strong brand reputation and diverse product portfolios further establish these manufacturers as key players in the smart building domain. These technologies are designed to optimize energy use, reducing costs, and minimizing environmental impact. Other studies also reflect the role of AI and the IoT in enhancing energy efficiency in smart buildings by enabling precise control and optimization of building systems, leading to significant cost reductions and minimized environmental impacts [41,42]. The products within this market are not only user-friendly but also highly convenient, offering intuitive interfaces and seamless integration into everyday life. Advanced control apps with sophisticated functionalities enhance the user experience by providing comprehensive control over building systems. These strengths reflect the sector's capability to drive forward with significant advancements in building management and efficiency.

Weaknesses, as revealed by the matrix, include challenges related to the interoperability of different technological systems, which can hinder the seamless integration essential for optimal smart building operations. The complexity of the installation processes is also marked as a weakness, potentially limiting wider adoption and impacting user experience negatively. Another significant issue is the limited focus on energy flexibility and storage solutions. For example, another study discusses the challenges of integrating energy storage solutions in smart buildings, noting the need for advanced energy management systems to enhance energy flexibility [43]. This gap leaves room for improvement in managing variable energy demands and storing energy efficiently. Additionally, there is a lack of integration of health functionalities into a single product, which can limit the holistic appeal of smart building solutions. This point is seconded by other researchers, indicating a lack of focus on health functionalities in existing smart building solutions, and highlights the need for incorporating such features to enhance occupant well-being [44]. The market also struggles with underdeveloped maintenance and fault prediction algorithms. Simplifying installation processes through plug-and-play designs, modular systems, and pre-configured setups can lower barriers for users and promote wider adoption. Additionally, providing robust training programs and professional support for complex installations could address challenges in commercial applications.

Opportunities emerge from a growing global emphasis on sustainability and energy efficiency. The sector is well-positioned to capitalize on governmental incentives aimed at green building initiatives, as well as advancements in related technologies that can enhance product offerings. Other studies also characterize the smart building sector by rapid technological advancements, particularly in integrating AI, machine learning, and IoT for energy management and user comfort [45]. Increasing awareness among consumers and businesses about the benefits of smart building solutions presents a significant opportunity for market expansion. Furthermore, there is a rising demand for comfort-increasing products as occupants seek more comfortable and livable environments. Providing enhanced information to occupants through interfaces, graphical dashboards, real-time monitoring, and detailed reports can significantly improve user satisfaction and engagement. The integration of IoT gateways is another promising opportunity, which some studies have also defined as a key game changer, through its enabling efficient data collection and real-time processing, enhancing both energy management and occupant comfort [46,47]. As IoT technology becomes more prevalent, integrating it into smart building solutions can create more connected, intelligent, and responsive environments [42].

Threats to the sector include the pace of technological evolution. Increased customer expectations, driven by rapid technological advancements and the proliferation of smart technologies, mean that market players must continuously innovate to stay relevant. A review of the research trends in the design and implementation of smart technologies in buildings also states that the rapid advancement of technologies such as AI, IoT, and automation in smart buildings presents a continuous challenge for companies to innovate and adapt. The need for ongoing technological upgrades and the integration of new systems can strain resources and require significant investment [48]. The market is highly competitive, with numerous players offering similar solutions, which can drive down prices and squeeze profit margins. Regulatory changes, particularly concerning data privacy and environmental standards, pose potential risks to operations and market strategy. Additionally, the threat of market saturation and the impact of economic fluctuations necessitate strategic foresight to navigate these external pressures effectively.

6. Conclusions

With the goal of illuminating the development status, current dynamics, and future trends of the smart building solution market, this paper first selected a range of prominent manufacturers and their products. The 19 identified products from 15 different manufacturers were categorized into six product categories. For the analysis, CAM was developed to evaluate products and manufacturers regarding their service impacts, investments, market penetration, business models, and value propositions. The framework and evaluation of the service impacts were based on the Smart Readiness Indicator (SRI) methodology.

The selected products and manufacturers were thoroughly investigated using manufacturers' websites, technical sheets, online newspaper articles, blog posts, and LinkedIn profiles. The gathered information was analyzed and synthesized into the CAM. Using color-coded visualization techniques, strengths in the market were identified in the areas of energy efficiency, comfort, and convenience. However, weaknesses were found in the integration of flexibility and storage functionalities. Additionally, health, well-being, and accessibility were identified as weak spots in the smart building solution market. There is development potential in further elaborating interoperability and fault detection algorithms as well as improvement of the information flow to building occupants. To further foster and enrich the study, it is recommended to provide a development of industry standards, such as open APIs and communication protocols, to enable seamless integration across systems. Open-source platforms and collaborative efforts among stakeholders could further enhance compatibility and foster innovation in smart building solutions.

Building on the CAM results, a SWOT analysis of the smart building solution market was conducted. Among the selected manufacturers and products, there is a noticeable dominance in the product categories of smart thermostats, smart radiator valves, and HVAC control, indicating the current focus of the most prominent manufacturers. The market is technically innovative (e.g., AI integration) but also highly competitive. Uncertain regulations and government incentives create uncertainty for manufacturers. Concerning business models, a trend towards subscription and contracting can be observed, particularly in the commercial and industrial sector. Approximately 60% of the manufacturers considered in this paper are actively investing in expanding their partner networks, reflecting the rapid developments in the market.

To conclude, the analysis conducted through the CAM and SWOT analysis offers a more holistic understanding of the smart building solutions sector. This exploration reveals a vibrant industry at the forefront of technological innovation, driven by a commitment to sustainability and efficiency. Despite facing challenges, the sector is ripe with opportunities for growth, particularly as global demand for sustainable infrastructure intensifies. Additionally, integrating smart technologies for electric vehicle charging within smart building ecosystems presents a promising revenue for future energy flexibility and management on the building and district level. The strategic insights underscore the importance of agility and foresight in navigating the competitive landscape. As research in this domain advances, future studies could delve deeper into the development of standardized frameworks and advanced methodologies to address interoperability challenges within the different impact categories.

While the study provides insights into the geographical distribution and application of products, it does not delve deeply into sales volumes or user demographics due to data availability constraints. Future research could leverage survey-based methods or partnerships with manufacturers to access anonymized datasets. Collaborations with industry stakeholders or access to third-party market reports may help address these limitations. Moreover, an accurate evaluation of manufacturers' claims requires real-life testing or controlled experiments, which were beyond the scope of this study. Future research could focus on comparative testing of products under standardized conditions, such as lab conditions or pilot implementations, to ensure a more objective validation of performance claims. It should also explore long-term environmental impacts using life cycle assessments and address social aspects like accessibility and equity to ensure alignment with sustainability goals.

As the sector continues to evolve, these insights can serve as a resource for different stakeholders, guiding strategic decisions and fostering innovation in the pursuit of smarter, more sustainable buildings. Future market developments should explore enhancing energy flexibility and storage technologies, integrating health-related functionalities into unified platforms, and establishing standardized protocols to improve interoperability across smart building systems.

Author Contributions: Conceptualization, N.M. and R.S.; methodology, N.M.; software, N.M. and N.S.; validation, N.M., N.S., R.S. and D.M.; formal analysis, N.M. and N.S.; investigation, N.S.; resources, D.M.; data curation, N.S.; writing—original draft preparation, N.M. and N.S.; writing—review and editing, N.M., N.S., R.S. and D.M.; visualization, N.M. and N.S.; supervision, N.M. and R.S.; project administration, N.M. and R.S.; funding acquisition, R.S. and D.M. All authors have read and agreed to the published version of the manuscript.

Funding: This study was funded by the European Commission's Horizon 2020 Research Programme under Grant Agreement Number 101023666.

Data Availability Statement: Data will be made available on request.

Conflicts of Interest: The authors declare that they have no known competing financial interests or personal relationships that could have appeared to influence the work reported in this paper.

References

1. Abdul Manap, N. *The Role of Smart Technologies in Reducing Energy Consumption*; Northern University of Malaysia: Kuala Lumpur, Malaysia, 2024. [CrossRef]
2. European Commission. In focus: Energy Efficiency in Buildings. Available online: https://commission.europa.eu/news/focus-energy-efficiency-buildings-2020-02-17_en (accessed on 17 July 2024).
3. European Commission. Regulation (EU) 2021/1119 of the European Parliament and of the Council of 30 June 2021 establishing the framework for achieving climate neutrality and amending Regulations (EC) No 401/2009 and (EU) 2018/1999 ('European Climate Law'). *Off. J. Eur. Union* **2021**, *243*, 1–17.
4. Deutsche Energie-Agentur. Dena—Gebäudereport 2024—Zahlen, Daten, Fakten zum Klimaschutz im Gebäudebestand. 2023. Available online: https://www.dena.de/infocenter/dena-gebaeudereport-2024/ (accessed on 23 October 2023).
5. European Network of Transmission System Operators for Electricity and Frontier Economics Ltd. Review of Flexibility Platforms—A Report Prepared by Frontier Economics for ENTSO-E. October 2021. Available online: https://eepublicdownloads.azureedge.net/clean-documents/SOC%20documents/SOC%20Reports/210957_entso-e_report_neutral_design_flexibility_platforms_04.pdf (accessed on 23 October 2023).
6. International Energy Agency. *World Energy Outlook 2023*; IEA: Paris, France, 2023.
7. González-Martín, J.; Kraakman, N.J.R.; Pérez, C.; Lebrero, R.; Muñoz, R. A state–of–the-art review on indoor air pollution and strategies for indoor air pollution control. *Chemosphere* **2021**, *262*, 128376. [CrossRef] [PubMed]
8. Özdamar Seitablaiev, M.; Umaroğulları, F. Thermal Comfort and Indoor Air Quality. *Int. J. Sci. Res. Innov. Technol.* **2018**, *5*, 90–109.
9. Lawrence, T.M.; Boudreau, M.-C.; Helsen, L.; Henze, G.; Mohammadpour, J.; Noonan, D.; Patteeuw, D.; Pless, S.; Watson, R.T. Ten questions concerning integrating smart buildings into the smart grid. *Build. Environ.* **2016**, *108*, 273–283. [CrossRef]
10. Hurtado, L.A.; Nguyen, P.H.; Kling, W.L. Agent-based control for building energy management in the smart grid framework. In Proceedings of the IEEE PES Innovative Smart Grid Technologies, Europe, Istanbul, Turkey, 12–15 October 2014; pp. 1–6. [CrossRef]
11. Ahmad, A.; Khan, J.Y. Real-Time Load Scheduling, Energy Storage Control and Comfort Management for Grid-Connected Solar Integrated Smart Buildings. *Appl. Energy* **2020**, *259*, 114208. [CrossRef]
12. Farzaneh, H.; Malehmirchegini, L.; Bejan, A.; Afolabi, T.; Mulumba, A.; Daka, P.P. Artificial Intelligence Evolution in Smart Buildings for Energy Efficiency. *Appl. Sci.* **2021**, *11*, 763. [CrossRef]
13. Cano-Suñén, E.; Martínez, I.; Fernández, Á.; Zalba, B.; Casas, R. Internet of Things (IoT) in Buildings: A Learning Factory. *Sustainability* **2023**, *15*, 12219. [CrossRef]
14. Al-Obaidi, K.M.; Hossain, M.; Alduais, N.A.M.; Al-Duais, H.S.; Omrany, H.; Ghaffarianhoseini, A. A Review of Using IoT for Energy Efficient Buildings and Cities: A Built Environment Perspective. *Energies* **2022**, *15*, 5991. [CrossRef]
15. Bitkom, e.V. Klimaschutz und Energieeffizienz Durch Digitale Gebäudetechnologien. 2021. Available online: https://www.bitkom.org/sites/main/files/2021-11/211111_st_klimaschutz-und-energieeffizienz.pdf (accessed on 15 July 2024).

16. European Commission. Fit for 55: Delivering on the Proposals. Available online: https://commission.europa.eu/strategy-and-policy/priorities-2019-2024/european-green-deal/delivering-european-green-deal/fit-55-delivering-proposals_en (accessed on 15 April 2024).
17. Verbeke, S.; Aerts, D.; Reynders, G.; Ma, Y.; Waide, P. Final Report on the Technical Support to the Development of a Smart Readiness Indicator for Buildings: Final Report. European Commission. 2020. Available online: https://data.europa.eu/doi/10.2833/41100 (accessed on 5 December 2023).
18. Nacer, A.; Delahoche, L.; Marhic, B. Smart Home, Smart HEMS, Smart heating: An Overview of the Latest Products and Trends. In Proceedings of the 2017 6th International Conference on Systems and Control (ICSC), Batna, Algeria, 7–9 May 2017. [CrossRef]
19. Lomas, K.J.; Oliveira, S.; Warren, P.; Haines, V.J.; Chatterton, T.; Beizaee, A.; Prestwood, E.; Gething, B. Do domestic heating controls save energy? A review of the evidence. *Renew. Sustain. Energy Rev.* **2018**, *93*, 52–75. [CrossRef]
20. Afram, A.; Janabi-Sharifi, F. Theory and applications of HVAC control systems—A review of model predictive control (MPC). *Build. Environ.* **2014**, *72*, 343–355. [CrossRef]
21. BeeBryte. Our Energy Optimization Solution—Unlock Unprecedented Energy Performance! Available online: https://www.beebryte.com/solution/ (accessed on 23 October 2023).
22. DABBEL—Automation Intelligence GmbH. DABBEL AI—Driving Decarbonization in Real Estate. Available online: https://www.dabbel.eu/ (accessed on 23 October 2023).
23. MeteoViva GmbH. MeteoViva—Smart Buildings as a Service. Available online: https://meteoviva.com/en/ (accessed on 23 October 2023).
24. Esrafilian-Najafabadi, M.; Haghighat, F. Occupancy-based HVAC control systems in buildings: A state-of-the-art review. *Build. Environ.* **2021**, *197*, 107810. [CrossRef]
25. Kleiminger, W.; Mattern, F.; Santini, S. Predicting household occupancy for smart heating control: A comparative performance analysis of state-of-the-art approaches. *Energy Build.* **2014**, *85*, 493–505. [CrossRef]
26. Louis, J.-N.; Caló, A.; Pongrácz, E. Smart Houses for Energy Efficiency and Carbon Dioxide Emission Reduction. In Proceedings of the Fourth International Conference on Smart Grids, Green Communications and IT Energy-aware Technologies, Chamonix, France, 20–25 April 2014.
27. Enel X, S.r.l. HOMIX, a Single Solution for a Smart Home. Available online: https://corporate.enelx.com/en/stories/2021/03/homix-smart-home-system (accessed on 23 October 2023).
28. Tado GmbH. Tado°—A Smart Solution for Maintaining the Perfect Temperature in Your Home. Available online: https://www.tado.com/gb-en (accessed on 15 July 2024).
29. Netatmo. With the Smart Thermostat, Save Energy Without Sacrificing Comfort. Available online: https://www.netatmo.com/en-gb/smart-thermostat (accessed on 15 July 2024).
30. Google LLC. Nest Learning Thermostat. Google Store. Available online: https://store.google.com/us/product/nest_learning_thermostat_3rd_gen?hl=en-US (accessed on 15 July 2024).
31. Robert Bosch Smart Home GmbH. Bosch Smart Home Top Ratings and Other Awards. Available online: https://www.bosch-smarthome.com/uk/en/smart-home-explained/outstanding-solutions/ (accessed on 15 July 2024).
32. Robert Bosch Smart Home GmbH. Contact, Help and Information on Bosch Smart Home. Available online: https://www.bosch-smarthome.com/uk/en/service/ (accessed on 15 July 2024).
33. MeteoViva GmbH. Revolutionary, Smart, and Effective. Reduce CO_2 Emissions by Up to 40 per Cent. Available online: https://meteoviva.com/en/meteoviva-climate-2/ (accessed on 15 July 2024).
34. Evogy, S.r.l. Ottimizzazione Predittiva in Tempo Reale: Simon Optimizer. Available online: https://www.evogy.it/monitoraggio-energetico-per-le-aziende-simon-optimizer (accessed on 15 July 2024).
35. Lux Products Corporation. Our KONO Smart Thermostat Was Made with You in Mind. With KONO Smart, We've Made It Personal. Available online: https://luxproducts.com/kono-4/ (accessed on 15 July 2024).
36. Robert Bosch Smart Home GmbH. Room Thermostat II 230 V (Underfloor Heating). Available online: https://www.bosch-smarthome.com/uk/en/products/devices/room-thermostat-for-underfloor-heating/ (accessed on 23 October 2023).
37. European Commission. Industry 5.0—What This Approach Is Focused On, How It Will Be Achieved and How It Is Already Being Implemented. Available online: https://research-and-innovation.ec.europa.eu/research-area/industrial-research-and-innovation/industry-50_en (accessed on 16 July 2024).
38. Ożadowicz, A. Generic IoT for Smart Buildings and Field-Level Automation—Challenges, Threats, Approaches, and Solutions. *Computers* **2024**, *13*, 45. [CrossRef]
39. Sleem, A.; Elhenawy, I. Survey of Artificial Intelligence of Things for Smart Buildings: A closer outlook. *J. Intell. Syst. Internet Things* **2023**, *8*, 63–71. [CrossRef]
40. Kumar, A.; Sharma, S.; Goyal, N.; Singh, A.; Cheng, X.; Singh, P. Secure and energy-efficient smart building architecture with emerging technology IoT. *Comput. Commun.* **2021**, *176*, 207–217. [CrossRef]

41. Wan, J.; Yang, J.; Wang, Z.; Hua, Q. Artificial Intelligence for Cloud-Assisted Smart Factory. *IEEE Access* **2018**, *6*, 55419–55430. [CrossRef]
42. Solanki, S.M. Industry 4.0 and Smart Manufacturing: Exploring the integration of advanced technologies in manufacturing. *Rev. Rev. Index J. Multidiscip.* **2023**, *3*, 36–46. [CrossRef]
43. Aliero, M.S.; Qureshi, K.N.; Pasha, M.F.; Ghani, I.; Yauri, R.A. Systematic Mapping Study on Energy Optimization Solutions in Smart Building Structure: Opportunities and Challenges. *Wirel. Pers. Commun.* **2021**, *119*, 2017–2053. [CrossRef]
44. Williams, J.; Lellouch, B.; Stein, S.; Vanderwel, C.; Gauthier, S. Low-Carbon Comfort Management for Smart Buildings. In Proceedings of the 2022 IEEE International Smart Cities Conference (ISC2), Pafos, Cyprus, 26–29 September 2022; pp. 1–5. [CrossRef]
45. Tanasiev, V.; Pluteanu, Ş.; Necula, H.; Pătraşcu, R. Enhancing Monitoring and Control of an HVAC System through IoT. *Energies* **2022**, *15*, 924. [CrossRef]
46. Charles Raja, S.; Vishnu Dharssini, A.C.; Nesmalar, J.J.D.; Karthick, T. Deployment of IoT-Based Smart Demand-Side Management System with an Enhanced Degree of User Comfort at an Educational Institution. *Energies* **2023**, *16*, 1403. [CrossRef]
47. Costa, A.A.; Lopes, P.M.; Antunes, A.; Cabral, I.; Grilo, A.; Rodrigues, F.M. 3I Buildings: Intelligent, Interactive and Immersive Buildings. *Procedia Eng.* **2015**, *123*, 7–14. [CrossRef]
48. Kim, D.; Yoon, Y.; Lee, J.; Mago, P.J.; Lee, K.; Cho, H. Design and Implementation of Smart Buildings: A Review of Current Research Trend. *Energies* **2022**, *15*, 4278. [CrossRef]

MDPI AG
Grosspeteranlage 5
4052 Basel
Switzerland
Tel.: +41 61 683 77 34

Energies Editorial Office
E-mail: energies@mdpi.com
www.mdpi.com/journal/energies

www.ingramcontent.com/pod-product-compliance
Lightning Source LLC
Chambersburg PA
CBHW061514210926
51850CB00008B/1915